INTRODUCTION TO NUMERICAL COMPUTATION

INTRODUCTION TO Numerical Computation

J. THOMAS KING

Professor of Mathematical Sciences
University of Cincinnati

McGRAW-HILL BOOK COMPANY

New York | St. Louis | San Francisco | Auckland | Bogotá | Hamburg
Johannesburg | London | Madrid | Mexico | Montreal | New Delhi
Panama | Paris | São Paulo | Singapore | Sydney | Tokyo | Toronto

To my parents, Jack and Ann

INTRODUCTION TO NUMERICAL COMPUTATION

234567890 DOCDOC 8987654

ISBN 0-07-034639-9

This book was set in Times Roman.
The editors were Peter R. Devine and Jo Satloff;
the designer was Nicholas Krenitsky;
the production supervisor was Marietta Breitwieser.
The drawings were done by Wellington Studios Ltd.
R. R. Donnelley & Sons Company was printer and binder.

Library of Congress Cataloging in Publication Data

King, J. Thomas.
    Introduction to numerical computation.

    Bibliography: p.
    Includes index.
    1. Numerical analysis—Data processing.   I. Title.
QA297.K564   1984   519.4′028′54   83-19531
ISBN 0-07-034639-9

CONTENTS

PREFACE

This book is concerned with the art and science of problem solving on a modern digital computer. More specifically it deals with the numerical solution of mathematical problems that arise as models of some phenomena in the physical sciences and engineering. This type of problem solving is commonly called scientific computing.

In writing the text, I have kept in mind its twofold objective: (1) to present simple and reasonably effective methods for the numerical solution of those problems that arise most frequently in applications and (2) to give an elementary exposition of the essential concepts in scientific computing—conditioning of a problem and its numerical method of solution, error bounds and estimation, order of accuracy, floating point arithmetic, computational pitfalls, and programming considerations. In my view it is better to treat some of the commonly occurring problems in depth rather than a large number of problems in a superficial manner.

The book is designed for a sophomore/junior level course in numerical methods. Typically this is a one-term (quarter or semester) course whose purpose is to introduce students to scientific computing. The material and its level of presentation have been chosen so as to be accessible to sophomores from a variety of disciplines, including computer science, engineering, mathematics, and the physical sciences. The only prerequisites for reading and understanding the text are a solid course in calculus and a working knowledge of a scientific programming language, preferably FORTRAN. Some knowledge of matrices and ordinary differential equations would be helpful but is not essential. The book is self-contained. This does not mean that it is a cookbook of numerical methods but rather that less emphasis is given to rigor and more to explanation and understanding. For each topic the underlying theory is there, but it is motivated and reinforced by examples and appropriate algorithmic descriptions. Proofs and derivations are given only when they are essential to the development and are accessible to the intended audience. Because the text is written for sophomores, several advanced topics of considerable importance have been omitted, including the matrix eigenvalue problem, optimization, and partial differential equations.

Much of the subject matter presented here has become standard in numerical methods books—the solution of a nonlinear equation, polynomial interpolation, and the elimination method for linear systems, to name a few. However, these traditional topics are presented here in the context of modern computing practice. For example, in addition to the usual discussion of Simpson's rule on a uniform mesh, an introduction to adaptive quadrature based on the Simpson formula is given. Other examples of modern procedures include a hybrid method for root finding, a variable step Runge–Kutta method, and basic cubic spline interpolation and approximation.

Most sections of the book terminate with a set of exercises. I have attempted to design a variety of problems (nearly 500 in total) that enable the student to gain valuable computational experience and to develop an appreciation of the necessity of studying the conditioning, obtainable accuracy, relative efficiency, and limitations of numerical methods. Many of the exercises can be solved using only pencil and paper, but the majority require the use of a digital computer. An essential component in learning the subject is the solution of a large number of computational exercises. The more difficult exercises are identified by an asterisk. In addition to the exercises the book contains several projects that require the student to write (or modify) and test a FORTRAN program based on the more sophisticated methods and ideas of the chapter. These projects are intended to introduce the student to state-of-the-art procedures and practice. Many of them require a substantial amount of time and/or effort and should be used as team projects.

All of the programs in the book are written in the WATFIV dialect of FORTRAN. In particular none of the structured features of FORTRAN 77 are used in these programs. This was a conscious decision on my part based on the belief that WATFIV is still the most readily available version of FORTRAN on most college campuses. These programs were all run on an AMDAHL machine that uses hexadecimal floating point arithmetic, chopping, and precision 6 in single precision; in double precision the precision is 14. The programs in the text are not meant to be of production quality. Indeed they were written with simplicity, readability, and correctness foremost in mind.

The reader is encouraged to familiarize himself or herself with one or more of the excellent scientific software packages available commercially. Most university computing centers possess one of these high quality packages. I mention only the two with which I am most familiar: the IMSL (International Mathematical and Statistical Library) package and the more specialized LINPACK-EISPACK library. Information on both libraries can be obtained by writing to IMSL Inc., 6th Floor, GNB Building, 7500 Bellaire Blvd., Houston, Texas 77036.

There is more than enough material in the book for a one-term course, and hence the instructor has considerable flexibility in choosing a syllabus. In my own one-quarter course taught to engineering students, I usually cover the material in Sections 1.1 to 1.4, 2.1 to 2.4, 3.1 to 3.3, 4.1 to 4.3, 5.1, 5.2.3, 5.3, 5.6, 6.1 to 6.3, and 7.1. Chapter 3, on linear systems, is a prerequisite only for parts of Section 4.3 and for Chapter 7. For a semester course, one could add more material on differential equations, spline interpolation, or least squares problems, depending on the audience. The entire book can be covered in a two-quarter sequence.

The books currently available for a sophomore/junior level numerical methods course can be grouped into two broad categories: traditional survey books and single-algorithm books. The traditional books are somewhat old-fashioned and resemble a rigorous numerical analysis book "toned down" for an audience that lacks mathematical maturity. Such books tend to be encyclopedic in that a large number of topics are covered but not in depth from either a theoretical or modern computational point of view. The single-algorithm books are strongly computer-oriented, with emphasis on one good algorithm for each major numerical problem. These books provide state-of-the-art codes that must be treated as black boxes. In my opinion, neither of these approaches is completely satisfactory; I believe that a better tack falls between the two. Certainly many of the traditional topics must be presented, but in a manner that is consistent with modern computing practice. Moreover, an introductory book should provide a rather complete explanation of numerical methods with an elementary analysis sufficient to explain why a method works. This is the type of book I have attempted to write.

I am indebted to my colleagues and reviewers for their helpful comments and criticisms of the manuscript: James L. Cornette, Iowa State University; Bruce H. Edwards, University of Florida; Richard S. Falk, Rutgers University; Herman Gollwitzer, Drexel University; Charles Groetsch, University of Cincinnati; Lee W. Johnson, Virginia Polytechnic Institute and State University; Lester Lipsky, University of Nebraska; Wendell H. Mills, Jr., Standard Oil Research and Development; Diego Murio, University of Cincinnati; Robert Plemmons, North Carolina State University; and Frank G. Walters, University of Missouri—Rolla. I thank Connie Spurlock, who did an excellent job of typing this and earlier versions of the manuscript. The University of Cincinnati provided a one-quarter release from my teaching duties, during which time a significant portion of the manuscript was written. The staff of McGraw-Hill have been most cooperative and helpful in this undertaking. Finally, I wish to thank my wife Susan, for her patience and understanding during the last three years.

J. Thomas King

CONCEPTS IN NUMERICAL PROBLEM SOLVING

1

Mathematics has traditionally been an intrinsic mode of expression and discovery in the physical sciences and engineering. The problems that arise in these fields have led to significant work in virtually every branch of the mathematical sciences. More recently mathematics has played an increasingly greater role in the solution of problems arising in such diverse fields as economics, medicine, psychology, and business administration, to name but a few. No doubt one of the primary reasons for this state of affairs is the development of the high speed electronic digital computer. Calculations that in the 1940s would have required months or years of human effort can now be performed by a computer in seconds. Consequently many intractable problems of 40 years ago are now routinely solved with the aid of a digital computer.

1.1 INTRODUCTION AND MATHEMATICAL PRELIMINARIES

In these various fields there arise mathematical problems that, in their complete form, cannot be solved by analytical means. Many such problems involve a myriad of interacting processes that make it difficult to decide which effects are important and which are insignificant. Consequently the investigator will, based on the principles of his or her discipline, neglect certain aspects in order to reduce the problem to a more manageable form. This simpler problem is then analyzed to give the investigator some qualitative insight into the original problem. The process of reduction to a simpler problem may be repeated several times under various simplifying assumptions. Such an approach is often not sufficient, however, especially when quantitative information is required. It is instructive to outline the various stages involved in numerical problem solving:

1 MATHEMATICAL MODELING In this formative stage the problem is phrased in the language of mathematics by identifying the dependent and independent variables and establishing the functional relationships between them. This process may result in a system of equations, a function that is to be minimized, an integral that is to be evaluated, an equation that involves an unknown function and one or more of its derivatives (a differential equation), and so forth.

2 SIMPLIFICATION Based on appropriate principles of the discipline, empirical evidence, and previous computational experience, the investigator then simplifies the model, if necessary, to a well-defined mathematical problem whose solution is expected to be practically attainable and to yield some specifically desired information.

3 NUMERICAL FORMULATION After the problem has been clearly identified and defined, one or more numerical methods are considered for its approximate solution. A preliminary human analysis of the method(s) is made regarding accuracy, efficiency, and stability. Specifically one attempts to answer, prior to any computation, the questions (a) Is the proposed numerical method sufficiently accurate? (b) Will the method require a reasonable amount of computer time and programming effort? (c) Is the method insensitive to small perturbations in the data? Assuming the answers to these questions are in the affirmative, an algorithm is developed for the implementation of the chosen method.

4 PROGRAMMING This stage consists in translating the algorithm into a computer program. Typically this program is written in a high level language such as FORTRAN. The program is tested and debugged to ensure its correctness.

5 EVALUATION In this phase several trial runs of the program are made for various sets of input data. The results of these trials are compared with empirical evidence and basic principles in order to assess the quality of the numerical solutions. If these results are judged satisfactory, then the program is normally refined with regard to efficiency, simplicity, and style. If warranted, the final version is carefully documented so that it is accessible to others. Should the results prove unsatisfactory, the process reverts to stage 2 or 3 for reexamination.

In this book we are most concerned with stages 3 and 4 of this process. Thus we start with a well-defined mathematical problem and consider one or more methods for its numerical solution. We endeavor to analyze and test these methods, with particular attention given to their efficiency, accuracy, and stability (conditioning). An integral part of the text consists in the computer implementation of these numerical methods.

Throughout the text we need to refer to several of the main results of

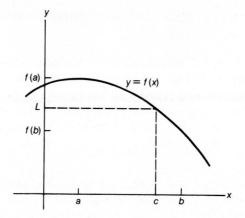

FIGURE 1.1

calculus. For the convenience of the reader we briefly review these basic mathematical results. We assume the reader is familiar with the essential ideas of calculus, such as functions, the concept of limits, continuity, differentiation, and integration, and especially with sequences (many numerical methods are iterative and thus result in a sequence of approximants). The theorems that follow can be found in any good calculus text.

THEOREM 1.1

INTERMEDIATE VALUE THEOREM   Suppose f is a continuous function on the closed interval $[a, b]$. If L is any number between $f(a)$ and $f(b)$, then there exists a point c in (a, b) such that $f(c) = L$.

This theorem is graphically illustrated in Figure 1.1.

EXAMPLE 1.1

Suppose we want to determine whether the equation $x^5 - 2x^3 = 7.2$ has a solution. If we define $f(x) = x^5 - 2x^3 - 7.2$, then we find $f(0) = -7.2$ and $f(2) = 8.8$. Since $f(0) < 0 < f(2)$, there is a point $c \in (0, 2)$ such that $f(c) = 0$. Therefore the given equation has at least one solution.

The approach given in Example 1.1 is frequently used to demonstrate that an equation has a solution. We formalize this observation in the following.

COROLLARY 1.1

Under the hypotheses of Theorem 1.1 and the condition $f(a)f(b) < 0$, there exists at least one point $c \in (a, b)$ such that $f(c) = 0$.

The next result states that a continuous function on a closed interval attains its maximum and minimum values.

THEOREM 1.2

EXTREME VALUE THEOREM Suppose f is continuous on $[a, b]$. Then there exist points $x_0, x_1 \in [a, b]$ such that $f(x_0) \le f(x) \le f(x_1)$ for all x in $[a, b]$.

We say that a function f is bounded on an interval I if there is a number $M > 0$ such that $|f(x)| \le M$ for all x in I. The number M is called a **bound** for f on I. If f is as in Theorem 1.2, we can determine a bound for f on $[a, b]$ by $M = \max\{|f(x_0)|, |f(x_1)|\}$. Thus a continuous function on a closed interval is always bounded.

EXAMPLE 1.2

We determine a bound for $g(x) = (x - a)(x - b)$ on $[a, b]$. Since g is continuous, we know that it is bounded. We find that $g'(x) = 2x - (a + b)$, and hence $x^* = (a + b)/2$ is the only critical point of g. Moreover we note that $g'(x) > 0$ for $x > x^*$ and $g'(x) < 0$ for $x < x^*$. It follows that g is decreasing to the left of x^* and increasing to the right of x^*. Since $g(a) = g(b) = 0$, we have $0 \ge g(x) \ge g(x^*) = -(b - a)^2/4$ for all $x \in [a, b]$. Therefore a bound for g on $[a, b]$ is given by $M = (b - a)^2/4$.

THEOREM 1.3

MEAN VALUE THEOREM Suppose f is continuous on $[a, b]$ and is differentiable on (a, b). Then there is a point $c \in (a, b)$ such that

$$f(b) - f(a) = f'(c)(b - a)$$

From Figure 1.2 we observe that the mean value theorem gives the existence of a point c where the tangent line to the graph of f is parallel to the secant line joining $(a, f(a))$ and $(b, f(b))$.

An important special case of Theorem 1.3 occurs when $f(a) = f(b)$. Then there is a point $c \in (a, b)$ for which $f'(c) = 0$. This result is called Rolle's theorem.

EXAMPLE 1.3

We show that the equation $x^3 - 3.1x + 2 = 0$ has exactly one solution in $(0, 1)$. Let $f(x) = x^3 - 3.1x + 2$; then we find $f(0) = 2 > 0$ and $f(1) = -0.1 < 0$. Hence by Corollary 1.1 the equation has at least one solution in $(0, 1)$. Let us suppose that the equation has two solutions in $(0, 1)$, say x_1 and x_2. Then by Rolle's theorem we must have, for some point c between x_1 and x_2,

$$f(x_1) - f(x_2) = 0 = f'(c)(x_1 - x_2)$$

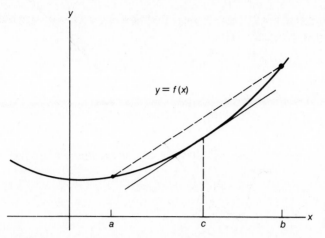

$y = f(x)$

FIGURE 1.2

and hence $f'(c) = 0$. However, $f'(x) = 3x^2 - 3.1 < 0$ for all $x \in (0, 1)$, which is a contradiction. Therefore the equation cannot have more than one solution in $(0, 1)$. Note that $f'(x) < 0$ implies that the graph of f is strictly decreasing on $(0, 1)$. Clearly a strictly decreasing graph cannot intersect the x axis at more than one point.

It should be apparent that if f satisfies the hypotheses of the mean value theorem and if $f'(x) \neq 0$ for all $x \in (a, b)$, then the graph of f can intersect the x axis at one point at most.

THEOREM 1.4

INTEGRAL MEAN VALUE THEOREM   Suppose f and g are continuous functions on $[a, b]$ and g does not change sign in $[a, b]$. Then there is a point $c \in [a, b]$ such that

$$\int_a^b f(x)g(x)\, dx = f(c) \int_a^b g(x)\, dx$$

EXAMPLE 1.4

Suppose f is continuous on $[a, b]$ and M is a bound for f on $[a, b]$. For $g(x) = (x - a)(x - b)$, we obtain a bound for $\int_a^b f(x)g(x)\, dx$. From Example 1.2 we have $g(x) \leq 0$ for all x in $[a, b]$. Hence by Theorem 1.4 we have

$$\left| \int_a^b f(x)(x - a)(x - b)\, dx \right| = \left| f(c) \int_a^b (x - a)(x - b)\, dx \right|$$

$$\leq M \int_a^b |(x - a)(x - b)|\ dx$$

Moreover we find that

$$\int_a^b |(x - a)(x - b)|\ dx = \int_a^b (x - a)(b - x)\ dx = \frac{(b - a)^3}{6}$$

and hence

$$\left| \int_a^b f(x)(x - a)(x - b)\ dx \right| \leq \frac{M(b - a)^3}{6}$$

The next theorem states that smooth functions can be approximated by polynomials.

THEOREM 1.5

TAYLOR'S THEOREM   Suppose f and its first $n + 1$ derivatives are continuous on (a, b) and x_0 is a point in (a, b). Then for any $x \in (a, b)$

$$f(x) = f(x_0) + f'(x_0)(x - x_0) + \cdots + \frac{f^{(n)}(x_0)}{n!}(x - x_0)^n + R_n(x)$$

where the remainder $R_n(x)$ is given by

$$R_n(x) = \frac{1}{n!} \int_{x_0}^x (x - t)^n f^{(n+1)}(t)\ dt$$

The polynomial $p_n(x) = \sum_{k=0}^n f^{(k)}(x_0)(x - x_0)^k / k!$ is called the nth **Taylor polynomial** for f about the point x_0. Thus

$$R_n(x) = f(x) - p_n(x)$$

gives the error in approximating $f(x)$ by $p_n(x)$. In order to obtain a bound for the error, we use Theorem 1.4 to give another form for $R_n(x)$. We note that the function $g(t) = (x - t)^n$ does not change sign on $[x_0, x]$, and hence

$$R_n(x) = \frac{1}{n!} f^{(n+1)}(c) \int_{x_0}^x (x - t)^n\ dt$$

$$= \frac{f^{(n+1)}(c)}{(n + 1)!}(x - x_0)^{n+1}$$

for some point c between x_0 and x.

Now suppose M_{n+1} is a bound for $f^{(n+1)}$ on (a, b); then we have

$$|R_n(x)| = \frac{|f^{(n+1)}(c)|}{(n + 1)!}|x - x_0|^{n+1} \leq \frac{M_{n+1}}{(n + 1)!}|x - x_0|^{n+1}$$

which furnishes a bound for the absolute value of the error $|f(x) - p_n(x)|$.

EXAMPLE 1.5

Use an appropriate Taylor polynomial to approximate $\sqrt{1.002}$ with an error of no more than 10^{-5}. We let $f(x) = (1 + x)^{1/2}$ and use Taylor's theorem with $x = 0.002$ and $x_0 = 0$. The derivatives of f are

$$f'(x) = \frac{1}{2(x + 1)^{1/2}}$$

$$f''(x) = \frac{-1}{2^2(x + 1)^{3/2}}$$

$$f'''(x) = \frac{1 \times 3}{2^3(x + 1)^{5/2}}$$

and in general

$$f^{(k)}(x) = (-1)^{k+1}2^{-k}[1 \times 3 \times 5 \times \cdots \times (2k - 1)](1 + x)^{1/2 - k}$$

Therefore for any $x \in (0, 0.002)$, we have

$$|f^{(k)}(x)| \leq 2^{-k}[1 \times 3 \times 5 \times \cdots \times (2k - 1)] = M_k$$

If we choose n so that $(0.002)^{n+1}M_{n+1}/(n + 1)! \leq 10^{-5}$, then $p_n(0.002)$ provides the required approximation to $\sqrt{1.002}$. It is easy to check that $n = 1$ is sufficient and $p_1(0.002) = 1.001$.

If Taylor's theorem is valid for all n and if $\lim_{n \to \infty} R_n(x) = 0$ for all x in (a, b), then

$$f(x) = \sum_{n=0}^{\infty} \frac{f^{(n)}(x_0)}{n!}(x - x_0)^n$$

which is called the **Taylor series expansion** for f about x_0. Moreover, if g is a function defined by a power series

$$g(x) = \sum_{n=0}^{\infty} a_n(x - x_0)^n \qquad x, x_0 \in (a, b)$$

then it follows that $a_n = g^{(n)}(x_0)/n!$; that is, a power series is its own Taylor series.

EXAMPLE 1.6

Find the Taylor series expansion for $f(x) = (1 - x)^{-1}$ about $x_0 = 0$. By simple multiplication we find

$$(1 - x)(1 + x^2 + x^3 + \cdots + x^n) = 1 - x^{n+1}$$

and therefore, if $x \neq 1$,

$$\sum_{k=0}^{n} x^k = \frac{1 - x^{n+1}}{1 - x} = \frac{1}{1 - x} - \frac{x^{n+1}}{1 - x}$$

For $|x| < 1$, we have $\lim_{n \to \infty} x^{n+1}/(1 - x) = 0$ and hence

$$\frac{1}{1 - x} = \lim_{n \to \infty} \sum_{k=0}^{n} x^k = \sum_{k=0}^{\infty} x^k \qquad |x| < 1$$

This series is called the **geometric** series and is the Taylor series expansion for $f(x) = (1 - x)^{-1}$ about $x_0 = 0$.

EXERCISES

1.1 Use Theorem 1.1 to show that the equation $x^5 + 3x = 9$ has a solution in $(1, 2)$.

*1.2 Suppose f is continuous on $[a, b]$ and differentiable with $f'(x) \neq 0$ for all $x \in (a, b)$. Show that there can be at most one point c in (a, b) such that $f(c) = 0$.

1.3 Use exercise 1.2 to show that the problem of exercise 1.1 has exactly one solution in $(1, 2)$.

1.4 Suppose $|f'(x)| \leq \alpha$ for all x and then use Theorem 1.3 to show that $|f(x_1) - f(x_2)| \leq \alpha |x_1 - x_2|$ for all x_1, x_2.

1.5 Let $p_n(x)$ denote the nth Taylor polynomial for $f(x) = \sin x$ about $x_0 = 0$. Determine an integer n such that $p_n(0.5)$ differs from $\sin(0.5)$ by no more than 10^{-4}. Hint: Choose n so that $|R_n(0.5)| \leq 10^{-4}$.

1.6 Find the Taylor series expansion for $f(x) = \cos x$ about $x_0 = \pi/4$. Use the first three terms to estimate $\cos(5\pi/16)$.

1.7 Use Example 1.6 to show that

$$\frac{1}{1 + t} = \sum_{n=0}^{\infty} (-1)^n t^n \qquad |t| < 1$$

1.8 Integrate the geometric series termwise to show that

$$\log|1 - x| = -\sum_{n=0}^{\infty} \frac{x^{n+1}}{n + 1} \qquad |x| < 1$$

1.9 Determine an appropriate Taylor polynomial for $\log|1 + x|$ about $x_0 = 0$ in order to find an approximate value of $\log 1.1$ that is accurate to within 10^{-3}. Hint: Replace x by $-x$ in the Taylor series of exercise 1.8. Then use the following estimate for alternating series. If

$$s = \sum_{n=0}^{\infty} (-1)^n a_n \qquad a_{n-1} > a_n > 0$$

then

$$\left| s - \sum_{n=0}^{N} (-1)^n a_n \right| < a_{N+1}$$

*1.10 Use the series of exercises 1.8 and 1.9 to show that

$$\log \left| \frac{1 + x}{1 - x} \right| = 2 \sum_{n=0}^{\infty} \frac{x^{2n+1}}{2n + 1} \qquad |x| < 1$$

Estimate log 1.1 by using the first three terms of this series with $x = \frac{1}{21}$. Compare with the value log 1.1 = 0.0953102, which is accurate to six digits.

1.11 Let $p_n(x)$ denote the nth Taylor polynomial about $x_0 = 0$ for e^x. Determine N such that $|p_N(x) - e^x| \le 10^{-5}$ for all x in $[0, 1]$.

1.2 SOURCES OF ERROR

We begin by describing some typical problems that frequently arise in science and engineering, together with the major sources of error involved in their numerical solution.

Problem A
Given a function f defined on $[a, b]$, find the value of

$$I(f) = \int_a^b f(x)\, dx \qquad\qquad (1.1)$$

complicated Integrals
impractical or im poss

Problem B
Given a continuous function g, find one or more points x such that

$$g(x) = 0 \qquad\qquad (1.2)$$

Problem C
Given a function h and a point x, find the value $h(x)$.

In many situations one cannot solve these problems by analytical means and hence is forced to find an approximate solution by some numerical method. In problem A the integrand f may be so complicated that it is impractical or impossible to apply the methods of calculus. In fact there are simple functions, such as $f(x) = e^{x^2}$, where the analytical methods of calculus fail to yield a solution. In problem B all but the simplest sort of function g require the use of a numerical method. Consider, for example, the function $g(x) = x - \tan x$. How does one find the point x that is closest to 10 and satisfies $x - \tan x = 0$? A typical example of problem C is that of evaluating an analytic function given by a power series. Recall from calculus the series for the exponential function:

$$e^x = \sum_{n=0}^{\infty} \frac{x^n}{n!}$$

which converges for all x. How does one compute the value of e^{16} or e^{-16} to six digits?

For problem A a standard class of numerical methods may be defined as follows. Let $a = x_0 < x_1 < \cdots < x_n = b$ and consider the formula

$$I_n(f) = \sum_{i=0}^{n} w_i f(x_i) \qquad\qquad (1.3)$$

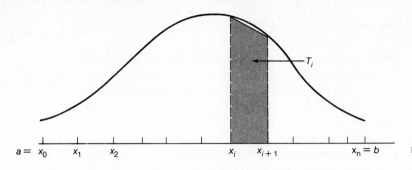

$$a = x_0 \quad x_1 \quad x_2 \quad\quad\quad x_i \quad x_{i+1} \quad\quad\quad x_n = b$$

FIGURE 1.3

where the **weights** $\{w_i\}_{i=0}^n$ are real constants. The quantity $I_n(f)$ is easily computed and is presumably a good approximation to $I(f)$ if the **nodes** $\{x_i\}_{i=0}^n$ and the weights are suitably chosen.

To be more specific, consider the familiar trapezoidal rule. The area of the trapezoid T_i in Figure 1.3 is

$$[f(x_i) + f(x_{i+1})] \frac{\Delta x_i}{2}$$

where $\Delta x_i = x_{i+1} - x_i$. By summing the areas of the trapezoids on each sub-interval $[x_i, x_{i+1}]$, we obtain an approximation to the value of $I(f)$. Thus if Δx_i is small for each i, we expect

$$T_n(f) = \sum_{i=0}^{n-1} \frac{[f(x_i) + f(x_{i+1})] \Delta x_i}{2} \tag{1.4}$$

to be a reasonably good approximation to $I(f)$. We examine the accuracy of this approximation more closely in Chapter 5. Note that $T_n(f)$ may be rewritten in the form of (1.3) with

$$w_0 = \frac{\Delta x_0}{2} \quad\quad w_n = \frac{\Delta x_{n-1}}{2}$$

and for $i = 1, 2, \ldots, n-1$

$$w_i = \frac{\Delta x_i + \Delta x_{i-1}}{2}$$

The difference between the exact solution of problem A and the exact value of $T_n(f)$ is called the **discretization** error. Thus for the trapezoidal rule the discretization error is given by $I(f) - T_n(f)$.

A second major source of error arises in terminating an infinite sequence or series after a finite number of terms. This is commonly called **truncation** error. For example, consider the numerical method for the computation of the square root of 3:

$$x_{n+1} = \frac{x_n + 3/x_n}{2} \quad\quad n = 0, 1, \ldots \tag{1.5}$$

where x_0 is an initial guess. This is a special case of Newton's method, which is discussed in Chapter 2. For any $x_0 > 0$, the sequence determined by (1.5) converges to $\sqrt{3}$. If we choose $x_0 = 2$, then we use (1.5) with $n = 0$ to compute $x_1 = 1.75$. Once x_1 has been determined, we use (1.5) with $n = 1$ to compute $x_2 = 1.73214$ and continue in this manner to generate a sequence $\{x_n\}$. Of course, we must terminate this sequence at some stage, say N, and then the truncation error is $\sqrt{3} - x_N$. The point $\sqrt{3} = \hat{x}$ may be characterized as that positive number that satisfies $\hat{x}^2 - 3 = 0$. Thus the determination of the square root of 3 is a problem of type B with $g(x) = x^2 - 3$.

In the preceding discussion we used the terminology "numerical method" as a description of some means of obtaining an approximate solution to a given problem. To be more precise, we make the following definition. A **numerical method** is an unambiguous description of the relationship between the input data of a given problem and the desired results (output) of the computation. An **algorithm** for a numerical method is a complete description of well-defined operations, arithmetic and logical, through which each admissible set of input data is transformed, in a finite number of steps, into output data.

In the calculation of the square root of 3, the input data consists of the function $g(x) = x^2 - 3$ and the initial guess x_0. The algorithm for the numerical method (1.5) could be given in terms of a flowchart; however, we shall use the following notation:

Do $n = 0, 1, \ldots,$ NMAX:
$$x_{n+1} \leftarrow \frac{x_n + 3/x_n}{2}$$

where $\leftarrow$ indicates value assignment, NMAX denotes a **termination criterion** (a means of selecting x_N), and $\llcorner$ indicates the range of the loop. We shall typically state algorithms in this concise form, which is a natural and convenient basis for writing a program in a high level language such as FORTRAN.

EXAMPLE 1.7

The following is a simple FORTRAN program for computing the square root of 3 based on the previous algorithm. Here we have chosen to terminate the iteration after eight steps.

```
C     A SIMPLE NEWTON METHOD PROGRAM FOR THE
C     COMPUTATION OF THE SQUARE ROOT OF THREE
      X = 15.
      WRITE(6,100)
100   FORMAT(' ',4X,'ITERATION',15X,'VALUE'//)
      DO 1 I=1,8
      X=(X+3./X)/2.
      WRITE(6,101)I,X
101   FORMAT(' ',7X,I2,15X,E14.7)
1     CONTINUE
      STOP
      END
```

ITERATION	VALUE
1	0.7599999E 01
2	0.3997368E 01
3	0.2373930E 01
4	0.1818829E 01
5	0.1734120E 01
6	0.1732052E 01
7	0.1732050E 01
8	0.1732050E 01

In Example 1.7 we specified the termination criterion to mean "stop after eight iterations." This choice is somewhat arbitrary for our simple example. In general the termination criterion depends on the nature of the problem and the algorithm being used. We consider the choice of appropriate termination criteria more carefully in Chapter 2.

A final source of error is the fact that numerical methods are typically not solved exactly. Computers, hand calculators, and human beings have only a finite computation capability. Consider the simple computation of dividing 4 by 3. Human beings can write down and computers can store only a finite number of digits, say 1.333333 in this example. This restriction necessitates the use of approximate arithmetic, and the resultant error is called **roundoff** error.

In the application of a numerical method at least one of these primary types of error is usually present, and in many methods all three types occur.

One of the primary functions of a numerical analyst is the derivation of appropriate error bounds for the numerical method (algorithm) under consideration. If v denotes a value for which we have an approximation $\tilde{v}$, then the **absolute error** is $v - \tilde{v}$, that is,

Absolute error = true value − approximate value

This is not to be confused with the absolute value of the error, which is given by $|v - \tilde{v}|$. A more appropriate measure of error takes the size of v into account. The **relative error** is $(v - \tilde{v})/v$.

EXAMPLE 1.8

Let $v = 0.137 \times 10^{-7}$, $\tilde{v} = 0.879 \times 10^{-6}$, $w = 0.137 \times 10^{8}$, and $\tilde{w} = 0.1374 \times 10^{8}$. Then the absolute error in $\tilde{v}$ and $\tilde{w}$ is

$$v - \tilde{v} = -0.8653 \times 10^{-6}$$

$$w - \tilde{w} = -0.4 \times 10^{5}$$

Although $|v - \tilde{v}|$ is much less than $|w - \tilde{w}|$, it is clear that $\tilde{w}$ is a better approximation to w than $\tilde{v}$ is to v. Indeed, we find the relative error to be $(v - \tilde{v})/v \simeq -63$ and $(w - \tilde{w})/w \simeq -0.003$.

In order to choose an effective numerical method or to decide whether the output of a given method is satisfactory, it is convenient to have a rigorous bound for the error or at least an estimate, not necessarily rigorous, for the size of the error. These error bounds or estimates are basically of two types. In one type a bound or estimate is obtained wherein all quantities are computable from the results (output) of the numerical method. This is referred to as an **a posteriori estimate** or **bound.**

EXAMPLE 1.9

In the method for approximating $\sqrt{3} = \hat{x}$ by (1.5), it can be shown that (disregarding roundoff), for any initial guess $x_0 \in [2, L]$,

$$|\hat{x} - x_{n+1}| \leq \alpha |x_{n+1} - x_n| \qquad \text{where } \alpha = \frac{1}{2}\left(1 - \frac{3}{L^2}\right)$$

This furnishes an a posteriori error bound because the right-hand side of the inequality is computed from the output of the method. For $L = 15$, we have $\alpha \simeq 0.494$, and hence from Example 1.7 we get, for $n = 6$,

$$|\sqrt{3} - 1.732050| \leq 0.494 |1.732050 - 1.732052|$$

$$\leq 0.988 \times 10^{-6}$$

In an a posteriori error analysis we are attempting to answer the question, How can the accuracy of the computed solution be checked? In the second type of analysis we address the question, What is the error in the proposed numerical method? In this approach one uses mathematical reasoning, without the aid of computational experience, to obtain an **a priori bound** or **estimate** of the error.

EXAMPLE 1.10

Consider the trapezoidal rule applied to $I(f) = \int_0^1 f(x)\, dx$ where the nodes are equally spaced, say $\Delta x_i = h$ for $i = 0, 1, \ldots, n - 1$. In Chapter 5 we establish that

$$|I(f) - T_n(f)| \leq \frac{M_2 h^2}{12}$$

where M_2 is a bound for f'' on $[0, 1]$. This gives an a priori error bound for the trapezoidal rule approximation since the right-hand side of the inequality is independent of the output. Suppose, for example, that $f(x) = e^{x^2}$. Then a simple calculation gives $f''(x) = 2e^{x^2}(1 + 2x^2)$, and hence for all $x \in [0, 1]$ we have

$$|f''(x)| \leq 2e(1 + 2) < 18 = M_2$$

Once M_2 is known, the significance of the a priori bound $|I(f) - T_n(f)| \leq$

becomes apparent: for a given accuracy requirement we can select an appropriate value of h ($= 1/n$) so as to guarantee the desired accuracy. For instance, if we require that $T_n(f) = I(f) \pm 10^{-2}$, then $h = \frac{1}{13}$ is sufficient.

EXERCISES

1.12 Calculate the square root of 3 to five decimal places using the method in equation (1.5). Use $x_0 = 3$ and compare with the correct value, $\sqrt{3} = 1.7320508\ldots$. Which do you think is the primary type (truncation, discretization, or roundoff) of error in your computations?

1.13 Compute the approximate value of $\int_0^1 e^{-x}\, dx$ by using the trapezoidal rule with $\Delta x_i = 0.1$ for each i. Retain only five decimal digits throughout your computations. Which do you think is the dominant source of error in this calculation, discretization or roundoff? Why?

1.14 Use a hand calculator to compute the value of the expression $4(1.632 + 0.188)/3$. Retain four digits throughout the computation. Is your computed value exact? If not, what types of error are present?

1.15 In this section we discuss the primary sources of error in numerical problem solving. Can you think of any other sources of error in solving problems on a digital computer?

1.16 In Example 1.5 we found that

$$|\sqrt{1.002} - p_n(0.002)| \le \frac{M_{n+1}(0.002)^{n+1}}{(n+1)!}$$

where M_{n+1} is a bound for the $(n+1)$th derivative of $f(x) = (1+x)^{1/2}$ on $[0, 0.002]$. Does this inequality furnish an a priori or an a posteriori error bound? Explain.

*1.17 Let $S_n(x) = \sum_{k=0}^{n} x^k$ and $S(x) = (1-x)^{-1}$; then $S(x) = \lim_{n \to \infty} S_n(x) = \sum_{k=0}^{\infty} x^k$ (Example 1.6).

(a) Show that

$$|S_n(x) - S(x)| \le \frac{|x|^{n+1}}{1 - |x|} \qquad \text{for } |x| < 1$$

Hint: $|a - b| \ge |a| - |b|$ for all a, b.

(b) Considering $S_n(x)$ as an approximation to $S(x)$, which type of error bound is given in part (a)?

(c) Show that $S_{n+1}(x) - S_n(x) = (x - 1)[S_n(x) - S(x)]$ and hence

$$|S_n(x) - S(x)| \le \frac{|S_{n+1}(x) - S_n(x)|}{1 - |x|}$$

(d) Is the inequality of part (c) an a posteriori error bound? Explain.

*1.18 Suppose we measure a and b experimentally to get approximations $\tilde{a}$ and $\tilde{b}$. How much relative error in $\tilde{a}$ and $\tilde{b}$ can be tolerated if we want the relative error in the product $\tilde{a}\tilde{b}$, $\rho = (ab - \tilde{a}\tilde{b})/ab$, to be less than 10^{-4}? Assume that $ab \ne 0$.

1.19 Modify the program of Example 1.7 so the iteration continues until successive iterates differ by no more than 10^{-6}. Make several runs of your program for various initial guesses, say $x_0 = 10, 15, 20, 25$. Print the iterations as in Example 1.7.

1.20 Write a simple FORTRAN program to compute the approximate value of $I(f) = \int_0^1 e^{-x^2}\, dx$ by using the trapezoidal rule with equally spaced points. Use the a priori error bound of Example 1.10 to select $h\,(= 1/n)$ so that

$$|I(f) - T_n(f)| \le 10^{-4}$$

Run your program with your choice of h and compare with the correct value, 0.74682413....

1.21 Use the program of exercise 1.20 to compute several approximations to $\int_0^1 e^{-x^2}\, dx$, say for $h = \frac{1}{8}, \frac{1}{10}, \frac{1}{15}, \frac{1}{20}, \frac{1}{30}, \frac{1}{40}$, and $\frac{1}{50}$. Print the error in each case. Does the error decrease as h does? If not, explain why not. Hint: The error in the computation has two components—one due to roundoff and the other to discretization. The overall error depends on which of these errors is dominant.

1.3 FLOATING POINT ARITHMETIC

Floating point numbers

In order to discuss roundoff error in a meaningful way, we need to examine, in some detail, the manner in which numbers are stored and manipulated when arithmetic operations are performed on a modern digital computer. In FORTRAN computation most numbers are in floating point form. This form is very similar to scientific notation.

A t-digit **floating point** number in **base** β has the form

$$x = \pm(.b_1 b_2 \dots b_t)\beta^e$$

where $\beta \ge 2$ is an integer and $0 \le b_i \le \beta - 1$ with each b_i an integer. The signed number e is called the **exponent**, t is the **precision**, b_i is called a **digit**, and $\pm(.b_1 b_2 \dots b_t)$ is called the **mantissa**. As a real number the floating point number x is given by

$$x = \pm\left(\sum_{j=1}^{t} b_j \beta^{-j}\right)\beta^e \tag{1.6}$$

On most computers $\beta = 2$, 8, or 16. Moreover, the exponent is required to satisfy $L \le e \le U$ for some constants L and U determined by the machine's hardware.

EXAMPLE 1.11

For $\beta = 2$ and $t = 5$, the real number $5\frac{3}{4}$ is given by the binary floating point number $(.10111) \times 2^3$. This may be verified by using (1.6) as follows:

$$(.10111) \times 2^3 = (2^{-1} + 2^{-3} + 2^{-4} + 2^{-5})2^3$$
$$= 2^2 + 1 + 2^{-1} + 2^{-2}$$
$$= 4 + 1 + \tfrac{1}{2} + \tfrac{1}{4} = 5\tfrac{3}{4}$$

EXAMPLE 1.12

For $\beta = 16$, the digits are 0, 1, 2, 3, 4, 5, 6, 7, 8, 9, A, B, C, D, E, F. Here B represents the base 10 number 11 and F represents the base 10 number 15. We

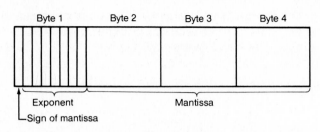

FIGURE 1.4

can denote this simply by writing $B_{16} = 11_{10}$ and $F_{16} = 15_{10}$. This is called the **hexadecimal** number system. Consider the floating point number $(.B108) \times 16^3$. This hexadecimal floating point number is written as a real number using (1.6) as follows:

$$(.B108) \times 16^3 = (11 \times 16^{-1} + 16^{-2} + 8 \times 16^{-4}) \times 16^3$$

$$= 11 \times 16^2 + 16 + \tfrac{8}{16} = 2832\tfrac{1}{2}$$

On an IBM 370 machine, floating point arithmetic is performed in base 16. A real constant is stored in a full **word** of memory. A word consists of four adjacent bytes, and each **byte** consists of eight bits. A **bit** is capable of storing a single **binary** (base 2) digit, that is, 0 or 1. We may visualize a word of memory as shown in Figure 1.4.

Every binary integer can be expressed as a hexadecimal integer by grouping the bits into groups of four from the right and replacing each group by the equivalent hexadecimal digit.

EXAMPLE 1.13

The binary integer 101101011 is converted to its hexadecimal equivalent as follows. Group the bits as $101101011 = 0001\ 0110\ 1011$; then

$$1011_2 = 2^0 + 2^1 + 2^3 = 11_{10} = B_{16}$$

$$0110_2 = 2^1 + 2^2 = 6_{10} = 6_{16}$$

$$0001_2 = 2^0 = 1_{10} = 1_{16}$$

Thus we see that $101101011_2 = 16B_{16}$.

Conversely, every hexadecimal digit can be expressed as a four-bit binary integer, that is, one half-byte of memory. The following is a table of the hexadecimal digits and their binary equivalents:

1	2	3	4	5	6	7	
0001	0010	0011	0100	0101	0110	0111	

8	9	A	B	C	D	E	F
1000	1001	1010	1011	1100	1101	1110	1111

From the table it is easy to see, for instance, that $1D9_{16} = 000111011001_2$.

Thus the memory configuration of Figure 1.4 allows the storage of six hexadecimal digits for the mantissa. The largest such mantissa is .FFFFFF, wherein each half-byte consists of the binary string 1111. In order to allow for a signed exponent, excess-64 notation is used for the storage of the exponent. By this we mean that 64 is added to e and this quantity, called the **characteristic**, is stored in the seven rightmost bits of byte 1. These seven bits allow for a range of 0 to 127_{10}. This corresponds to an exponent range of -64 to 63. Hence in FORTRAN on the IBM 370 the largest floating point number is .FFFFFF $\times 16^{63} \simeq 0.72 \times 10^{76}$ and the smallest positive number is .100000 $\times 16^{-64}$. Nonzero floating point numbers are stored in **normalized form**, which means that the first digit to the right of the base point is nonzero; in other words, $b_1 \neq 0$.

The computer translates a given real number y to a t-digit, base β floating point number, call it fl(y), by either rounding or chopping. In **chopping**, fl(y) is taken to be the normalized floating point number whose mantissa agrees with that of y to t digits. In **rounding**, fl(y) is the normalized floating point number that is closest to y with a rule for ties. In either case the roundoff error is $y - $ fl(y).

EXAMPLE 1.14

Consider a base 10, precision 3 machine that uses chopping with $-L = U = 5$. The following is a table of some real numbers and their normalized floating point representations:

Real Number	Floating Point Number
1638	$.163 \times 10^4$
$21\frac{1}{3}$	$.213 \times 10^2$
$\frac{1}{30}$	$.333 \times 10^{-1}$
101,037	Overflow

If the machine uses rounding, we have

1638	$.164 \times 10^4$
$\frac{2}{3}$	$.667 \times 10^0$

EXAMPLE 1.15

Consider a machine that does arithmetic in base 10, precision 3, $-L = U = 5$, and uses chopping. Suppose we want to execute the algorithm

$$x \leftarrow 1.632$$

$$v \leftarrow 0.189$$

$$z \leftarrow \tfrac{4}{3}(x + v)$$

$$w \leftarrow 4\left(\frac{x + v}{3}\right)$$

If the arithmetic were performed exactly, the values of z and w would be the same. Let us examine the actual values on our hypothetical machine. First we store x and v on our machine. Since $t = 3$, the value for x is stored as $.163 \times 10^1$ and that for v is stored exactly as $.189 \times 10^0$. Let us assume that our machine computes each operation exactly but stores only the normalized floating point result of the operation. Then we have $x + v = .181 \times 10^1$, whereas the exact sum is 1.819. For the computation of z we first divide 4 by 3 to get $.133 \times 10^1$, which we then multiply by $.181 \times 10^1$ to get $z = .240 \times 10^1$. For w we get $(x + v)/3 = .603 \times 10^0$, and hence $w = .241 \times 10^1$.

In Example 1.15 we note that the two methods of evaluation produce different answers, neither of which is equal to the exact value of 2.428. In both methods we encounter roundoff error in simply storing 1.632 on the machine. Moreover, the first method introduces additional roundoff error due to the division and multiplication, whereas these operations are exact in the second approach. This shows that the usual associative rule $a(bc) = (ab)c$ of real arithmetic does not hold for floating point computation.

Another difficulty arises when we attempt the computation $.863 \times 10^5 + .256 \times 10^5$. In this case the exact result is $.1119 \times 10^6$. However, we have $U = 5$, and so we get **exponent overflow** and the computation cannot be performed on our fictitious machine. **Underflow** results when the exponent is less than -5 on the machine of Example 1.15.

The roundoff error is best measured relative to the size of the exact value. We define for $y \neq 0$ the relative roundoff error in $\mathrm{fl}(y)$ by $\rho = [y - \mathrm{fl}(y)]/y$. We may rewrite this expression as $\mathrm{fl}(y) = (1 - \rho)y$. Let us consider the size of the relative roundoff error. For rounding, $\mathrm{fl}(y)$ differs from y by at most $\beta/2$ in the $(t + 1)$st digit to the right of the base point. Thus if the mantissa of $\mathrm{fl}(y)$ is m and the exponent is e, then $y = (m + \eta)\beta^e$ with $|\eta| \leq \beta^{-t}/2$. Therefore we have $y - \mathrm{fl}(y) = (m + \eta)\beta^e - m\beta^e = \eta\beta^e$ and hence

$$\frac{|\mathrm{fl}(y) - y|}{|y|} = \frac{|\eta|}{|m + \eta|} \leq \frac{\frac{1}{2}\beta^{-t}}{|m + \eta|}$$

However, $\mathrm{fl}(y)$ is normalized so that $|m + \eta| \geq \beta^{-1}$, and hence the relative roundoff error is bounded by $\beta^{1-t}/2$. For chopping, $\mathrm{fl}(y)$ differs from y by at most β in the $(t + 1)$st digit. We summarize this in the following theorem.

THEOREM 1.6

For every nonzero number y within the floating point range of the system

$$\frac{|\mathrm{fl}(y) - y|}{|y|} \leq \begin{cases} \frac{1}{2}\beta^{1-t} & \text{rounding} \\ \beta^{1-t} & \text{chopping} \end{cases}$$

We note that for $t = 6$ and $\beta = 16$ we have the estimate $\beta^{1-t} = 16^{-5} \simeq 10^{-6}$. If a chopping machine uses base β, precision t floating point arithmetic, then $\beta^{1-t} = u$ is called the **machine unit**. For a rounding machine, $u = \beta^{1-t}/2$. The quantity u is, in view of Theorem 1.6, an appropriate unit for relative error. In performing arithmetic on a machine, one cannot in general expect the relative error in the computation of a given quantity to be less than u.

Suppose x is a real number and $\tilde{x}$ is an approximation to x. We say that $\tilde{x}$ is **correct** to p base β **digits** if $x = (.b_1 b_2 \ldots)\beta^e$ and $\tilde{x} = (.\tilde{b}_1 \tilde{b}_2 \ldots)\beta^f$ with $f = e$, $b_1 \neq 0$, $\tilde{b}_1 \neq 0$, and $|.b_1 b_2 \ldots - .\tilde{b}_1 \tilde{b}_2 \ldots| \leq \frac{1}{2}\beta^{-p}$; that is, in normalized form the exponents are the same and the mantissas differ by no more than half a unit in the pth digit. Thus if $x = \frac{2}{30}$, then $\tilde{x} = .0667$ is correct to three decimal digits since $x = .6666\ldots \times 10^{-1}$ and $\tilde{x} = .667 \times 10^{-1}$ with $|.667 - .6666\ldots| \leq 0.5 \times 10^{-3}$. Similarly -2.999 is correct to three decimal digits as an approximation to $x = -3$.

Propagation of roundoff error

We next consider the effect that a finite precision has on the basic arithmetic operations. Let us denote by ω any one of the operations $+, -, \div, \cdot$. Let $\bar{\omega}$ represent the corresponding floating point operations $\oplus, \ominus, /, *$. If x and y are two floating point numbers, we want to examine the relative error in $x\bar{\omega}y$. A completely rigorous analysis requires a clear understanding of the manner in which arithmetic is performed on a given machine. For our purposes it is sufficient to define $\bar{\omega}$ by

$$x\bar{\omega}y = \mathrm{fl}(x\omega y)$$

that is, $x\bar{\omega}y$ is the floating point number that represents the exact result. On an actual machine this is not always the case, but in general it is very close to the real computation. Moreover, this definition allows us to easily examine the roundoff error in arithmetic operations. We assume throughout that no overflow or underflow occurs. By virtue of Theorem 1.6 we have

$$|x\bar{\omega}y - x\omega y| \leq u|x\omega y| \tag{1.7}$$

Suppose that x and y are not known exactly; they may be the results of previous computations. How does the error in x and y propagate as the result of an arithmetic operation? Specifically we suppose that $\tilde{x} = (1 - \sigma)x$ and $\tilde{y} = (1 - \eta)y$, where $\tilde{x}$ and $\tilde{y}$ are floating point numbers stored in memory. Had these previous computations been exact, then $\eta = \sigma = 0$. Let us examine how $\tilde{x}\bar{\omega}\tilde{y}$ compares with $x\omega y$.

We begin by considering floating point multiplication. Let ρ denote the relative error in multiplication so that $\tilde{x} * \tilde{y} = (1 - \rho)\tilde{x} \cdot \tilde{y}$. In view of (1.7) we have $|\rho| \leq u$. Moreover, it is reasonable to presume that the relative error in x and y is much less than unity, which we denote by $|\sigma| \ll 1$ and $|\eta| \ll 1$. We find that

$$\tilde{x} * \tilde{y} = (1 - \rho)(1 - \sigma)(1 - \eta)x \cdot y$$

and hence

$$x \cdot y - \tilde{x} * \tilde{y} = [1 - (1 - \rho)(1 - \sigma)(1 - \eta)]x \cdot y$$

Therefore if the overall relative error is denoted by τ, we see that

$$\tau = \frac{x \cdot y - \tilde{x} * \tilde{y}}{x \cdot y}$$

$$= 1 - (1 - \rho)(1 - \sigma)(1 - \eta)$$

$$= \rho + \sigma + \eta - \rho\sigma - \rho\eta - \sigma\eta + \sigma\rho\eta$$

As a result of the assumption that $|\sigma| \ll 1$ and $|\eta| \ll 1$, the third-order term $\sigma\rho\eta$ and the second-order terms like $\sigma\eta$ are negligible and hence

$$|\tau| \leq |\sigma| + |\rho| + |\eta| + |\rho\sigma| + |\rho\eta| + |\sigma\eta| + |\sigma\rho\eta|$$

$$\simeq |\sigma| + |\rho| + |\eta|$$

Thus the overall relative error in floating point multiplication is essentially bounded by the sum of the relative errors in the operands $\tilde{x}$ and $\tilde{y}$ and the relative error in $*$.

In a similar manner it can be shown that the overall relative error in $\tilde{x}/\tilde{y}$ is essentially bounded by $|\sigma| + |\eta| + |\gamma|$, where γ denotes the relative error in floating point division, that is,

$$\frac{|x \div y - \tilde{x}/\tilde{y}|}{|x \div y|} \simeq |\sigma| + |\eta| + |\gamma| \tag{1.8}$$

Next we analyze floating point addition where x and y are of the same sign. We consider the case where x and y are positive; the case where both are negative is handled by writing $x + y = -[(-x) + (-y)]$. In this context we let ρ denote the relative error in addition, that is, $\tilde{x} \oplus \tilde{y} = (1 - \rho)(\tilde{x} + \tilde{y})$. Then

$$\tilde{x} \oplus \tilde{y} = (1 - \rho)(\tilde{x} + \tilde{y}) = (1 - \rho)[(1 - \sigma)x + (1 - \eta)y]$$

$$= (1 - \rho)[x + y - (\sigma x + \eta y)]$$

However for $x, y > 0$, we have

$$\min\{\sigma, \eta\}(x + y) \leq \sigma x + \eta y \leq \max\{\sigma, \eta\}(x + y)$$

suppose $\sigma < n$
the $(\sigma x + \sigma y) \leq (\sigma x + ny) \leq (nx + ny)$

and hence if we define ϕ by $(x + y)\phi = \sigma x + \eta y$, then $|\phi| \leq \max\{|\sigma|, |\eta|\}$. It follows that $\tilde{x} \oplus \tilde{y} = (1 - \rho)(1 - \phi)(x + y)$ and therefore

$$\left| \frac{(x + y) - (\tilde{x} \oplus \tilde{y})}{x + y} \right| = |\rho + \phi + \rho\phi| \leq |\rho| + |\phi| + |\rho\phi|$$

$$\simeq |\rho| + |\phi|$$

since $|\phi| \ll 1$ is implied by the assumption that $|\sigma| \ll 1$ and $|\eta| \ll 1$. We have established that the overall relative error in $\tilde{x} \oplus \tilde{y}$ with x and y of the same sign is approximately the sum of the relative error in $\oplus$ and the largest relative error in the operands.

EXAMPLE 1.16

Suppose we want to find the sum of the convergent series

$$\frac{9}{8} + \frac{3}{4} \sum_{n=1}^{\infty} \frac{1}{n(n + 1)(n + 2)(n + 0.5)}$$

This series converges quite rapidly, and an accurate approximation (correct to five digits) to this sum can be found by computing the value of the finite series

$$\frac{9}{8} + \frac{3}{4} \sum_{n=1}^{40} \frac{1}{n(n+1)(n+2)(n+0.5)}$$

Consider the following program, which computes this finite series in two different orders:

```
C   SUM:BACKWARD SUMMATION
C   SSUM:FORWARD SUMMATION
        SUM=0.
        SSUM=9./8.
        DO 1 I=1,40
          N=41-I
          X=FLOAT(N)
          Y=FLOAT(I)
          T=X*(1.+X)*(2.+X)*(.5+X)
          U=Y*(1.+Y)*(2.+Y)*(.5+Y)
          T=.75/T
          U=.75/U
          SUM=SUM+T
          SSUM=SSUM+U
    1   CONTINUE
        SUM=SUM+9./8.
        WRITE(6,100) SUM
        WRITE(6,101) SSUM
  101   FORMAT(' ','THE FORWARD SUM IS',2X,E14.7)
  100   FORMAT(' ','THE BACKWARD SUM IS',2X,E14.7)
        STOP
        END
```

```
THE BACKWARD SUM IS    0.1227407E 01
THE FORWARD SUM IS     0.1227389E 01
```

In Example 1.16 the computed value of the series depended on the order of summation. How do we explain the difference in the computed sums? The next example should give the reader some insight into the difficulty.

EXAMPLE 1.17

Suppose we use the hypothetical machine of Example 1.14 to add the numbers .0438, .0693, and 13.2. Then

$$(.438 \times 10^{-1} \oplus .693 \times 10^{-1}) \oplus .132 \times 10^2 = .133 \times 10^2$$

whereas

$$(.132 \times 10^2 \oplus .693 \times 10^{-1}) \oplus .438 \times 10^{-1} = .132 \times 10^2$$

The exact sum is 13.3131, and the relative errors in the calculations are 0.00098 and 0.0085, respectively.

When adding two normalized floating point numbers of different orders of magnitude, that is, having different exponents, there is more likelihood of roundoff error due to chopping than when the numbers are of the same order of magnitude. Moreover, this error is propagated if additional sums are performed. Consequently, as a general rule the programmer should (when feasible) arrange the computation so that numbers of the same sign are summed in increasing order of magnitude.

For floating point addition of numbers of opposite sign there is no simple bound for the overall relative error.

EXAMPLE 1.18

Consider a chopping machine with base 10 and precision 4. Suppose $x = .1238$, $\tilde{x} = .1237$, $y = .1232$, and $\tilde{y} = .1236$. Then $x - y = 6 \times 10^{-4}$ and $\tilde{x} \ominus \tilde{y} = .1000 \times 10^{-3}$. A simple calculation shows that the overall relative error is about .83, whereas the relative error in $\tilde{x}$ is 10^{-3} and that in $\tilde{y}$ is -4×10^{-3}. Note that the floating point subtraction was performed exactly; however, the overall relative error is roughly 10^3 times as large as the relative error in the operands. The difficulty can be overcome only by having more accuracy in $\tilde{x}$ and $\tilde{y}$.

[handwritten: ABSOLUTE ERROR]

The loss of accuracy that results when two nearly equal numbers are subtracted is called **subtractive cancellation**. In Example 1.18, $\tilde{x}$ and $\tilde{y}$ are correct to three decimal digits but $\tilde{x} - \tilde{y} = .1000 \times 10^{-3}$ has no correct digits. This loss of correct digits through subtractive cancellation can, in some instances, be avoided. Suppose, in the course of some computation, it is required to evaluate the expression $1 - \cos x$. For $x \simeq 0$, a straightforward evaluation results in subtractive cancellation. One means of avoiding this difficulty is to make use of the identity

$$2 \sin^2 \left(\frac{t}{2}\right) = 1 - \cos t$$

and another approach consists in using the Taylor series expansion for $\cos x$ about $x = 0$:

$$1 - \cos x = 1 - \sum_{n=0}^{\infty} (-1)^n \frac{x^{2n}}{(2n)!}$$

$$= - \sum_{n=1}^{\infty} (-1)^n \frac{x^{2n}}{(2n)!}$$

For $x \simeq 0$, we have $1 - \cos x \simeq x^2/2!$, or $\simeq x^2/2! - x^4/4!$, or $\simeq x^2/2! - x^4/4! + x^6/6!$, and so on. The choice of how many terms in the series to use for the evaluation depends on the required accuracy and the precision of the machine.

In FORTRAN one is permitted to write a sum, difference, product, or quotient without inserting parentheses. The justification for $*$ or $/$ is that these operations are approximately correct, that is, the relative error is good. For example, it can be shown that $x * (y * z) = (1 + \delta)[(x * y) * z]$ with $|\delta| \leq 2u/(1 - u)$. For floating point addition (subtraction) the justification for this liberty is not clear. The sum $(x \oplus y) \oplus z$ may be quite different from $x \oplus (y \oplus z)$.

EXAMPLE 1.19

On the machine of Example 1.14 we compute the sums $x \oplus (y \oplus z)$ and $(x \oplus y) \oplus z$, where $x = -y = .936 \times 10^5$ and $z = -.985 \times 10^2$. We have $x \oplus y = 0$ and $y \oplus z = y$, so that

$$x \oplus (y \oplus z) = 0 \qquad \text{and} \qquad (x \oplus y) \oplus z = z$$

We used (1.7) as the basis of an analysis of roundoff error propagation in the basic arithmetic operations. The relative error in operation ω is given by $\rho = (x\omega y - x\bar{\omega}y)/x\omega y$ or, equivalently, by

$$x\bar{\omega}y = (1 - \rho)x\omega y \tag{1.9}$$

where x and y are two normalized floating point numbers and $|\rho| \leq u$. We can interpret (1.9) to mean that the floating point computation is the result of the corresponding exact computation wherein the operand(s) has been perturbed by an amount that does not exceed u. This point of view is the foundation of a subject called **backward error analysis**. In this analysis one applies, in a given algorithm, this interpretation backward step by step. Using this procedure, it has been shown for many algorithms (some of which are very complicated) that the output of the algorithm in the presence of roundoff error is the exact result of a problem of the same type in which the input data has been perturbed by some multiple of u. A simple example should suffice to illustrate the main idea.

EXAMPLE 1.20

Consider the computation of $x^2 - y^2$, where x and y are normalized floating point numbers. We use the algorithm

$$z \leftarrow x \oplus y$$

$$w \leftarrow x \ominus y$$

$$v \leftarrow z * w$$

We have from (1.9) that $z = (1 - \sigma)(x + y)$ and $w = (1 - \eta)(x - y)$ for some σ, η that satisfy $|\sigma|, |\eta| \leq u$. Thus for some τ such that $|\tau| \leq u$, we find

$$v = (1 - \tau)z \cdot w = (1 - \tau)(1 - \sigma)(1 - \eta)(x^2 - y^2)$$

$$\equiv (1 - \gamma)(x^2 - y^2)$$

where $\gamma = -\tau - \sigma - \eta + \tau\sigma + \tau\eta + \sigma\eta - \tau\sigma\eta$ satisfies $|\gamma| \ll 1$. Therefore if we define $\tilde{x} = (1 - \gamma)^{1/2}x$ and $\tilde{y} = (1 - \gamma)^{1/2}y$, then v can be interpreted as the output of the same algorithm applied to $\tilde{x}$ and $\tilde{y}$ with the floating point operations replaced by their exact arithmetic counterparts.

The point of backward error analysis is that the problem of roundoff error estimation in an algorithm is reduced to estimating the result of perturbations of the input data to the problem itself.

EXERCISES

1.22 Using chopping, write the following as normalized decimal (base 10) floating point numbers with precision 5:

$$1\tfrac{2}{3} =$$

$$\tfrac{56}{10,000} =$$

$$-\tfrac{3}{100} =$$

$$10^{-5} =$$

1.23 Suppose the real number x is stored on the hypothetical machine of Example 1.14. If this results in $\mathrm{fl}(x) = .371 \times 10^{-4}$, how large can $|x - \mathrm{fl}(x)|$ be? If $\mathrm{fl}(y) = .371 \times 10^4$, how large can $|y - \mathrm{fl}(y)|$ be? How large can the relative errors be?

1.24 Suppose x and y are positive real numbers. Assuming no overflow or underflow occurs, determine a bound for the overall relative error in computing $x^2 + xy$ on the machine of Example 1.14. Caution: Be sure to include the error that results when x and y are simply stored in memory.

1.25 On a base β, precision t machine how many correct digits does $\tilde{\mathrm{fl}}(x)$ have? Does your answer depend on whether the machine uses chopping or rounding? Explain.

1.26 Suppose x and y are normalized floating point numbers in base β with precision t. Assuming no underflow or overflow, how many base β digits are correct in

(a) $x * y$

(b) x / y

(c) $x \oplus y$ with $x, y > 0$

1.27 Let $\tilde{x}$ be a normalized floating point number in base β with precision t. Suppose $\tilde{x}$, considered as an approximation to $x \neq 0$, has $r\ (\leq t)$ correct digits. Find a bound for $|x - \tilde{x}|/|x|$ in terms of the machine unit.

*1.28 Let x be a normalized hexadecimal floating point number of precision 6. Let $\tilde{x}$ be the normalized decimal (base 10) floating point representation of x with chopping and precision 7. How many correct decimal digits does $\tilde{x}$ have?

1.29 For addition and subtraction, show that the bound for the absolute error in the result is given by the sum of the bounds for the absolute errors in the operands.

1.30 Write a program to implement the floating point sums

$$w \leftarrow (x \oplus y) \oplus z$$

$$v \leftarrow x \oplus (y \oplus z)$$

Find x, y, z so that $|w - v| > 10^{60}$. How large can you make $|w - v|/|v|$?

1.31 Consider the floating point algorithm

$$w \leftarrow x^2 \ominus y^2$$

$$z \leftarrow (x \ominus y) * (x \oplus y)$$

implemented on a hypothetical machine that uses chopped floating point arithmetic with base 10 and precision 4. Find w and z for $x = 9 - 10^{-3}$ and $y = 9 + 10^{-3}$. Make sure you chop the result of each operation.

1.32 If u is the machine unit, what is the value of fl$(1 + u)$? What is the smallest positive floating point number ε such that fl$(1 + \varepsilon) > 1$? Determine the approximate value of ε on your computer by using the following algorithm:

$$\varepsilon \leftarrow 1$$

Do $n = 1, 2, \ldots,$ NMAX;

$\quad\quad \varepsilon \leftarrow \varepsilon/2$

$\quad\quad \delta \leftarrow \varepsilon + 1$

$\quad\quad$ If $\delta \leq 1$ then terminate loop

1.33 Convert to decimal (base 10) representations: (a) 101_8, (b) 32_{16}, (c) 1101101_2, (d) $F4A_{16}$.

1.34 Convert 431_{10} to the equivalent binary, octal, and hexadecimal numbers.

1.35 Subtractive cancellation results when evaluating $f(x) = e^x - 1 - x$ for $x \simeq 0$. One means of avoiding this difficulty is to rearrange the formula for f as $f(x) = [e^{2x} - (1 + x)^2]/(e^x + 1 + x)$. Another method is to use the series expansion for e^x to get $f(x) = \sum_{n=2}^{\infty} x^n/n!$. Then for $x \simeq 0$, we have $f(x) \simeq x^2/2 + x^3/6$. Use four decimal digit, chopped floating point arithmetic to evaluate $f(0.01)$ by each of the two alternatives. How do these computed values compare with the use of the original formula for f?

1.36 Consider the following program:

```
      N=63
      A=101.
      X=A
      B=104.
      H=(B-A)/FLOAT(N)
      WRITE(6,10)
10    FORMAT(' ',7X,'METHOD1',7X,'METHOD2',8X,'POINT'//)
      DO 1 J=1,N
      X=X+H
      XX=FLOAT(J)*H+A
      Y=EXP(X)
      YY=EXP(XX)
      IF(J/3*3.NE.J) GO TO 1
      WRITE(6,11) Y,YY,XX
11    FORMAT(' ',3X,E14.7,3X,E14.7,2X,F7.3)
1     CONTINUE
      STOP
      END
```

METHOD1	METHOD2	POINT
0.8428873E 44	0.8429130E 44	101.143
0.9722914E 44	0.9723507E 44	101.286
0.1121562E 45	0.1121665E 45	101.429
0.1293750E 45	0.1293928E 45	101.571
0.1492373E 45	0.1492623E 45	101.714
0.1721489E 45	0.1721830E 45	101.857
0.1985780E 45	0.1986234E 45	102.000
0.2290646E 45	0.2291276E 45	102.143
0.2642317E 45	0.2643124E 45	102.286
0.3047978E 45	0.3049002E 45	102.429
0.3515918E 45	0.3517260E 45	102.571
0.4055696E 45	0.4057370E 45	102.714
0.4678348E 45	0.4680417E 45	102.857
0.5396589E 45	0.5399145E 45	103.000
0.6225102E 45	0.6228331E 45	103.143
0.7180809E 45	0.7184753E 45	103.286
0.8283242E 45	0.8288043E 45	103.429
0.9554921E 45	0.9560903E 45	103.571
0.1102184E 46	0.1102908E 46	103.714
0.1271397E 46	0.1272270E 46	103.857
0.1466587E 46	0.1467640E 46	104.000

From the output, determine which method is better for generating a table of values. Explain why one method is to be preferred over the other. Hint: Consider how round-off error is propagated. (In order to conserve space, we have printed only every third value.)

1.37 How would you use the quadratic formula to find the solutions of $x^2 - 55x + 1 = 0$ so as to avoid subtractive cancellation?

1.38 In Example 1.16 modify the program as follows. Replace the statement SSUM = 9./8. by the statement SSUM = 0. and insert the statement SSUM = SSUM + 9./8. following the CONTINUE statement. Run the modified program and compare the resultant forward sum with the output given in Example 1.16. Note: The backward sum is more accurate.

***1.39** Perform a backwards error analysis on the floating point computation $z \leftarrow x * y \oplus w$, where x, y, w are positive normalized floating point numbers. Hint: Express z as $\tilde{x} \cdot \tilde{y} + \tilde{w}$, where $\tilde{x}$ is a perturbed value for $\tilde{x}$, and so on.

***1.40** Demonstrate the validity of (1.8). Hint: Use the geometric series and neglect higher order terms.

1.4 CONDITIONING AND WELL-POSED PROBLEMS

In this section we consider some ill-conceived computations wherein a bad algorithm can produce a poor answer to a well-posed problem. We say that a problem is **well-posed** if (1) for each admissible set of input data the problem has a solution (existence), (2) for each admissible set of input data the problem has at most one solution (uniqueness), and (3) a sufficiently small perturbation of the input data results in a small change in the solution (continuous dependence on data).

EXAMPLE 1.21

Consider the problem of calculating the definite integral of a function on the interval $[a, b]$. We take the set of admissible input data to consist of those functions that are continuous on the interval $[a, b]$. By the fundamental theorem of calculus the problem satisfies the existence and uniqueness requirements. To see that the problem's solution depends continuously on the data, we proceed as follows. Suppose f is a given continuous function and $g = f + f_\delta$, where f_δ is continuous and $|f_\delta(x)| \le \delta$ for all x in the interval. Then by properties of the integral

$$\left| \int_a^b g(x)\, dx - \int_a^b f(x)\, dx \right| = \left| \int_a^b f_\delta(x)\, dx \right|$$

$$\le \int_a^b |f_\delta(x)|\, dx \le \delta(b - a)$$

Thus a small perturbation in the data f_δ produces in the solution a change that is no larger than $\delta(b - a)$.

The statement that a problem is well-posed is actually a relative one. Suppose we consider two problems of the same type, say the evaluation of

$$I_1(f) = \int_0^1 f(x)\, dx \quad \text{and} \quad I_2(g) = \int_0^{100} g(x)\, dx$$

Then if $F(x) = f(x) + f_\delta(x)$ and $G(x) = g(x) + g_\delta(x)$ with $|f_\delta(x)|, |g_\delta(x)| \le \delta$ for all x in $[0, 1]$ and $[0, 100]$, respectively, we have, by Example 1.21, for $\delta = 10^{-6}$,

$$|I_1(F) - I_1(f)| \le 10^{-6} \quad \text{and} \quad |I_2(g) - I_2(G)| \le 10^{-4}$$

Both problems are well-posed, but problem $I_1(f)$ is "more" well-posed than problem $I_2(g)$.

EXAMPLE 1.22

Consider the problem of evaluating $f(x) = \tan x$ for x near $\pi/2$, say $1.5 \le x < \pi/2$. The problem is uniquely solvable, but what about continuous dependence on data? Consider $x_0 = 1.5$ and $x_1 = 1.5001$; then to seven digits we find $\tan x_0 = 14.10150$ and $\tan x_1 = 14.12143$; that is, a perturbation of 10^{-4} results in a change of about 10^{-2} in the solution. For $x_2 = 1.500001$, we get $\tan x_2 = 14.10162$; that is, a perturbation of 10^{-6} results in a solution change of about 10^{-4}. This problem is well-posed, but nonetheless the solution exhibits a sensitivity to perturbation in the data.

In the next example we exhibit a problem that is also sensitive to perturbations in the data, but the nature of the sensitivity is completely different from that of Example 1.22.

EXAMPLE 1.23

We show that the problem of differentiation of a function at a point is not well-posed. As the set of admissible data we consider those functions that are differentiable at a point, say x_0. Let g be such a function and define $G(x) = g(x) + \delta \sin(x/\delta^2)$. Then $|G(x) - g(x)| \leq \delta$, but $G'(x) = g'(x) + \delta^{-1} \cos(x/\delta^2)$; thus $|G'(x) - g'(x)| = \delta^{-1} |\cos(x/\delta^2)|$. It should be clear that the derivatives may differ by a large amount as a result of a small perturbation in the data; in fact, for this perturbation the difference in derivatives grows as δ decreases in size.

For well-posed problems we anticipate that a reasonable algorithm should produce good answers, whereas for ill-posed problems we suspect that any algorithm may result in poor-quality answers. From the viewpoint of numerical computation the lack of continuous dependence on data for a problem typically results in severe difficulties in its numerical solution because of the inevitable presence of roundoff errors.

Let us examine the standard approach suggested in most calculus texts for the evaluation of the exponential function. From Taylor's theorem we recall that for all x

$$e^x = \sum_{n=0}^{N} \frac{x^n}{n!} + R_N(x)$$

where $R_N(x) = e^{\xi} x^{N+1}/(N+1)!$ for some ξ between 0 and x. Since $\lim_{N \to \infty} R_N(x) = 0$, the value of $S_N(x) = \sum_{n=0}^{N} x^n/n!$ should be a good approximation to e^x for a suitable choice of N. Consider the following simple WATFIV FORTRAN code for the computation, with $x = 16$.

EXAMPLE 1.24

```
C         AN EXAMPLE OF POWER SERIES COMPUTATION
C         OF THE EXPONENTIAL FUNCTION..SUMMATION
C         PROCEEDS UNTIL THE NEXT TERM ADDED TO
C         THE SUM DOES NOT CHANGE THE VALUE
          READ,X
          WRITE(6,500)
   500    FORMAT(' ',6X,'SUM',9X,'NTH TERM'//)
          S=1.
          Y=1.
          DO 1 K=1,100
            N=K
            Y=X*Y/FLOAT(K)
            SUM=S
            S=S+Y
            IF(SUM.EQ.S) GO TO 2
```

```
   1 CONTINUE
   2 WRITE(6,100) S,N
 100 FORMAT(' ', E14.7,5X,I4)
     STOP
     END
```

```
    SUM              NTH TERM

0.8886089E 07          41
```

Let us examine how the computations proceed in this example. At each stage the terms added to the sum S are of the form $t_n = x^n/n!$. For $x = 16$, the t_n's grow quite rapidly until $n = 17$; thereafter they decrease in size. Working in base 16, precision 6, we retain only the leading seven decimal digits of each t_n (Y in the program). Thus instead of calculating t_n, we have an approximation, say $\tilde{t}_n = 16^n/n! + \varepsilon_n$, and the final value of S is $S = S_N(16) + \sum_{n=0}^{N} \varepsilon_n$, where N is determined by the logical IF statement in the program. Here we have ignored the effect of roundoff due to floating point addition. Of course the ε_n's are small relative to t_n, and so $\sum_{n=0}^{N} \varepsilon_n$ is small relative to $S_N(x)$. In fact, it can be shown that $S = (1 + \sigma)S_N(16)$, where $|\sigma| \le 2(N - 1) \times 16^{-5}$, and we expect the program to produce a good approximation to e^{16}. The value printed for $X = 16$. is 0.8886089E 07, and the correct (to seven digits) value is 0.8886110E 07.

However, if we use the same program for $X = -16.$, we find the value printed is $S = -0.3549890E - 02$, whereas the correct value is 0.1125352E - 06. The computed value is not even of the correct sign! In order to explain the difficulty, we note that $t_8 = 16^8/8! = 1456.355556\ldots$ and only the first seven digits (actually six hexadecimal digits) are retained, so that $\varepsilon_8 \gg e^{-16}$. Moreover, the roundoff error propagates throughout the summation. Even though the t_n's alternate in sign, the sum of the ε_n's is large, in magnitude, relative to the final value of S. Using this approach, the only remedy is to use more precision. The problem of evaluation of e^{-x} for $x = 16$ is well-posed; the program given here simply reflects a bad choice of algorithm.

It has no doubt occurred to the reader that one could compute a reasonable approximation to e^{-16} by inversion of $S_N(16)$. This is certainly preferable to the approach given above. For a hexadecimal machine a better method is as follows. Let $y = e^{w \log 16} = 16^w$, where $y = e^x$ is the value to be computed. This determines w; that is, $w = x/\log 16$. Now write w as $w = I - F$, where I is an integer and $0 < F \le 1$. Then $y = 16^{-F}16^I$ so that I is the exponent of the result and 16^{-F} is the mantissa. Therefore one need only compute $16^{-F} = e^{-F \log 16}$ for $0 < F \le 1$. This type of approach is used in the FORTRAN built-in function subprogram EXP.

As another example of the difficulties that arise from an ill-conceived algorithm, we consider the evaluation of the hyperbolic sine function. We remark that in early versions of FORTRAN this function was not in the library of built-in functions and so the user had to compute it. Recall that

$$\sinh x = \tfrac{1}{2}(e^x - e^{-x}) \tag{1.10}$$

If $x \simeq 0$, we have $e^x \simeq e^{-x}$ and subtractive cancellation results in a large relative error. More specifically, suppose we have a means of evaluating the exponential function that gives seven correct decimal digits for any x within the range of the floating point number system. For $x = .1234567 \times 10^{-5}$, we have $e^x \simeq 1.000001$ and $e^{-x} \simeq 0.9999988$ and formula (1.10) yields sinh x $\simeq 0.0000011$. However, for such a small value of x the approximation sinh x $\simeq x$ is correct to almost 12 decimal digits. Thus the value given by (1.10) is correct only to one digit (first nonzero digit). A simple calculation shows that the relative error is roughly 0.1.

In general we say that the evaluation of $f(x)$ is well-conditioned if a small relative error in x results in a small relative error in $f(x)$. This is not a precise mathematical definition as we have not specified the meaning of "small." If $f(x) \neq 0$, then we may express the notion of well-conditioning by the statement

$$\frac{|x - \tilde{x}|}{|x|} \text{ is small implies } \frac{|f(x) - f(\tilde{x})|}{|f(x)|} \text{ is small}$$

Let us assume that f is continuously differentiable on some interval containing x and $\tilde{x}$. Then, by the mean value theorem, we have

$$f(x) - f(\tilde{x}) = f'(\xi)(x - \tilde{x})$$

for some point ξ between x and $\tilde{x}$. If $|x - \tilde{x}|$ is sufficiently small, then $f'(\xi) \simeq f'(x)$ by continuity of f'. It follows that $f(x) - f(\tilde{x}) \simeq f'(x)(x - \tilde{x})$ and hence

$$\frac{f(x) - f(\tilde{x})}{f(x)} \simeq x \left[\frac{f'(x)}{f(x)} \right] \left[\frac{x - \tilde{x}}{x} \right] \tag{1.11}$$

We may interpret (1.11) as follows. The quantity $m = xf'(x)/f(x)$ gives a measure of the amount by which the relative error in $\tilde{x}$, $\sigma = (x - \tilde{x})/x$, is magnified to yield the relative error in $f(\tilde{x})$.

Let us apply these ideas to the function $f(x) = $ sinh x. The magnification factor is x cosh $x/$sinh x. For $x \simeq 0$, we have cosh $x \simeq x/$sinh $x \simeq 1$ so that the relative error in $f(\tilde{x})$ is nearly the same as that in $\tilde{x}$. Therefore the problem of evaluation of sinh x for $x \simeq 0$ is well-conditioned. How do we explain the large relative error that results for $x = .1234567 \times 10^{-5}$? In view of this analysis we cannot blame the problem; rather, it is the method of evaluation (algorithm) that is at fault. As was previously observed, the difficulty in (1.10) is subtractive cancellation for $x \simeq 0$. In order to alleviate this difficulty, we use the power series expansion for the exponential function:

$$\text{sinh } x = \frac{1}{2} \left[\sum_{n=0}^{\infty} \frac{x^n}{n!} - \sum_{n=0}^{\infty} (-1)^n \frac{x^n}{n!} \right]$$

$$= x + \frac{x^3}{3!} + \frac{x^5}{5!} + \cdots = \sum_{n=0}^{\infty} \frac{x^{2n+1}}{(2n+1)!}$$

which is valid for all x. For $x \simeq 0$, one of the approximations sinh $x \simeq x$, sinh $x \simeq x + x^3/3!$, or sinh $x \simeq x + x^3/3! + x^5/5!$ should be sufficiently accurate and does not result in any significant subtractive cancellation.

Previously we gave a definition of a well-conditioned function evaluation. This is simply a special case of the idea of a **well-conditioned problem**. We say that a problem is well-conditioned if a small relative error in the input data results in a small relative error in the solution. In certain problems we can quantify this conditioning concept by defining the **condition** (or condition number) of the problem. Loosely speaking, the condition is the **magnification factor** by which the relative error in the input data is multiplied to give the relative error in the solution. For instance, the condition of the problem of evaluating a continuously differentiable function f at x is $m = xf'(x)/f(x)$.

In summary, a problem is well-conditioned if its solution is insensitive to small perturbations in the input data. If the solution is sensitive to small perturbations of the input data, we call the problem **ill-conditioned**.

We may also use these terms to describe an algorithm. What does it mean to say that an algorithm is well-conditioned? In the environment of floating point computation, with finite precision, roundoff errors are inevitable during the execution of an algorithm. A central concern of the user of an algorithm is the growth of roundoff error. If the roundoff error grows at an unwieldy rate, then the algorithm is ill-conditioned; the algorithm is well-conditioned if the roundoff error propagation is relatively small. An algorithm can be well-conditioned for some input data and ill-conditioned for others. In Example 1.24 the algorithm is well-conditioned for $x = 16$ but ill-conditioned for $x = -16$. The algorithm given by (1.10) is ill-conditioned for $x \simeq 0$ but well-conditioned for $x > 1$.

We have seen two examples of the situation whereby an ill-conditioned algorithm results in poor accuracy: the evaluation of e^{-x} by truncation of the power series for the exponential function and the evaluation of $\sinh x$ for small x by use of (1.10). However, poor accuracy can also result from the nature of the problem itself; that is, it can happen that for any algorithm the output is very sensitive to perturbations in the input data. Consider the problem of evaluation of a continuously differentiable function f at some point x. Suppose A is an algorithm for this problem, in which backwards error analysis is applicable. We may think of A as a black box:

$$x \longrightarrow \boxed{\quad A \quad} \longrightarrow y$$

Input   Output

As a result of roundoff error the output is of the form $y = (1 - \rho)f(x)$. By backwards error analysis the application of A may also be viewed as

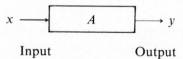

$$\tilde{x} = (1 - \sigma)x \longrightarrow \boxed{\quad f \quad} \longrightarrow y \qquad - \text{BACKWARDS}$$

for some perturbation σx of the input data. Thus $y = f(\tilde{x}) = (1 - \rho)f(x)$ and y is the result of evaluating f at a perturbed point $\tilde{x}$. Moreover, by (1.11) we may relate ρ and σ:

$$\rho = \frac{f(x) - f((1 - \sigma)x)}{f(x)} \simeq \frac{\sigma x f'(x)}{f(x)}$$

The quantity σ is a measure of the conditioning of the algorithm A, and, as was observed before, $m = xf'(x)/f(x)$ is a measure of the conditioning of the problem. It is important to realize that these two measures are independent of one another; σ can be large even if the magnification factor m is small and vice versa.

EXAMPLE 1.25

Consider the algorithm for the evaluation of $\sinh x$ given by

$$z \leftarrow \frac{1}{2}\left[E(x) - \frac{1}{E(x)} \right]$$

where $E(x)$ denotes an approximation to e^x that satisfies $E(x) = (1 - \rho)e^x$ with $|\rho| \le u$. The approximation $E(x)$ could be the output of the FORTRAN built-in function EXP. Let $\tilde{x} = (1 - \sigma)x$, where $\sigma = -(1/x)\log(1 - \rho)$; then for $f(x) = \sinh x$, we have $z = f(\tilde{x})$ and σ gives a measure of the conditioning of the algorithm. If $x = 0.1234 \times 10^{-8}$ and $u = 10^{-5}$, we have $|\sigma| \simeq (1/x)|\log(1 - u)| \simeq 8104$, whereas $m \simeq 1$. In this case the problem is well-conditioned but the algorithm is ill-conditioned.

EXAMPLE 1.26

Here we given an illustration of a well-conditioned algorithm for an ill-conditioned problem. The problem is to evaluate $f(x) = \tan x$ for $x \simeq 1.55$. As we saw in Example 1.22, this problem is sensitive to perturbations in x. The algorithm for the evaluation is

$$z \leftarrow \frac{S(x)}{C(x)}$$

where $S(x)$, $C(x)$ denote approximate values of $\sin x$ and $\cos x$, respectively, with $S(x) = (1 - \rho)\sin x$, $C(x) = (1 - \gamma)\cos x$, $|\rho| \ll 1$, and $|\gamma| \ll 1$. Let $\tilde{x} = (1 - \sigma)x$, where

$$\sigma = 1 - \frac{1}{x}\arctan\left(\frac{1 - \rho}{1 - \gamma}\tan x \right)$$

Then $z = f(\tilde{x})$ and, since $(1 - \rho)/(1 - \gamma) \simeq 1$, we have $|\sigma| \ll 1$, whereas $m = x\sec x/\sin x \simeq x/\cos x$ and for $x = 1.55$ we have $m = 74.5$.

In solving a problem by a numerical method, it is certainly advantageous if the problem and the algorithm for the numerical method are well-conditioned. However, this is not the only concern of the problem solver. Another primary consideration is efficiency. Can the algorithm produce an approximation with good relative error in a reasonable amount of computer time? For the novice problem solver there is a temptation to view the computer as an all-powerful device that can handle computations without regard to their length or complexity.

We give an example to illustrate the futility of using the power of a modern digital computer without a human analysis of the efficiency of the proposed method of solution.

EXAMPLE 1.27

The problem is to evaluate the sum of the infinite series $\sum_{n=1}^{\infty} 1/n(n + 0.5)$ with an error of at most 10^{-6}. The brute-force approach for this problem consists in writing (and using) a program to compute $\sum_{n=1}^{N} 1/n(n + 0.5)$ for some large value of N. Let us examine how large N should be by considering the given series as a special case of $f(x) = \sum_{n=1}^{\infty} 1/n(n + x)$. Certainly the given series $f(0.5)$ converges no faster than does the series $f(1)$. The series $f(1)$ is a telescopic series, and hence it is not difficult to estimate its truncation error. We have

$$f(1) - \sum_{n=1}^{N} \frac{1}{n(n + 1)} = \sum_{n=N+1}^{\infty} \frac{1}{n(n + 1)}$$

$$= \sum_{n=N+1}^{\infty} \left(\frac{1}{n} - \frac{1}{n + 1}\right)$$

$$= \lim_{m \to \infty} \sum_{n=N+1}^{m} \left(\frac{1}{n} - \frac{1}{n + 1}\right)$$

$$= \lim_{m \to \infty} \left(\frac{1}{N + 1} - \frac{1}{m + 1}\right)$$

$$= \frac{1}{N + 1} \tag{1.12}$$

Thus we need 1 million terms in the sum in order to evaluate $f(1)$ with a truncation error of at most 10^{-6}. Let us use this to estimate the truncation error in the given series $f(0.5)$. We have

$$f(0.5) - \sum_{n=1}^{N} \frac{1}{n(n + 0.5)} = \sum_{n=N+1}^{\infty} \frac{1}{n(n + 0.5)}$$

$$\leq 2 \sum_{n=N+1}^{\infty} \frac{1}{n(n + 1)}$$

$$\leq \frac{2}{N + 1} \tag{1.13}$$

where we have used the inequality $1/n(n + 0.5) \leq 2/n(n + 1)$ and (1.12). From (1.13) we estimate that 2 million terms in the sum are required to evaluate $f(0.5)$ with a truncation error of at most 10^{-6}. Moreover, this presumes that no roundoff error is present. Let us reformulate the problem in terms of another series, one that converges much faster than the one defining $f(0.5)$. From (1.12) we have

$$f(1) = \lim_{N \to \infty} \frac{1}{(N + 1)} + \lim_{N \to \infty} \sum_{n=1}^{N} \frac{1}{n(n + 1)}$$

$$= \lim_{N \to \infty} \sum_{n=1}^{N} \left(\frac{1}{n} - \frac{1}{n + 1}\right) = \lim_{N \to \infty} \left(1 - \frac{1}{N + 1}\right) = 1$$

Therefore

$$f(x) - f(1) = \sum_{n=1}^{\infty} \frac{1}{n(n+x)} - \sum_{n=1}^{\infty} \frac{1}{n(n+1)}$$

$$= \sum_{n=1}^{\infty} \frac{1}{n}\left(\frac{1}{n+x} - \frac{1}{n+1}\right) = \sum_{n=1}^{\infty} \frac{1-x}{n(n+1)(n+x)}$$

so that

$$f(x) = 1 + (1-x)\sum_{n=1}^{\infty} \frac{1}{n(n+1)(n+x)} \tag{1.14}$$

This formulation is an improvement as the series in (1.14) converges more rapidly than does the original series for $f(x)$. Let us apply this technique again to find a series for $f(x)$ that converges faster still. We evaluate $f(2)$ from (1.14) and use the equation $1/[n(n+1)(n+2)] = \frac{1}{2}[(1/n) - 2/(n+1) + 1/(n+2)]$ to get

$$f(2) = 1 - \sum_{n=1}^{\infty} \frac{1}{n(n+1)(n+2)}$$

$$= 1 - \frac{1}{2} \lim_{N \to \infty} \sum_{n=1}^{N} \left(\frac{1}{n} - \frac{2}{n+1} + \frac{1}{n+2}\right)$$

$$= 1 - \frac{1}{2} \lim_{N \to \infty} \left(\frac{1}{2} + \frac{1}{N+2} - \frac{1}{N+1}\right) = 1 - \frac{1}{4} = \frac{3}{4}$$

Then we find

$$f(x) + (1-x)f(2) = (2-x) + (1-x)\sum_{n=1}^{\infty} \frac{1}{n(n+1)}\left(\frac{1}{n+x} - \frac{1}{n+2}\right)$$

$$= (2-x) + (1-x)(2-x)\sum_{n=1}^{\infty} \frac{1}{n(n+1)(n+2)(n+x)}$$

and finally

$$f(x) = \frac{1}{4}(5-x) + (1-x)(2-x)\sum_{n=1}^{\infty} \frac{1}{n(n+1)(n+2)(n+x)}$$

For $x = 0.5$ we have

$$f(0.5) = \frac{9}{8} + \frac{3}{4} \sum_{n=1}^{\infty} \frac{1}{n(n+1)(n+2)(n+0.5)} \tag{1.15}$$

which converges much more rapidly than does the original series for $f(0.5)$. In fact, the truncation error after N terms for the series in (1.15) is roughly $1/[3(N+1)^3]$, so that approximately 70 terms should be sufficient for the desired accuracy. See Example 1.16, where 40 terms were used in the sum. The exact value of the sum is $4(1 - \log 2) = 1.2274112\ldots$. For most numerical problems human analysis and computer power are required, and neither is likely to succeed without the other.

As another example of the need for human evaluation of the efficiency of a proposed numerical method, we consider the following. Suppose it is required to design a function subprogram for the natural logarithm. We recall the series expansion (exercise 1.8)

$$\log|1 - x| = -\sum_{n=0}^{\infty} \frac{x^{n+1}}{n+1}$$

which is valid for $-1 \le x < 1$. For example, with $x = -1$, we could estimate log 2 by using the finite series

$$\log 2 \simeq \sum_{n=0}^{N} \frac{(-1)^n}{n+1}$$

Since this is an alternating series, the truncation error after $N + 1$ terms is bounded by $1/(N + 2)$. Thus we need approximately 10,000 terms to find the value of log 2 correct to four decimal digits. Clearly this is not an efficient method. In exercise 1.10 we established another series for the logarithm, namely,

$$\log \frac{1 + x}{1 - x} = 2 \sum_{n=0}^{\infty} \frac{x^{2n+1}}{2n+1} \qquad |x| < 1 \tag{1.16}$$

For $x = \frac{1}{3}$, we get the series

$$\log 2 = 2 \sum_{n=0}^{\infty} \frac{1}{3^{2n+1}(2n+1)} \tag{1.17}$$

Although this is not an alternating series, we get a "ballpark" estimate of the truncation error after $N + 1$ terms by the first truncated term:

$$\log 2 - 2 \sum_{n=0}^{N} \frac{1}{3^{2n+1}(2n+1)} \simeq \frac{2}{3^{2N+3}(2N+3)}$$

For $N = 4$, we have $2/[(2N + 3)3^{2N+3}] \simeq 10^{-6}$, and we estimate that five terms of the series will give log 2 correct to five decimal digits (see exercise 1.53).

EXERCISES

1.41 Evaluate $\int_0^1 1/(1 + x^2)\, dx$ and $\int_0^1 [1/(1 + x^2) + e^x/10^6]\, dx$. By how much do the integrals differ in value? Find a bound for the difference of the integrands.

1.42 Interpret the output of the following program:

```
H=.5
U=ALOG(2.)
DO 1 J=1,10
  H=H/4.
  V=ALOG(2.+H)
  DER=(V-U)/H
```

```
        ERR=ABS(DER-.5)
        WRITE(6,10) H,DER,ERR
10      FORMAT(' ',3(2X,E14.7))
1    CONTINUE
     STOP
     END
```

```
0.1250000E 00     0.4849973E 00     0.1500273E-01
0.3125000E-01     0.4961357E 00     0.3864288E-02
0.7812500E-02     0.4990311E 00     0.9689331E-03
0.1953125E-02     0.4997559E 00     0.2441406E-03
0.4882813E-03     0.5000000E 00     0.0000000E 00
0.1220703E-03     0.5000000E 00     0.0000000E 00
0.3051758E-04     0.5000000E 00     0.0000000E 00
0.7629395E-05     0.5000000E 00     0.0000000E 00
0.1907349E-05     0.5000000E 00     0.0000000E 00
0.4768372E-06     0.0000000E 00     0.5000000E 00
```

Hint: Recall the definition of the derivative of $\log x$ at $x = 2$. Also keep in mind how roundoff error influences DER.

1.43 Determine whether the problem of evaluating f at x is well-conditioned for each of the following:

(a) $f(x) = x - \sin x$ $x \simeq 0$
(b) $f(x) = x/\log x$ $x \simeq 1, x \neq 1$
(c) $f(x) = 1 - \cos x$ $x \simeq 2\pi$
(d) $f(x) = \tan x$ $x \simeq 3\pi/2, x \neq 3\pi/2$
(e) $f(x) = \sin x/x$ $x \simeq 0, x \neq 0$

*1.44 Can a problem be well-posed but not well-conditioned? Give an example to support your answer.

1.45 Determine well-conditioned algorithms for the evaluations in parts (a) and (e) of exercise 1.43. Test your algorithms by writing a FORTRAN program to perform the evaluations for $x = 0.1, 0.01, 0.0001, 0.000001$.

1.46 Modify the program of Example 1.16 to sum the first 70 terms of the series. Compare the computed backward sum with the exact value of $4(1 - \log 2)$.

1.47 Consider the function $f(x) = 1 - (1 - x^2)^{1/2}$ for $x \simeq 0$.

(a) Is $y \leftarrow 1 - (1 - x^2)^{1/2}$ a good algorithm for evaluation, assuming that the square root is available with a good relative error? Explain.
(b) Rearrange the formula for f in order to avoid subtractive cancellation.
(c) Use Taylor's theorem to find a polynomial (of degree 2) approximation for f that is accurate for $x \simeq 0$.
(d) Is the problem of evaluating f well-conditioned for x near 0? Justify your answer.
(e) Is the problem of evaluating f well-posed?
(f) For $x = 0.1$, use chopped floating point arithmetic with precision 3 to compute the value of f by (1) the algorithm of part (a), (2) the algorithm that results from part (b), and (3) the result of part (c). Which gives the most accurate approximation? Compare the relative errors in these three methods. If precision 5 arithmetic is used, how do the relative errors compare? Which algorithm is best conditioned for x near 0?

1.48 For a convergent alternating series, say $\sum_{n=0}^{\infty} (-1)^n a_n = A$ with $a_{n-1} > a_n > 0$ for all n, it is known that $|A - \sum_{n=0}^{N} (-1)^n a_n| < a_{N+1}$. How many terms are required to estimate log 1.1 from the series log $1.1 = \sum_{n=0}^{\infty} 10^{-n}(-1)^n/(n + 1)$ with a relative error of at most 10^{-6}? Do you think that this is an efficient means of evaluating log 1.1?

1.49 It is known that $\pi^2/6 = \sum_{n=1}^{\infty} 1/n^2$. One could find the approximate sum of the series by computing $S_N = \sum_{n=1}^{N} 1/n^2$ for some value of N. Write and use a FORTRAN program to compute S_N for $N = 10,000$ by computing the sum in (a) decreasing order, that is, $[(1 + \frac{1}{4}) + \frac{1}{9}] + \frac{1}{16}...$, and (b) increasing order. Explain why (b) produces a more accurate approximation. Which algorithm is better conditioned? Note that a more efficient means of summing the series is furnished by the rapidly convergent series of Example 1.27.

1.50 Subroutine NATLOG uses (1.16) to compute the natural logarithm. Try several points Y and compare ALOG(Y) with the values produced by NATLOG.

```
      SUBROUTINE NATLOG(Y,XLOG,IFLAG)
C   Y:POINT AT WHICH LOGARITHM IS EVALUATED
C   XLOG: VALUE OF LOG FUNCTION AT Y
C   IFLAG:SIGNALS MODE OF RETURN; 1 NORMAL, -1 XLOG
C       IS UNDEFINED, 0 CONVERGENCE IS NOT OBTAINED
C       WITH 200 TERMS OF THE SERIES
      IFLAG=-1
      IF(Y.LE.0.) RETURN
      IFLAG=-IFLAG
      XLOG=0.
      IF(Y.EQ.1.) RETURN
      YY=(Y-1.)/(1.+Y)
      SGN=SIGN(1.,YY)
      YY=ABS(YY)
      DO 1 K=1,200
        J=2*K-1
        TERM=YY**J/FLOAT(J)
        TEST=XLOG
        XLOG=XLOG+TERM
        IF(XLOG.EQ.TEST) GO TO 2
    1 CONTINUE
      IFLAG=0
    2 XLOG=2.*SGN*XLOG
      RETURN
      END
```

1.51 Verify the choice of $\tilde{x}$ given in Example 1.25; that is, show that $f(\tilde{x}) = z$.

1.52 Repeat exercise 1.51 for Example 1.26.

1.53 Use a pocket calculator to sum the first five terms, in increasing order of magnitude, in (1.17). Compare with the value log $2 = 0.693147180....$

Project 1

The purpose of this project is to design a subprogram for the evaluation of the hyperbolic sine function. Recall that in early versions of FORTRAN this was not in the library of built-in functions.

1 Show that $\sinh x > x$ for $x > 0$. Show that $\sinh (-x) = -\sinh x$ for all x.

2 Let $p_5(x) = 1 + x^3/3! + x^5/5!$; then by Taylor's theorem

$$\sinh x - p_5(x) = (\sinh c) \frac{x^6}{6!}$$

for some point c between 0 and x. Thus if $x \neq 0$

$$\left| \frac{\sinh x - p_5(x)}{\sinh x} \right| = \left| \frac{\sinh c}{\sinh x} \right| \frac{|x|^6}{6!} < \left| \frac{\sinh x}{x} \right| \frac{|x|^6}{6!}$$

Use this inequality to show that, for $0 < x \leq 0.25$, the relative error in $p_5(x)$ is bounded by 10^{-6}.

3 Assume that e^x is available from a library subprogram with a relative error of no more than 0.5×10^{-6}, say $E(x) = (1 - \rho)e^x$ with $|\rho| \leq 10^{-6}/2$. For $|x| > 0.25$, show that the algorithm

$$z \leftarrow \tfrac{1}{2}\left[E(x) - \frac{1}{E(x)} \right]$$

produces an approximation to $\sinh x$ such that

$$\left| \frac{z - \sinh x}{\sinh x} \right| \leq |\rho| \left| 1 - \frac{\rho}{1 - \rho} \frac{1}{e^{2x} - 1} \right| \leq 10^{-6}$$

Note: This inequality ignores any roundoff error resulting from division or subtraction.

4 Use the results of (2) and (3) to design an algorithm for the evaluation of $\sinh x$ whose relative error is bounded by 10^{-6} for all x. Hint: Use the FORTRAN function EXP(X) for $E(x)$.

5 Write a FORTRAN subprogram called HSIN to implement the algorithm of (4). Test HSIN by comparison with the built-in function SINH for several values of x, say $x = 0.0001, 0.001, 0.01, 0.1, 0.2, 0.25$ and $x = 0.3, 0.5, 0.7, 0.9, 1, 5, 10$.

NOTES AND COMMENTS

Section 1.1

Our discussion of mathematical models and numerical problem solving gave only a brief introduction to this field of study. For a more detailed discussion at an elementary level, see Malkevitch and Meyer (1974) and Noble (1967). A more advanced treatment is given in Haberman (1977), Maki and Thompson (1973), and Andrews and McLone (1976).

Section 1.3

One of the better sources for the analysis of floating point arithmetic and error propagation is Sterbenz (1974). See also the books by Wilkinson (1963) and Knuth (1969). For additional discussion of backward error analysis, consult Conte and deBoor (1980) or Ralston and Rabinowitz (1978).

Section 1.4

Our discussion of conditioning is similar to that given in Dahlquist and Björck (1974). For a more detailed presentation of methods for function evaluation, see Fike (1968). The book by Rice (1971) addresses most of the essential questions in the design of programs for mathematical computation.

NONLINEAR EQUATIONS

2

This chapter is concerned with numerical methods for solving an equation of the form

$$f(x) = 0 \qquad (2.1)$$

where f is a given continuous function. If x is a solution of (2.1), we say that x is a **root** (or zero) of f. From a practical viewpoint equation (2.1) is usually impossible to solve exactly. For example, a simple sketch of the curves $y = -x$ and $y = \log x$ shows that the equation $x + \log x = 0$ has a unique positive solution, say x^*. However, there are no known methods for finding the exact value of x^*. Even for simple functions like polynomials the determination of the roots is difficult. For a polynomial of degree 2 we can use the familiar quadratic formula (see exercise 1.37). If the polynomial has degree 3 or 4, it is still possible to find the roots by formula, but unfortunately the determination of the roots by analytical means is generally impossible if the degree is greater than 4. Therefore, in order to solve a nonlinear equation, we must typically rely on some numerical method for an approximate solution.

These approximation methods are iterative in nature; that is, starting with one or more initial (educated) guesses for a solution, they produce a sequence $\{x_n\}$ that presumably converges to a root. There are basically two types of iterative methods. In one type the convergence is guaranteed for any admissible initial guesses; this is called a **globally convergent** method. A **locally convergent** method is one in which the initial guesses must be sufficiently close to the desired root (solution) in order to be guaranteed of convergence. Locally convergent methods usually converge more rapidly than do globally

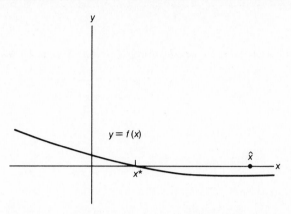

FIGURE 2.1

convergent ones. Hence, an appropriate algorithm is one in which the iteration is started with a globally convergent method and then switched to a locally convergent method when the most recent approximation is sufficiently close to the root.

In attempting to find an approximate root of $f(x) = 0$ by a numerical method, we seek a point $\hat{x}$ such that $|f(\hat{x})|$ is small and/or $|x^* - \hat{x}|/|x^*|$ is small, where x^* is the desired root. If the approximate root is known to satisfy $|f(\hat{x})| \leq \delta$ only for some small number $\delta > 0$, then $\hat{x}$ need not be close to the desired root. This can be seen in Figure 2.1. The difficulty here is that the graph of f is nearly flat near x^*. Indeed, by the mean value theorem we have for some point ξ between $\hat{x}$ and x^*

$$-f(\hat{x}) = f(x^*) - f(\hat{x}) = f'(\xi)(x^* - \hat{x})$$

Thus we find

$$x^* - \hat{x} = \frac{-f(\hat{x})}{f'(\xi)} \tag{2.2}$$

and for $|f'(\xi)| \ll 1$ the difference $|x^* - \hat{x}|$ may be quite large even though $|f(\hat{x})|$ is small. Clearly it is preferable, though not always possible, to find $\hat{x}$ such that $|x^* - \hat{x}|/|x^*|$ is small.

2.1 ROOT BRACKETING METHODS—GLOBAL CONVERGENCE

Bisection method

The simplest approach for finding an approximate root of f is based on the intermediate value theorem. If f is continuous on $[a, b]$ with $f(a)f(b) < 0$, then, by Corollary 1.1, $[a, b]$ contains a root of f. If w is any point in (a, b), then we may ask, Is $f(a)f(w) < 0$? If the answer is yes, then (a, w) contains a root. If, on the other hand, $f(a)f(w) > 0$, then (w, b) contains a root. [If $f(w)f(a) = 0$, we're finished.] In either case we find an interval, smaller than $[a, b]$, that contains a root of f. Based on this observation, we consider two methods that differ only in the manner in which the **test point** w is generated.

The first method generates the test point by always choosing the midpoint of the most recent interval. This is called the **bisection method**.

ALGORITHM 2.1

BISECTION METHOD Given a continuous function f on $[a, b]$ with $f(a)f(b) < 0$, let $a_1 = a$ and $b_1 = b$.

Do $n = 1, 2, \ldots,$ NMAX:

$$w \leftarrow \frac{a_n + b_n}{2}$$

If $f(a_n)f(w) < 0$ then $a_{n+1} \leftarrow a_n, b_{n+1} \leftarrow w$

If $f(a_n)f(w) > 0$ then $a_{n+1} \leftarrow w, b_{n+1} \leftarrow b_n$

At the nth stage of the algorithm we produce an interval $[a_n, b_n]$ that brackets a root of f. If we terminate this iteration at $[a_N, b_N]$ and take $x_N = (a_N + b_N)/2$, then there is a root x^* of f such that $|x^* - x_N| \le |b_N - a_N|/2$. Since each interval is half the length of its predecessor, we have

$$|x^* - x_N| \le \frac{|b_N - a_N|}{2} \le \frac{|b_{N-1} - a_{N-1}|}{2^2} \le \cdots \le \frac{|b - a|}{2^N}$$

and we have the a priori absolute error bound

$$|x^* - x_N| \le \frac{|b - a|}{2^N} \tag{2.3}$$

The bound (2.3) suggests a suitable termination criterion. If ε denotes a specified absolute error tolerance, then we terminate the iteration at stage N, where N is the least integer such that $|b - a| \le 2^N \varepsilon$. A better termination criterion takes the size of x^* into account. If ε denotes a relative error tolerance, then we iterate until

$$|b_N - a_N| \le \frac{\varepsilon|b_N + a_N|}{2} \tag{2.4}$$

In this case we have for x_N the bound

$$|x^* - x_N| \le \frac{|b_N - a_N|}{2} \le \frac{\varepsilon|b_N + a_N|}{4} = \frac{\varepsilon|x_N|}{2}$$

and hence

$$\frac{|x^* - x_N|}{|x^*|} \le \frac{\varepsilon}{2}\left|\frac{x_N}{x^*}\right| \tag{2.5}$$

In (2.5) the term $|x_N|/|x^*|$ is usually less than 2 for $\varepsilon < 1$ (see exercise 2.13), and so the termination criterion (2.4) guarantees that the relative error is less than ε.

The bisection method, when properly used, is guaranteed to produce a sequence of approximants that converge to a root of f; that is, it is a globally convergent method. We note, however, that the method uses little information

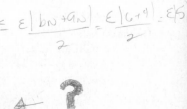

about f, namely, only its sign. Let us summarize our results for the bisection method.

THEOREM 2.1

Let x_n denote the midpoint of the nth interval $[a_n, b_n]$, which is generated by the bisection method applied to f. If f is continuous on $[a, b] = [a_1, b_1]$ and $f(a)f(b) < 0$, then there is a point x^* in (a, b) such that

(i) $f(x^*) = 0$

(ii) $\lim_{n \to \infty} x_n = x^*$

(iii) If $0 < \varepsilon < 1$ and $|b_N - a_N| \le \varepsilon |b_N + a_N|/2$, $a_N b_N > 0$, then $|x^* - x_N|/|x^*| \le \varepsilon$

EXAMPLE 2.1

The problem is to find the unique root of $f(x) = 2x^3 - 5x - 1$ in the interval $[1, 2]$. We make use of the following subprograms:

```
      SUBROUTINE BYSEKT(A,B,XMID,TOL)
      WRITE(6,400)
 400  FORMAT(' ',3X,'LEFT ENDPOINT',3X,'RIGHT ENDPOINT')
      FA=F(A)
   1  XMID=(A+B)/2.
      FMID=F(XMID)
      TEST=FA*FMID
      IF(TEST.LE.0.) GO TO 2
      A=XMID
      FA=FMID
      GO TO 3
   2  B=XMID
   3  IF((B-A).LE.TOL) GO TO 4
      WRITE(6,500)A,B
 500  FORMAT(' ',2(2X,E14.7))
      GO TO 1
   4  XMID=(B+A)/2.
      RETURN
      END
C
      REAL FUNCTION F(X)
      F=2.*X**3-5.*X-1.
      RETURN
      END
```

These subprograms, when used with the following main program,

```
      A=1.
      B=2.
      TOL=1.E-5
      CALL BYSEKT(A,B,X,TOL)
```

```
      ERROR=(B-A)/2.
      WRITE(6,101)X,ERROR
101   FORMAT('-','ROOT IS',2X,F8.6,2X,'ERROR BOUND =',E14.7)
      STOP
      END
```

result in the output

```
LEFT ENDPOINT       RIGHT ENDPOINT
0.1500000E 01       0.2000000E 01
0.1500000E 01       0.1750000E 01
0.1625000E 01       0.1750000E 01
0.1625000E 01       0.1687500E 01
0.1656250E 01       0.1687500E 01
0.1671875E 01       0.1687500E 01
0.1671875E 01       0.1679688E 01
0.1671875E 01       0.1675781E 01
0.1671875E 01       0.1673828E 01
0.1672852E 01       0.1673828E 01
0.1672852E 01       0.1673340E 01
0.1672852E 01       0.1673096E 01
0.1672974E 01       0.1673096E 01
0.1672974E 01       0.1673035E 01
0.1672974E 01       0.1673004E 01
0.1672974E 01       0.1672989E 01
```

```
ROOT IS  1.672985  ERROR BOUND = 0.3814697E-05
```

BYSEKT is a rather naive subroutine for the bisection algorithm. We point out some possible sources of difficulty in the program and suggest suitable improvements. The subroutine has several shortcomings, some of which can be serious. Perhaps the most obvious flaw in the subroutine is the absolute error termination criterion in statement 3. Subroutines are usually written for repeated use on several problems, and it is cleary advisable to terminate the iteration with a relative error criterion. Another difficulty with the program is that no test or safeguard is provided for the situation wherein the user supplies an unreasonably small tolerance TOL. If TOL is to small, then the iteration will never terminate even if a relative error criterion is used. A more subtle difficulty is present in the test for a sign change. The use of FA*FMID may result in underflow during the execution of the subroutine. As the root is approached, the value of the function is close to 0 and the product of two such numbers is likely to produce exponent underflow. As we are interested only in the sign of the product, it is preferable to make use of the FORTRAN built-in function SIGN. A final criticism is the subroutine's failure to protect against the uninformed user. Specifically we suggest that an initial check that $f(a)f(b) < 0$ be made and that ABS(B−A) be used in place of (B−A) in statement 3. Notice what happens if B is less than A in subroutine BYSEKT.

The following subroutine incorporates these improvements:

```
      SUBROUTINE BISECT(A,B,TOL,IFLAG,ERR,XMID)
C  A,B:ENDPOINTS OF INTERVAL
C  TOL:RELATIVE ERROR TOLERANCE;TOL=0.0 IS OK
C  ERR:ABSOLUTE ERROR ESTIMATE
C  XMID:MIDPOINT OF FINAL INTERVAL;APPROX. ROOT
C  IFLAG:SIGNALS MODE OF RETURN;1 IS NORMAL;-1 IF THE
C          VALUES F(A) & F(B) ARE OF THE SAME SIGN
      IFLAG=1
      FA=F(A)
      SFA=SIGN(1.,FA)
      TEST=SFA*F(B)
      IF(TEST.LE.0.) GO TO 1
      IFLAG=-IFLAG
      RETURN
    1 IF(B.GT.A) GO TO 2
      TEMP=A
      A=B
      B=TEMP
    2 ERR=B-A
      XMID=(A+B)/2.
C   DETERMINE THE APPROXIMATE MACHINE UNIT
      UNIT=1.
    3 UNIT=.5*UNIT
      U=UNIT+1.
      IF(U.GT.1.) GO TO 3
C   PROTECT AGAINST UNREASONABLE TOLERANCE
      TOL1=UNIT+TOL
    4 ERR=ERR/2.
C   CHECK THE TERMINATION CRITERION
      TOL2=TOL1*ABS(A+B)/4.
      IF(ERR.LE.TOL2) RETURN
      FMID=F(XMID)
C   TEST FOR SIGN CHANGE & UPDATE ENDPOINTS
      IF(SFA*FMID.LE.0.) GO TO 5
      A=XMID
      FA=FMID
      SFA=SIGN(1.,FA)
      XMID=XMID+(B-A)/2.
      GO TO 4
    5 B=XMID
      XMID=XMID-(B-A)/2.
      GO TO 4
      END
```

When BISECT is used with the same function subprogram as in Example 2.1, and the main program is

```
A=1.
B=2.
TOL=1.E-8
```

```
      CALL BISECT(A,B,TOL,IFLAG,ERROR,X)
      IF(IFLAG.GT.0) GO TO 1
      WRITE(6,300)
300   FORMAT(' ','F HAS THE SAME SIGN AT THE ENDPOINTS')
      STOP
  1   WRITE(6,400)X,ERROR
400   FORMAT(' ','ROOT =',2X,E14.7,1X,'WITH ERROR BOUND=',E14.7)
      STOP
      END
```

the resulting output is

```
ROOT =    0.1672981E 01 WITH ERROR BOUND= 0.2384186E-06
```

In this application of BISECT we chose a tolerance of 10^{-8} for the relative error. This specification is too small for single-precision computations on an IBM machine; however, this poses no difficulty for BISECT since TOL1 $\simeq$ UNIT. In fact the choice TOL $= 0.$ is an acceptable specification; its use results in the determination of the root to machine precision. We make use of the approximate machine unit in determining a tolerance TOL1 whose value is $\simeq$ UNIT if TOL is too small. If, on the other hand, TOL is large, then TOL1 $\simeq$ TOL. The execution-time determination of the machine unit allows this subroutine to be safely used on more than one machine. We note that subroutine BISECT never evaluates F more than once at the same point; rather, the value is saved for subsequent assignment. Note also that the new test point XMID is of the form: new value = old value + correction term.

False position methods

We anticipate that a method that uses more information about f, such as its value at a point, should converge at a faster rate. Such a method, which is based on the intermediate value theorem, is the method of false position. In this approach the test point w is taken to be the point of intersection of the x axis and the secant line joining $(a_n, f(a_n))$ and $(b_n, f(b_n))$, as shown in Figure 2.2. A simple computation shows that $w = [a_n f(b_n) - b_n f(a_n)]/[f(b_n) - f(a_n)]$. Note that for $|f(b_n)| > |f(a_n)|$ we have that $|w - a_n| < |w - b_n|$; that is, the test point is closer to a_n than to b_n. Moreover, we expect that the root should be closer to a_n, and this is the rationale for the choice of the test point.

ALGORITHM 2.2

FALSE POSITION METHOD   Given a continuous function f on $[a, b]$ with $f(a)f(b) < 0$, let $a_1 = a, b_1 = b.$

Do $n = 1, 2, ..., $ NMAX:

$$w \leftarrow \frac{a_n f(b_n) - b_n f(a_n)}{f(b_n) - f(a_n)}$$

If $f(a_n)f(w) < 0$ then $a_{n+1} \leftarrow a_n, b_{n+1} \leftarrow w$

If $f(a_n)f(w) > 0$ then $a_{n+1} \leftarrow w, b_{n+1} \leftarrow b_n$

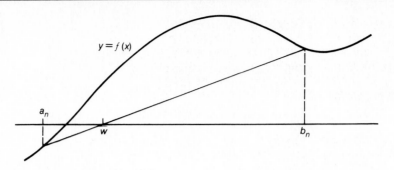

FIGURE 2.2

EXAMPLE 2.2
_____

In the following we use a straightforward implementation of the false position
method for the function $f(x) = x^2 - e^{-x}$ on [0, 1]. The function f has exactly
one positive root. After 10 iterations we find the root is approximately
0.7031875.

```
C   A SIMPLE FALSE POSITION PROGRAM
        A=0.
        B=2.5
        FA=F(A)
        FB=F(B)
        SFA=SIGN(1.,FA)
        DO 3 N=1,10
          TOP=A*FB-B*FA
          DIFF=FB-FA
          X=TOP/DIFF
          FX=F(X)
          TEST=SFA*FX
          IF(TEST.LE.0.) GO TO 1
          A=X
          FA=FX
          GO TO 2
    1     B=X
          FB=FX
    2     WRITE(6,100)N,A,N,B,X
  100     FORMAT(' ','A(',I2,')=',F9.7,2X,'B(',I2,')=',
        *F9.7,2X,'X=',F9.7)
    3   CONTINUE
        STOP
        END
C
        REAL FUNCTION F(X)
        F=X**2-EXP(-X)
        RETURN
        END
```

(handwritten annotations:) IF(FA .LT.0) THEN SFA = -1 END IF   END DO

```
A( 1)=0.3487765  B( 1)=2.5000000  X=0.3487765
A( 2)=0.5348169  B( 2)=2.5000000  X=0.5348169
A( 3)=0.6258942  B( 3)=2.5000000  X=0.6258942
A( 4)=0.6683712  B( 4)=2.5000000  X=0.6683712
```

```
A( 5)=0.6877114   B( 5)=2.5000000   X=0.6877114
A( 6)=0.6964189   B( 6)=2.5000000   X=0.6964189
A( 7)=0.7003192   B( 7)=2.5000000   X=0.7003192
A( 8)=0.7020621   B( 8)=2.5000000   X=0.7020621
A( 9)=0.7028403   B( 9)=2.5000000   X=0.7028403
A(10)=0.7031875   B(10)=2.5000000   X=0.7031875
```

As in the bisection method, the false position method produces, at each stage, an interval that contains a root of f. However, it need not be true that $b_n - a_n \rightarrow 0$ as $n \rightarrow \infty$. To see that this is the case, it is enough to consider the method applied to an increasing (or decreasing) function whose graph is concave up.

Figure 2.3 shows the first four iterations of the false position method applied to an increasing function. In this case the right endpoint remains fixed throughout the iteration and we have, for all n, that

$$b_n - a_n \lessapprox b_n - x^* = b - x^*$$

The situation depicted in Figure 2.3 is exactly that which occurs in Example 2.2, where the function $f(x) = x^2 - e^{-x}$ is increasing and its graph is concave up (that is, $f'' > 0$). The reader should, by use of sketches, convince herself or himself that a similar situation prevails if the graph is concave down or if the function is decreasing (with either concavity). This is a serious shortcoming of the false position method, as almost all functions are eventually one of these types in the proximity of a root. Even though $b_n - a_n$ need not approach 0 as $n \rightarrow \infty$, it is apparently true, for the function of Figure 2.3, that $\lim_{n \rightarrow \infty} a_n = x^*$ and the method produces a sequence that converges to the root. We state, without proof, the following convergence result.

THEOREM 2.2

Suppose f is continuous on $[a, b]$ with $f(a)f(b) < 0$ and let $a_1 = a$, $b_1 = b$. Let $\{x_n\}$ denote the sequence of test points generated by the false position method. Then there is a point x^* in (a, b) such that $f(x^*) = 0$ and $\lim_{n \rightarrow \infty} x_n = x^*$.

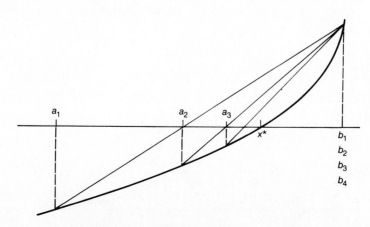

FIGURE 2.3

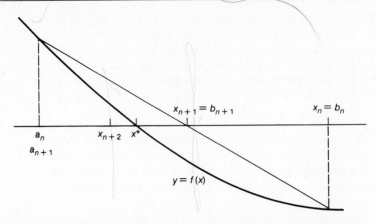

FIGURE 2.4

From our observations regarding Figure 2.3, it is clear that a termination criterion such as the ones suggested for the bisection method are inadequate for false position and may result in an infinite loop. Therefore, we must view the method as one that produces an approximate root $\hat{x}$ in the sense that $|f(\hat{x})|$ is small. Recall the difficulty with error estimation we observed in connection with Figure 2.1. The false position method has few redeeming qualities, and its use is to be avoided.

Rather than dismiss the false position method entirely, we suggest some appropriate modifications. The essential difficulty with the method is that, eventually, successive test points, say x_n and x_{n+1}, fall on the same side of the root (as in Figure 2.3). What we require is a different means of determining the next test point, x_{n+2}, so that $[x_{n+1}, x_{n+2}]$ contains the root. The first such method that comes to mind is bisection.

Referring to Figure 2.4, we choose $x_{n+2} = (a_{n+1} + x_{n+1})/2$. If $f(x_{n+2})f(x_{n+1}) < 0$, we set $a_{n+2} = x_{n+2}$, $b_{n+2} = b_{n+1}$ and proceed as in the false position method. If $f(x_{n+2})f(x_{n+1}) > 0$, we bisect again, that is, $x_{n+3} = (a_{n+1} + x_{n+2})/2$, until we find a test point that falls to the other side of the root.

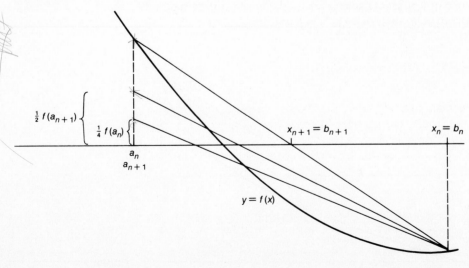

FIGURE 2.5

A better approach for modifying the test point is illustrated in Figure 2.5. We have $f(x_{n+1})f(x_n) > 0$, and hence the left endpoint remains fixed. Now take x_{n+2} to be the point of intersection of the line joining $(a_{n+1}, f(a_{n+1})/2)$ and $(b_{n+1}, f(b_{n+1}))$ with the x axis. If $f(x_{n+2})f(x_{n+1}) < 0$, we continue as in the false position method. Otherwise we repeat the process of halving the value at a_{n+1} until we find a test point that falls to the other side of the root. In Figure 2.5 we find success after two steps of this modification.

ALGORITHM 2.3

MODIFIED FALSE POSITION METHOD Given a continuous function f on $[a, b]$ with $f(a)f(b) < 0$, let $a_1 = a$, $b_1 = b$.

$$A \leftarrow f(a_1)$$

$$B \leftarrow f(b_1)$$

$$x_1 \leftarrow (Ab_1 - Ba_1)/(A - B)$$

If $Af(x_1) < 0$ then $a_2 \leftarrow a_1$, $b_2 \leftarrow x_1$, $B \leftarrow f(x_1)$

If $Af(x_1) > 0$ then $a_2 \leftarrow x_1$, $b_2 \leftarrow b_1$, $A \leftarrow f(x_1)$

Do $n = 2, 3, \ldots,$ NMAX:

> $x_n \leftarrow (Ab_n - Ba_n)/(A - B)$
>
> If $Af(x_n) < 0$ then $a_{n+1} \leftarrow a_n$, $b_{n+1} \leftarrow x_n$, $B \leftarrow f(x_n)$
>
> > If $f(x_{n-1})f(x_n) > 0$ then $A \leftarrow A/2$
>
> If $Af(x_n) > 0$ then $a_{n+1} \leftarrow x_n$, $b_{n+1} \leftarrow b_n$, $A \leftarrow f(x_n)$
>
> > If $f(x_{n-1})f(x_n) > 0$ then $B \leftarrow B/2$

A suitable termination criterion for this algorithm is as follows. Let ε_1 and ε_2 denote specified tolerances. Then we terminate the iteration if

$$|f(x_n)| \leq \varepsilon_1 \quad \text{or if} \quad \frac{|x_n - x_{n-1}|}{|x_n|} \leq \varepsilon_2$$

The rationale for the second part of the termination criterion is that there is little justification for continuing the iteration if successive iterates are nearly equal (in a relative sense). Moreover, if the iterates x_n and x_{n-1} are close to x^*, then $|x_n - x_{n-1}|$ is an a posteriori estimate for the absolute value of the error.

EXAMPLE 2.3

For the same problem as in Example 2.2 we apply subroutine MFP, which implements the modified false position method:

```
      EXTERNAL F
      READ,A,B
      TOL1=1.E-7
      TOL2=0.
      NMAX=50
      CALL MFP(A,B,F,ROOT,VALUE,TOL1,TOL2,NMAX)
      WRITE(6,100)A,B
  100 FORMAT(' ','ENDPOINTS ARE',2(2X,F9.7)/)
      WRITE(6,110)ROOT,VALUE
  110 FORMAT(' ','ROOT=',2X,F9.7,2X,'F(ROOT)=',2X,E14.7)
      STOP
      END
C
      SUBROUTINE MFP(A,B,F,XNEW,FXN,TOL1,TOL2,NTOL)
C  XNEW IS THE APPROXIMATE ROOT
C  ENDPOINTS OF INTERVAL ARE A AND B
C  F IS EXTERNAL FUNCTION SUPPLIED BY USER
C  FXN IS F(XNEW)
C  TOL1 IS TOLERANCE FOR FUNCTION VALUE
C  TOL2 IS TOLERANCE FOR RELATIVE DIFFERENCE
C      BETWEEN SUCCESSIVE ITERATES
C  NTOL IS MAX. NUMBER OF ITERATIONS ALLOWED
      UNIT=1.
    1 UNIT=.5*UNIT
      U=1.+UNIT
      IF(U.GT.1.) GO TO 1
C  PROTECT AGAINST UNREASONABLE TOLERANCE
      TOL2=TOL2+UNIT
C  INITIALIZATION
      FA=F(A)
      SFA=SIGN(1.,FA)
      FB=F(B)
      TOP=A*FB-B*FA
      DIFF=FB-FA
      XOLD=TOP/DIFF
      FXO=F(XOLD)
      TEST=SFA*FXO
      IF(TEST.GT.0.) GO TO 2
      B=XOLD
      FB=FXO
      GO TO 3
    2 A=XOLD
      FA=FXO
    3 CONTINUE
C  BEGIN ITERATION
      DO 6 K=1,NTOL
        TOP=A*FB-B*FA
        DIFF=FB-FA
        XNEW=TOP/DIFF
        XDIFF=ABS(XNEW-XOLD)
        XOLD=XNEW
        RERR=XDIFF/ABS(XNEW)
```

```
C   CHECK RELATIVE ERROR CRITERION
          IF(RERR.LE.TOL2) RETURN
          FXN=F(XNEW)
          SIZE=ABS(FXN)
C   CHECK SIZE OF FUNCTION VALUE
          IF(SIZE.LE.TOL1) RETURN
          SFXN=SIGN(1.,FXN)
          TEST2=SFXN*FXO
          TEST=SFA*FXN
C   UPDATE ENDPOINT VALUES
          IF(TEST.GT.0.) GO TO 4
          B=XNEW
          FB=FXN
          IF(TEST2.GT.0.) FA=FA/2.
          GO TO 5
    4     A=XNEW
          FA=FXN
          IF(TEST2.GT.0.) FB=FB/2.
    5     FXO=FXN
    6     CONTINUE
          WRITE(6,101)
  101     FORMAT(' ',/'MAX. ITERATIONS ACHIEVED'/)
          RETURN
          END
C
          REAL FUNCTION F(X)
          F=X**2-EXP(-X)
          RETURN
          END

    ENDPOINTS ARE  0.7034672  0.7034684

    ROOT=  0.7034670  F(ROOT)= - 0.4172325E-06
```

This program gives essentially the same root as in Example 2.2; however, the final interval has length 1.2×10^{-6} rather than length $\simeq 1.8$ produced by the false position method. (We used the same initial interval as in Example 2.2.) Thus MFP produces a more accurate approximate root than do 10 iterations of the false position method.

EXERCISES

2.1   For each of the following, determine the number of positive roots:

(a) $f(x) = x + \cos x$
(b) $f(x) = x - \cos x$
(c) $f(x) = e^{-x} + \sin x$
(d) $f(x) = x^3 - 6x^2 + 11x - 6$

2.2   Determine b such that $(0, b)$ contains a root of $f(x) = 1 - x - e^{-2x}$.

2.3   How many solutions does the equation $2x + \tan x = 0$ have? Can you determine any of them exactly? Can you determine any others approximately by graphical means?

2.4   The function $f(x) = x^3 - x - 1$ has exactly one root in $(1, 2)$. Verify this. Determine the number of iterations required in the bisection method to determine this root with an absolute error of no more than 10^{-5}.

2.5   Apply subroutine BISECT to find the positive root of $f(x) = x^2 - e^{-x}$. Modify BISECT to count and print the number of bisections, call it NUM, required for a given TOL. Use TOL $= 0$. and TOL $= 1.E - 04$ and compare the results.

2.6   In BISECT is it necessary to update SFA after each iteration? Justify your answer.

2.7   Compare BISECT and MFP for rate of convergence as follows. Use the test problem $f(x) = x^2 - e^{-x}$ on the interval $[0, 15]$. Modify both subprograms to count and print the number of iterations required. Use TOL $= 0$. in BISECT and TOL2 $= 0$. in MFP. Print the final endpoints for both methods.

2.8   Use the programs of Examples 2.2 and 2.3 to find the root x^* ($\simeq 0.5$) of $\log x + x = 0$, with a relative error of no more than 10^{-6}. Use the same initial endpoints in both methods. Print the number of iterations required in each method.

2.9   Use subroutine MFP to find the unique root of $f(x) = \sin x - e^{-x}$ in $[0, 1]$. Use TOL2 $= 0$. and TOL2 $= 1.E - 03$. Compare the number of iterations required in each run.

 2.10   Repeat exercise 2.7 for the function $f(x) = 2x^3 - 5x - 1$ on $[-2, 15]$.

2.11   Rerun the program of Example 2.2 with 20 iterations instead of 10. Compare with Example 2.3.

2.12   The initialization phase of subroutine MFP consists of one iteration of false position in order to obtain the first values of FX0 and TEST2. This could also be accomplished by simply treating one of the initial endpoints as XOLD. Modify MFP to implement this approach. Test your version of MFP on the problem of Example 2.3.

*2.13   Show that part (iii) of Theorem 2.1 is valid. Hint: Let $v = \min\{|a_N|, |b_N|\}$; then show that (a) $|x^*| \geq v$ and (b) $|x_N| \leq v/(1 - \varepsilon/2)$ and combine these with (2.5).

*2.14   Does BISECT always produce an approximate root with relative error $\simeq$ TOL? Hint: See part (iii) of Theorem 2.1.

2.15   Suppose it is required to find the root of $f(x) = e^{-x} - \cos x$ that is closest to 50. If BYSEKT is used on an IBM 370, would it be reasonable to specify TOL $= 1.E - 06$? Explain your answer.

2.16   Modify the program of Example 2.2 so the iteration terminates only if ABS(FX).LE.TOL has the value TRUE. Is TOL $= 1.E - 09$ reasonable for the problem of exercise 2.15? Justify your answer.

2.17   Subroutines BYSEKT and BISECT share a shortcoming that is easy to correct. Suppose you want to use either subroutine in a main program for two different functions, say $F(x)$ and $G(x)$. This poses a serious difficulty because the function is not in the parameter list. Correct this defect in BISECT by adding a parameter F to the list. Caution: In using this new version of BISECT, you must declare F as an external subprogram; in other words, the first statement in the main program should be EXTERNAL F.

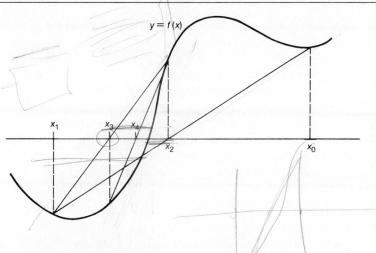

$y = f(x)$

FIGURE 2.6

2.2 LOCALLY CONVERGENT METHODS

The methods of the previous section have the desirable property that convergence is guaranteed on $[a, b]$ provided f is continuous and $f(a)f(b) < 0$. There is no requirement that $b - a$ be small to ensure convergence. In this section we consider two methods that generally require a good estimation of the location of the root prior to the iteration.

Secant method

The first method we consider is similar to the false position method, but here we do not require f to be of opposite signs at the initial points. Starting with two points on the graph of f, we generate the next point by the intersection of the secant line with the x axis. The new point, together with one of the two preceding points, is used in the same manner to determine the next point, and so on.

ALGORITHM 2.4

SECANT METHOD Given a continuous function f and two points x_0 and x_1,

Do $n = 1, 2, \ldots, \text{NMAX}$:

$$x_{n+1} \leftarrow \frac{x_{n-1} f(x_n) - x_n f(x_{n-1})}{f(x_n) - f(x_{n-1})}$$

The algorithm is illustrated in Figure 2.6. The point x_{n+1} is the one where the secant line joining $(x_{n-1}, f(x_{n-1}))$ and $(x_n, f(x_n))$ intersects the x axis. The secant method is subject to roundoff error difficulties because $f(x_{n-1})$ and $f(x_n)$ are not required to be of opposite signs. Thus the difference $f(x_n) - f(x_{n-1})$ is prone to subtractive cancellation as we approach the root. In the worst case it may happen that $f(x_{n-1}) = f(x_n)$ so that x_{n+1} is undefined, that is, the secant line is horizontal. However, let us rewrite the expression for x_{n+1} as

$$x_{n+1} = x_n - \frac{x_n - x_{n-1}}{f(x_n) - f(x_{n-1})} f(x_n) \tag{2.6}$$

correction term [handwritten annotation]

As we approach the root, the second term in (2.6) can be viewed as a correction term to x_n. Suppose, for example, that $x_n = 126.4315$ and the correction term is 0.1437614. Because of subtractive cancellation it may be that only the first four digits of the correction term are correct. The value of x_{n+1} ($= 126.4315 - 0.1437614$) is nonetheless correct to six digits. Thus only a few correct digits in the correction term are required as we approach the root. It is good practice, in computing approximations, to use the form

Improved value = approximate value $\pm$ correction term

Thus (2.6) is to be preferred over the form given in the secant algorithm. A suitable termination criterion for the secant method is the same as we suggested for the modified false position method.

In order for the secant method to converge, it is normally necessary that the initial guesses x_0 and x_1 be sufficiently close to the desired root.

THEOREM 2.3

Suppose $f(x^*) = 0$, $f'(x^*) \neq 0$, and f'' is continuous near x^*. Then there exists $\delta > 0$ such that, for x_0, x_1 in $[x^* - \delta, x^* + \delta]$, the sequence $\{x_n\}$ generated by the secant method converges to x^*.

The proof of this result is beyond the scope of the text and we omit it.

EXAMPLE 2.4

We give a simple program for the secant method applied to $f(x) = 2x^3 - 5x - 1$ on $[1, 2]$.

```
      TOL=1.E-6
      NMAX=15
      WRITE(6,100)
  100 FORMAT(' ',2X,'ITERATE',7X,'VALUE'/)
      XOLD=2.
      XNEW=1.
      FO=F(XOLD)
      DO 1 N=1,NMAX
        FN=F(XNEW)
        W=XNEW-FN*(XNEW-XOLD)/(FN-FO)
        XOLD=XNEW
        FO=FN
        XNEW=W
        TOLX=TOL*ABS(XNEW)
        TEST=ABS(XNEW-XOLD)
        WRITE(6,120)N,XNEW
```

```
120   FORMAT(' ',5X,I2,5X,E14.7)
         IF(TEST.LE.TOLX) GO TO 2
   1 CONTINUE
     WRITE(6,110)
 110 FORMAT(' ',/'MAX. NUMBER OF ITERATIONS REACHED')
     STOP
   2 WRITE(6,130)
 130 FORMAT(' ',2X,/'TERMINATION CRITERION SATISFIED')
     STOP
     END
C
     REAL FUNCTION F(X)
     F=2.*X**3-5.*X-1.
     RETURN
     END
```

```
 ITERATE        VALUE

    1       0.1444444E 01
    2       0.1984804E 01      3  on vax
    3       0.1616105E 01
    4       0.1660096E 01
    5       0.1673635E 01
    6       0.1672974E 01
    7       0.1672981E 01
    8       0.1672981E 01      2 on VAX
```

```
TERMINATION CRITERION SATISFIED
```

In this program we terminate the iteration if $|x_n - x_{n-1}| \le 10^{-6}|x_n|$. From the output it is seen that this occurs for $n = 8$ in our example.

Let us examine the truncation error in the secant method. Our aim is to find a formula for the error in the nth iterate. This formula will be used in the next section to assess the speed of convergence of the method. Let $e_n = x^* - x_n$ and assume that the secant method sequence is well-defined with $x_n \ne x_{n-1}$ for all n. After some manipulation we find

$$e_{n+1} = \frac{f(x_n)e_{n-1} - f(x_{n-1})e_n}{f(x_n) - f(x_{n-1})}$$

which we can rewrite as

$$e_{n+1} = e_n e_{n-1} \frac{f(x_n)/e_n - f(x_{n-1})/e_{n-1}}{f(x_n) - f(x_{n-1})} \tag{2.7}$$

Using the fact that $f(x^*) = 0$, we have

$$\frac{\dfrac{f(x_n)}{e_n} - \dfrac{f(x_{n-1})}{e_{n-1}}}{f(x_n) - f(x_{n-1})} = \frac{\dfrac{f(x_{n-1}) - f(x^*)}{x_{n-1} - x^*} - \dfrac{f(x_n) - f(x^*)}{x_n - x^*}}{f(x_n) - f(x_{n-1})} \tag{2.8}$$

Now let $F(x) = [f(x) - f(x^*)]/(x - x^*)$. Then by the mean value theorem we have, for some point ξ_n between x_{n-1} and x_n, that $F(x_{n-1}) - F(x_n) = F'(\xi_n)(x_{n-1} - x_n)$. A simple calculation gives

$$F'(x) = \frac{f'(x)(x - x^*) + f(x^*) - f(x)}{(x - x^*)^2}$$

Moreover, by Taylor's theorem, we have

$$f(x^*) = f(x) + f'(x)(x^* - x) + f''(\eta)\frac{(x^* - x)^2}{2}$$

and hence

$$f(x^*) - [f(x) + f'(x)(x^* - x)] = \frac{f''(\eta)(x^* - x)^2}{2}$$

where η is some point between x and x^*. Combining the last three equations gives

$$F(x_{n+1}) - F(x_n) = \frac{(x_{n-1} - x_n)f''(\eta_n)}{2}$$

which is the negative of the numerator on the right-hand side of (2.8). Thus (2.8) may be written as

$$\frac{f(x_n)/e_n - f(x_{n-1})/e_{n-1}}{f(x_n) - f(x_{n-1})} = \frac{-f''(\eta_n)}{2}\frac{x_{n-1} - x_n}{f(x_{n-1}) - f(x_n)} \tag{2.9}$$

The second factor on the right-hand side of (2.9) is equal to $1/f'(\gamma_n)$ by the mean value theorem, and we may combine (2.9) and (2.7) to get

$$e_{n+1} = -\frac{e_{n-1}e_n f''(\eta_n)}{2f'(\gamma_n)} \tag{2.10}$$

where η_n and γ_n are in the smallest interval that contains x^*, x_{n-1}, and x_n. The derivation of (2.10) assumes that f'' exists and that $f' \neq 0$ near x^*. If f'' is continuous, we have $f''(\eta_n) \simeq f''(x^*)$ and $f'(\gamma_n) \simeq f'(x^*)$ by continuity; hence $e_{n+1} \simeq -e_n e_{n-1} f''(x^*)/2f'(x^*)$. Thus the error in the nth iterate is proportional to the product of the errors in the previous two iterates.

Newton's method

The final method we consider is Newton's method, which may be viewed as a modification of the secant method. We assume that f is continuously differentiable. Then by the mean value theorem the correction term in (2.6) can be written as

$$\frac{x_n - x_{n-1}}{f(x_n) - f(x_{n-1})}f(x_n) = \frac{f(x_n)}{f'(\gamma_n)}$$

for some γ_n between x_{n-1} and x_n. Thus the secant iterates satisfy

$$x_{n+1} = x_n - \frac{f(x_n)}{f'(\gamma_n)}$$

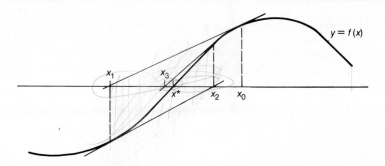

FIGURE 2.7

As we approach x^*, $x_n \simeq x_{n-1} \simeq x^*$, and so by continuity of f' we expect that $f'(\gamma_n) \simeq f'(x_n)$. Thus we have

$$x_{n+1} \simeq x_n - \frac{f(x_n)}{f'(x_n)} \tag{2.11}$$

The right-hand side of (2.11) depends only on x_n, and Newton's method results when we require the sequence $\{x_n\}$ to satisfy (2.11) with $\simeq$ replaced by $=$.

ALGORITHM 2.5

NEWTON'S METHOD   Given a continuously differentiable function f and a point x_0,

Do $n = 1, 2, \ldots$, NMAX:

$$x_{n+1} \leftarrow x_n - \frac{f(x_n)}{f'(x_n)}$$

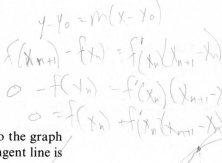

Newton's method is graphically illustrated in Figure 2.7.

The Newton iterate x_{n+1} is the point where the tangent line to the graph of f at $(x_n, f(x_n))$ intersects the x axis. Indeed, the equation of the tangent line is given by

$$y = y(x) = f(x_n) + f'(x_n)(x - x_n)$$

If x_{n+1} denotes the point where this line intersects the x axis, then $y(x_{n+1}) = 0$; that is,

$$0 = f(x_n) + f'(x_n)(x_{n+1} - x_n) \tag{2.12}$$

and solving for x_{n+1} gives $x_{n+1} = x_n - f(x_n)/f'(x_n)$.

Since Newton's method is not a root-bracketing method, a suitable termination criterion is to iterate until

$$|f(x_n)| \le \varepsilon_1 \qquad \text{or} \qquad |x_n - x_{n-1}| \le \varepsilon_2 |x_n|$$

where, as before, ε_1 and ε_2 denote specified tolerances. We note that Newton's method requires two function evaluations per iteration, namely, $f(x_n)$ and $f'(x_n)$. In addition, we require $f'(x_n) \ne 0$ for all n. For $f'(x^*) = 0$ or $f'(x^*) \simeq 0$, we expect convergence to be impaired (exercise 2.28).

EXAMPLE 2.5

Suppose it is desired to find the pth root of a number $A > 0$. That is, we seek the number $\hat{x}$ that satisfies $f(\hat{x}) = \hat{x}^p - A = 0$. We have $f'(x) = px^{p-1}$, and so Newton's method becomes

$$x_{n+1} = x_n - \frac{x_n^p - A}{px_n^{p-1}}$$

or equivalently

$$x_{n+1} = \frac{(p-1)x_n + A/x_n^{p-1}}{p}$$

EXAMPLE 2.6

Suppose we had a computer that could perform only addition, subtraction, and multiplication. How could we find the quotient $a \div b$? Newton's method furnishes a means of performing approximate division using only these three operations. We write $a \div b$ as $a \cdot (1 \div b)$ and let $f(x) = x^{-1} - b$. The Newton iterates for f are given by

$$x_{n+1} = x_n + (x_n^{-1} - b)x_n^2$$

$$= 2x_n - bx_n^2$$

Suppose $a = 4$ and $b = 3$. If $x_0 = 0.5$, we find $x_1 = 0.25$, $x_2 = 0.3125$, $x_3 = 0.33203125$, $x_4 = 0.333328247$, and so on. Then $4 \cdot x_4 = 1.333312\ldots \simeq \frac{4}{3}$.

EXAMPLE 2.7

```
C   A SIMPLE NEWTON PROGRAM
        TOL=1.E-6
        XO=5.0
        WRITE(6,100)
  100 FORMAT(' ',2X,'ITERATE',7X,'VALUE'/)
        DO 1 N=1,20
        DFO=DF(XO)
        IF(DFO.EQ.0.) GO TO 3
        XN=XO-F(XO)/DFO
        TEST=ABS(XN-XO)
        TOLX=TOL*ABS(XN)
        WRITE(6,110)N,XN
        IF(TEST.LE.TOLX) GO TO 2
        XO=XN
  110   FORMAT(' ',5X,I2,5X,E14.7)
    1 CONTINUE
        WRITE(6,111)
  111 FORMAT(' ',/'MAXIMUM NUMBER OF ITERATIONS')
        STOP
    2 WRITE(6,112)
```

```
112 FORMAT(' ',/2X,'TERMINATION CRITERION SATISFIED')
    STOP
  3 WRITE(6,113)
113 FORMAT(' ',/'DERIVATIVE=0,USE ANOTHER X0')
    STOP
    END
C

    REAL FUNCTION F(X)
    F=X**2-EXP(-X)
    RETURN
    END
C

    REAL FUNCTION DF(X)
    DF=2.*X+EXP(-X)
    RETURN
    END
```

ITERATE	VALUE
1	0.2502357E 01
2	0.1287422E 01
3	0.8028350E 00
4	0.7071615E 00
5	0.7034728E 00
6	0.7034674E 00
7	0.7034674E 00

TERMINATION CRITERION SATISFIED

In this program we terminate the iteration if $|x_n - x_{n-1}| \leq 10^{-6}|x_n|$. This occurs at $n = 7$, with the resultant approximate root of 0.7034674. Compare this with Examples 2.2 and 2.3.

Next we derive a formula for the truncation error in Newton's method. Suppose $f(x^*) = 0$, $f'(x^*) \neq 0$, and f'' is continuous near x^*. Let $\{x_n\}$ denote the Newton sequence and assume that $\lim_{n \to \infty} x_n = x^*$. By Taylor's theorem we have

$$0 = f(x^*) = f(x_n) + f'(x_n)(x^* - x_n) + f''(\xi_n)\frac{(x^* - x_n)^2}{2}$$

where ξ_n is between x^* and x_n. By (2.12) we can write this equation as

$$0 = f'(x_n)(x^* - x_{n+1}) + f''(\xi_n)\frac{(x^* - x_n)^2}{2}$$

Thus the error in the nth Newton iterate $e_n = x^* - x_n$ satisfies

$$e_{n+1} = \frac{-e_n^2 f''(\xi_n)}{2f'(x_n)}$$

and by continuity of f' and f'' we have $e_{n+1} \simeq -e_n^2 f''(x^*)/2f'(x^*)$ for n sufficiently large. Therefore,

$$|e_{n+1}| \simeq K|e_n|^2 \tag{2.13}$$

where $K = |f''(x^*)|/2|f'(x^*)|$. The estimate (2.13) says that the error in the $(n+1)$st iterate is proportional to the square of the error in the nth iterate. We return to (2.13) in the next section when discussing the order of convergence.

The previous error analysis can be refined in order to prove the following convergence result.

THEOREM 2.4

Suppose $f(x^*) = 0, f'(x^*) \neq 0$, and f'' is continuous near x^*. There exists $\delta > 0$ such that if x_0 is in $[x^* - \delta, x^* + \delta]$, then Newton's method converges to x^*.

Loosely speaking, this theorem says that Newton's method works if the initial guess x_0 is close enough to x^*. Clearly it is desirable to have a practically verifiable criterion under which we are guaranteed convergence for any admissible initial guess. One such criterion is given in the following theorem.

THEOREM 2.5

Suppose f'' is continuous on $[a, b]$ with (i) $f(a)f(b) < 0$, (ii) $f'(x) \neq 0$ for all x in $[a, b]$, (iii) f'' does not change sign in $[a, b]$, and (iv) $|f(a)/f'(a)|$, $|f(b)/f'(b)| < b - a$. Then for any x_0 in $[a, b]$ the Newton sequence converges to x^* and x^* is the only root of f in $[a, b]$.

Condition (i) guarantees that $[a, b]$ contains a root of f. By (ii) f is either strictly increasing ($f' > 0$) or strictly decreasing ($f' < 0$) and hence $[a, b]$ contains exactly one root of f. Condition (iii) says that the graph of f is either concave up ($f'' \geq 0$) or concave down ($f'' \leq 0$). The last condition is most easily examined by graphical means. In view of Figure 2.8, condition (iv) guarantees that if x_0 is chosen to be a or b, then x_1 is in (a, b). If x_0 is chosen to be b in Figure 2.8, then the Newton sequence is increasing, that is, $x_{n+1} > x_n$ for $n \geq 1$, and converges to x^*.

It is also interesting to note that if a and b are chosen as the initial guesses in the secant method, say $a = x_0$ and $b = x_1$, then the next secant iterate, x_2, is in (a, b). In fact the conditions of Theorem 2.5 are sufficient to guarantee the convergence of the secant method for any x_0, x_1 in $[a, b]$. The reader should verify this graphically using a function of the type depicted in Figure 2.8.

EXAMPLE 2.8

Consider the problem of finding the root of $f(x) = x^2 - 3$ in $[1, 4]$, that is, of finding $\sqrt{3}$. We have $f(1)f(4) = -26$, $f'(x) = 2x > 0$ for $x \in [1, 4]$, and

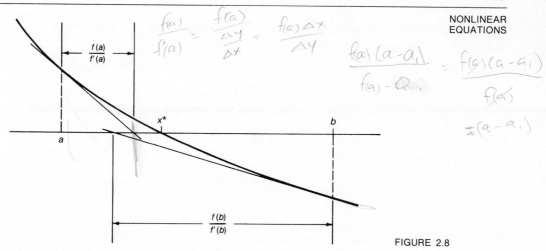

$\frac{f(a)}{f'(a)} = \frac{f(a)}{\frac{\Delta y}{\Delta x}} = \frac{f(a)\Delta x}{\Delta y}$

$\frac{f(a)(a-a_1)}{f(a) - \cancel{a_1}} = \frac{f(a)(a-a_1)}{f(a)}$

$= (a-a_1)$

FIGURE 2.8

$f''(x) = 2 \neq 0$ so that conditions (i), (ii), and (iii) of Theorem 2.5 are satisfied. In addition we see that

$$\left| \frac{f(1)}{f'(1)} \right| = 1 < 3 \qquad \text{and} \qquad \left| \frac{f(4)}{f'(4)} \right| = \frac{13}{8} < 3$$

and hence (iv) is satisfied as well. It follows that Newton's method and the secant method are guaranteed to converge for any starting values in [1, 4].

In Example 2.8 it was particularly easy to verify the hypotheses of Theorem 2.5; however, this is not always the case (see exercise 2.26).

EXERCISES

2.18 The program of Example 2.4 has a serious defect. Find it and make an appropriate modification of the program. Hint: Protect against division by 0.

2.19 By modifying the program of Example 2.4, write a subroutine called SECANT with parameter list (XOLD, XNEW, NTOL, TOL1, TOL2, X, IFLG, F) that accepts starting values (XOLD and XNEW), a limit on the maximum number of iterations (NTOL), a tolerance for $|F(XNEW)|$ as a termination requirement (TOL1), a tolerance for $|XNEW - XOLD|/|XNEW|$ as a termination requirement (TOL2), and an external function subprogram (F). The subroutine should return the approximate root in X and signal the mode of return in IFLG ($= 0$ if zero divide occurs, $= 1$ if TOL1 is satisfied, $= 2$ if TOL2 is satisfied). Carefully debug and test your program for the problems of Example 2.2 and 2.4. Be sure to document the subroutine by use of comments.

2.20 Suppose l is the secant line joining $(x_n, f(x_n))$ and $(x_{n-1}, f(x_{n-1}))$. Find the point of intersection of l and the x axis.

2.21 Use the secant method to determine the root of $x - \tan x = 0$ that is closest to 10. Your approximate root should have a relative error of at most 10^{-6}.

2.22 Use the program of Example 2.4 with TOL $= 1.E-06$ for the function

$f(x) = x^2 - e^{-x}$ with initial guesses $x_0 = 0$ and $x_1 = 1$. The unique positive root of f is $x^* \simeq 0.7034674$. Print the iterates and note the speed of convergence.

2.23 Use the program of Example 2.4 on the problem of exercise 2.8 and compare the results.

2.24 Use the ideas of Example 2.6 to write a simple program that calculates the ratio $2/7$ to six correct decimal digits without using the division operation.

2.25 It is required to find the approximate value of $\sqrt{7}$. For which starting values in Newton's method can you be assured of convergence? Hint: You may want to consider Theorem 2.5.

2.26 For $f(x) = x^3 - 5x - 3$, find an interval for which the conditions of Theorem 2.5 hold. Use the program of Example 2.7 to find the root in your interval.

*2.27 Frequently the problem of finding a solution of $f(x) = 0$ can be reformulated as that of finding a solution of $F(x) = x$ for some function F. Such a point is called a **fixed point** of F. If $f(x) = x^3 - 5x - 3$, then one choice for F is $F(x) = (5 + 3/x)^{1/2}$. Verify this.

(a) Determine at least two other choices of F for which solutions of $F(x) = x$ are roots of f and vice versa.

(b) The function f has exactly one root in $(2, 3)$. Consider the iteration

$$x_{n+1} = F(x_n) = (5 + 3/x_n)^{1/2}$$

If $x_0 = 2.5$, we find that $x_7 = 2.490863615\ldots$, with $|f(x_7)| \leq 0.5 \times 10^{-8}$. This is called **fixed point iteration**, and x_7 is the seventh fixed point iterate. Try your choices of F from (a) in a fixed point iteration with $x_0 = 2.5$. How do your iterations compare with the above fixed point iteration? Do they produce a convergent iteration?

2.28 We say that x^* is a root of **multiplicity** k of f if $f(x^*) = f'(x^*) = \cdots = f^{(k-1)}(x^*) = 0$, $f^{(k)}(x^*) \neq 0$. What is the multiplicity of $x^* = 0$ for $f(x) = e^{2x} - 1 - 2x - 2x^2$? Use the program of Example 2.7 for f with $x_0 = 0.5$ and observe the speed of convergence. Write a simple program for the following modification of Newton's method:

$$x_{n+1} = x_n - 3 \frac{f(x_n)}{f'(x_n)}$$

and observe the speed of convergence for $x_0 = 2.5$.

2.3 AN OVERVIEW

Our aim in this section is to make a comparison of the methods introduced in the previous two sections and to make some general comments about root finding. We begin by considering the conditioning of the problem.

Suppose $\{x_n\}$ is a sequence of approximations, obtained by any means, that converge to a root of f, say x^*. By the mean value theorem

$$f(x_n) - f(x^*) = f(x_n) = f'(\xi_n)(x_n - x^*)$$

for some ξ_n between x^* and x_n. Then $e_n = x^* - x_n$ satisfies

$$|e_n| = \frac{|f(x_n)|}{|f'(\xi_n)|}$$

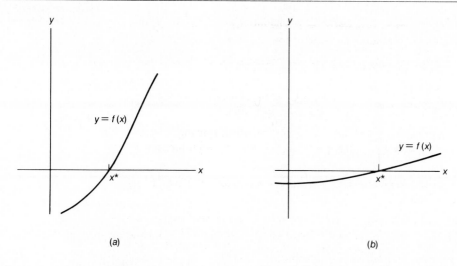

(a)

(b)

FIGURE 2.9
(a) Well-conditioned.
(b) Ill-conditioned.

If f' is continuous and $f'(x^*) \neq 0$, then for some interval $I_\delta = [x^* - \delta, x^* + \delta]$ we have $f'(x) \neq 0$ for all x in I_δ. Moreover, by the extreme value theorem there is a constant $m_\delta > 0$ such that $|f'(x)| \geq m_\delta$ for all x in I_δ. Since x_n converges to x^*, we have for n sufficiently large that x_n is in I_δ and hence ξ_n is in I_δ. Therefore we have the absolute error bound

$$|e_n| \leq \frac{|f(x_n)|}{m_\delta} \tag{2.14}$$

which is method-independent. If m_δ is small, then the problem of finding x^* is ill-conditioned (recall the discussion of Figure 2.1) as is depicted in Figure 2.9b.

In general the computation of multiple roots, that is, those for which $f'(x^*) = 0$, is ill-conditioned (exercise 2.28). If m_δ is known, then (2.14) may be used as an a posteriori error estimate for such methods as modified false position, secant, and Newton's.

Up to this point we have not mentioned the propagation of roundoff error in root finding methods. The reason for this omission is that roundoff error propagation is "nonexistent" for iterative methods. In an iterative method the determination of x_{n+1} depends only on x_n (and possibly x_{n-1}). This is the case for the methods presented in the text. We may regard x_n (and x_{n-1}) as a new initial guess for the iteration, and as such any roundoff errors in $x_{n-2}, x_{n-3}, \ldots$, are irrelevant. More specifically, consider an iterative method of the form $x_{n+1} = F(x_n)$. Let $\{\tilde{x}_n\}$ denote the computed sequence and δ_n denote the roundoff error in the computation of $F(\tilde{x}_n)$. Then

$$\tilde{x}_{n+1} = F(\tilde{x}_n) + \delta_n \qquad n \geq 0$$

[as an example take $F(x) = x - f(x)/f'(x)$, which corresponds to Newton's method], and we have

$$x^* - \tilde{x}_{n+1} = F(x^*) - F(\tilde{x}_n) - \delta_n$$

By the mean value theorem this may be written as

$$x^* - \tilde{x}_{n+1} = F'(\xi_n)(x^* - \tilde{x}_n) - \delta_n$$

or, equivalently,

$$[1 - F'(\xi_n)](x^* - \tilde{x}_{n+1}) = F'(\xi_n)(\tilde{x}_{n+1} - \tilde{x}_n) - \delta_n$$

Assume M is known such that $|F'(\xi_n)| \le M < 1$. Then we have

$$|x^* - \tilde{x}_{n+1}| \le \frac{M}{1 - M} |\tilde{x}_{n+1} - \tilde{x}_n| + \frac{|\delta_n|}{1 - M} \tag{2.15}$$

which furnishes an a posteriori error bound provided a bound for $|\delta_n|$ is known in terms of the machine unit. The first term on the right-hand side of (2.15) represents the truncation error, and the second term estimates the roundoff error. Note that the roundoff error depends only on δ_n and not on δ_k, $k < n$.

EXAMPLE 2.9

In Example 1.9 we gave the a posteriori error bound

$$|x_{n+1} - \sqrt{3}| \le \alpha |x_{n+1} - x_n| \qquad \alpha = \frac{1}{2}\left(1 - \frac{3}{L^2}\right)$$

for Newton's method, with $x_0 \in (2, L)$, applied to $f(x) = x^2 - 3$. This bound is obtained from (2.15) by ignoring roundoff error. Because $x_0 \in (2, L)$, it can be shown that $x_n \in [\sqrt{3}, L]$ for all $n \ge 1$. Also we have $F(x) = (x + 3/x)/2$, and hence $F'(x) = (1 - 3/x^2)/2$. It follows that $|F'(\xi_n)| \le \alpha$ for each n.

Next we introduce the order of convergence of a sequence. This is a measure of how rapidly a sequence converges. We say that a sequence $\{y_n\}$ converges to y with **order p** if

$$\lim_{n \to \infty} \frac{|y_{n+1} - y|}{|y_n - y|^p} = C_p > 0 \tag{2.16}$$

where C_p is called the asymptotic error constant. We refer to the convergence as **linear** if $p = 1$, **superlinear** if $1 < p < 2$, and **quadratic** if $p = 2$.

Loosely speaking, (2.16) means that for n sufficiently large, say $n \ge N$,

$$|y_{n+1} - y| \simeq C_p |y_n - y|^p$$

To gain an appreciation of the advantage of superlinear or quadratic convergence, let us suppose that $|y_n - y| = 10^{-k_n}|y|$ so that k_n essentially gives the number of correct decimal digits in the nth term. Then

$$10^{-k_{n+1}}|y| \simeq C_p 10^{-pk_n}|y|^p$$

and by taking logarithms we find

$$-k_{n+1} + \log_{10}|y| \simeq -pk_n + \log_{10} C_p|y|^p$$

or $\qquad k_{n+1} \simeq pk_n - \log_{10} C_p|y|^{p-1} \tag{2.17}$

If $C_p|y|^{p-1} \le 10$, then (2.17) says that the number of correct decimal digits in the $(n + 1)$st term is roughly p times that in the nth term.

For the secant method we can heuristically determine the order of convergence as follows. Let p denote the order; then for n sufficiently large we have by definition (2.16)

$$|e_{n+1}| \simeq C_p|e_n|^p \tag{2.18}$$

where $e_n = x^* - x_n$ and x_n is the nth secant iterate. Moreover, by (2.10) we have

$$|e_{n+1}| \simeq K|e_n||e_{n-1}|$$

where $K = |f''(x^*)|/2|f'(x^*)|$. It follows that

$$K|e_n||e_{n-1}| \simeq C_p|e_n|^p$$

and using (2.18) with n replaced by $n-1$ gives $C_p^{-1/p}|e_n|^{1/p} \simeq |e_{n-1}|$. Therefore, upon combining these expressions, we get

$$KC_p^{-1/p}|e_n|^{1+1/p} \simeq C_p|e_n|^p$$

and by equating the exponents of $|e_n|$ we find that $p = 1 + 1/p$ or, equivalently, $p^2 - p - 1 = 0$. Solving for p by the quadratic formula gives $p = (1 \pm \sqrt{5})/2$. Clearly the root $(1 - \sqrt{5})/2$ is less than 0 and can be ignored. Thus the secant method has order of convergence $(1 + \sqrt{5})/2 \simeq 1.618$ and the convergence is superlinear.

For Newton's method we recall (2.13), which says that the convergence is quadratic. However, if x^* is a multiple root, then it can be shown that the Newton sequence converges to x^* at a linear rate (exercise 2.28). It is possible to show that the order of convergence of the false position method is 1 and that for modified false position the order is approximately 1.442. On the basis of order the modified false position method is to be preferred among the root bracketing methods. For the locally convergent methods Newton's method converges fastest to a simple root; however, it requires two function evaluations per iteration. The evaluation of f' may require considerable additional computation time and in some problems may be impractical or impossible to find. In such cases the secant method is more efficient and should be used.

EXAMPLE 2.10

Consider the function $f(x) = x + 0.6 \cos x - 1$ on the interval $[0, 1]$. It is not difficult to see that f has a unique root x^* ($\simeq 0.5$) in $[0, 1]$. We use the programs of Examples 2.1, 2.2, 2.3, 2.4, and 2.7 on this problem so as to compare the rates of convergence. The bisection method required the most iterations, whereas the modified false position, secant, and Newton methods required no more than five iterations.

```
OUTPUT FROM BYSEKT,A=0.,B=1.,TOL=1.E-6


LEFT ENDPOINT    RIGHT ENDPOINT

0.0000000E 00    0.5000000E 00
0.2500000E 00    0.5000000E 00
```

```
0.3750000E 00    0.5000000E 00
0.4375000E 00    0.5000000E 00
0.4375000E 00    0.4687500E 00
0.4531250E 00    0.4687500E 00
0.4609375E 00    0.4687500E 00
0.4609375E 00    0.4648438E 00
0.4628906E 00    0.4648438E 00
0.4628906E 00    0.4638672E 00
0.4628906E 00    0.4633789E 00
0.4631348E 00    0.4633789E 00
0.4631348E 00    0.4632568E 00
0.4631958E 00    0.4632568E 00
0.4632263E 00    0.4632568E 00
0.4632263E 00    0.4632416E 00
0.4632263E 00    0.4632339E 00
0.4632301E 00    0.4632339E 00
0.4632320E 00    0.4632339E 00
```

ROOT IS 0.463233 ERROR BOUND = 0.4768372E-06

FALSE POSITION METHOD,A=0.,B=1.,14 ITERATIONS

A	B	TEST PT.
0.0000000E 00	0.5523484E 00	0.5523484E 00
0.0000000E 00	0.4770622E 00	0.4770622E 00
0.0000000E 00	0.4653462E 00	0.4653462E 00
0.0000000E 00	0.4635546E 00	0.4635546E 00
0.0000000E 00	0.4632818E 00	0.4632818E 00
0.0000000E 00	0.4632409E 00	0.4632409E 00
0.0000000E 00	0.4632341E 00	0.4632341E 00
0.0000000E 00	0.4632329E 00	0.4632329E 00
0.0000000E 00	0.4632328E 00	0.4632328E 00
0.0000000E 00	0.4632327E 00	0.4632327E 00
0.0000000E 00	0.4632325E 00	0.4632325E 00
0.0000000E 00	0.4632324E 00	0.4632324E 00
0.4632322E 00	0.4632324E 00	0.4632322E 00
0.4632322E 00	0.4632324E 00	0.4632324E 00

MODIFIED FALSE POSITION,A=0.,B=1.,TOL1=1.E-10,TO2=1.E-6

A	B	TEST PT.
0.0000000E 00	0.5523484E 00	0.4770622E 00
0.0000000E 00	0.4770622E 00	0.4541918E 00
0.4541918E 00	0.4770622E 00	0.4632778E 00
0.4541918E 00	0.4632778E 00	0.4632335E 00

ENDPOINTS ARE 0.4541918 0.4632778

ROOT= 0.4632335 F(ROOT)= 0.0000000E 00

SECANT METHOD,XOLD=0.,XNEW=1.,TOL=1.E-6

```
ITERATE        VALUE

    1       0.5523484E 00
    2       0.4441040E 00
    3       0.4638659E 00
    4       0.4632369E 00
    5       0.4632329E 00
    6       0.4632329E 00
```

TERMINATION CRITERION SATISFIED

NEWTON METHOD,XO=1.,TOL=1.E-6

```
ITERATE        VALUE

    1       0.3452451E 00
    2       0.4583765E 00
    3       0.4632236E 00
    4       0.4632323E 00
    5       0.4632323E 00
```

TERMINATION CRITERION SATISFIED

As was mentioned in the introduction, a reasonable general-purpose algorithm would use a bracketing method, such as bisection, to produce an initial approximation x_N to x^* (with a relative error of 10^{-2}, for instance). Then x_N and x_{N-1} could be used as starting points for several secant iterations to produce a refined approximation. One of the best algorithms currently available for root finding is due to Brent. This algorithm utilizes the bisection and secant methods in such a way as to retain the advantages of each, namely, certainty of bisection and speed of secant. The algorithm produces a small interval that contains the root, and if the function is well-behaved near the root, the order of convergence is superlinear. In the next section we discuss an algorithm that utilizes the bisection and Newton methods. The reader is asked to write a program for a simple version of the Brent algorithm in Project 3.

An adaptive hybrid method

In this section we give a subroutine called NEWBIS that utilizes the certainty of the bisection method and the speed of Newton's method. The basic idea is simple. We start with an interval $[a, b]$ in which f has a root. Initially we use the endpoint b to generate a Newton iterate, call it XN. If XN is in (a, b), we use it as the test point. Otherwise we use the midpoint as the test point. The test point is used (in the same manner as in bisection or false position) to produce a new, smaller interval that contains the root. The process is then repeated until the difference between successive test points is less than some prescribed tolerance (TOL) in a relative sense.

This hybrid approach is designed so that a sufficiently small interval is first produced by the bisection method and then used in Newton's method to produce the approximate root quickly and within the specified tolerance.

Subroutine NEWBIS is as follows:

```
        SUBROUTINE NEWBIS(A,B,F,DF,TOL,NMAX,X,IFLG)
C   A,B:ENDPOINTS OF INTERVAL WHICH CONTAINS ROOT
C   F,DF:EXTERNAL FUNCTION AND DERIVATIVE
C   X:APPROXIMATE ROOT
C   IFLG:SIGNALS MODE OF RETURN
        IF(A.LE.B) GO TO 1
        TEMP=A
        A=B
        B=TEMP
    1   FA=F(A)
        SFA=SIGN(1.,FA)
        FB=F(B)
        IFLG=1
        IF(SFA*FB.LE.0.) GO TO 3
    2   IFLG=-IFLG
        RETURN
    3   WRITE(6,300)
  300   FORMAT(' ',2X,'METHOD',8X,'VALUE'/)
        X=B
        FX=FB
C   BEGIN ITERATION
        DO 7 K=1,NMAX
        XN=X-FX/DF(X)
C   DETERMINE TEST POINT
        XTEST=XN
        IF(A.GT.XN.OR.B.LT.XN) XTEST=(A+B)/2.
        TOLX=TOL*ABS(XTEST)
        DIFF=ABS(XTEST-X)
        IF(XTEST.EQ.XN) GO TO 5
C   USE BISECTION FOR THE TESTPOINT
        WRITE(6,100)XTEST
  100     FORMAT(' ',2X,'BISECTION',3X,E14.7)
        GO TO 6
C   USE NEWTON FOR THE TEST POINT
    5     WRITE(6,101)XTEST
  101     FORMAT(' ',2X,'NEWTON',6X,E14.7)
    6     X=XTEST
C   CHECK TERMINATON CRITERION
        IF(DIFF.LE.TOLX)RETURN
        FX=F(X)
```

```
C   UPDATE INTERVAL
        IF(SFA*FX.GE.O.)A=X
        IF(A.EQ.X) GO TO 7
        B=X
  7   CONTINUE
        WRITE(6,200)
 200  FORMAT(' ',/'MAX. ITERATIONS REACHED'/)
        RETURN
        END
```

The following gives a simple run of NEWBIS.

EXAMPLE 2.11

```
      EXTERNAL F,DF
      A=1.8
      B=9.
      TOL=1.E-6
      NMAX=40
      CALL NEWBIS(A,B,F,DF,TOL,NMAX,X,IFLG)
      IF(IFLG.LT.0) GO TO 1
      WRITE(6,10)A,B
 10   FORMAT(' ','LEFT END=',2X,F9.7,2X,'RIGHT END=',F9.7/)
      WRITE(6,20)X
 20   FORMAT(' ','TERMINAL ITERATE=',2X,F9.7)
      STOP
  1   WRITE(6,11)
 11   FORMAT(' ','SAME SIGNS AT THE ENDPOINTS')
      STOP
      END
C
      REAL FUNCTION F(X)
      F=X**2-6.*X+6.75
      RETURN
      END
C
      REAL FUNCTION DF(X)
      DF=2.*X-6.
      RETURN
      END
```

METHOD	VALUE
NEWTON	0.6187500E 01
NEWTON	0.4946692E 01
NEWTON	0.4551251E 01
NEWTON	0.4500844E 01
NEWTON	0.4500000E 01
NEWTON	0.4500000E 01

LEFT END= 4.5000000 RIGHT END=4.5008440

TERMINAL ITERATE= 4.5000000

EXAMPLE 2.12

In the previous example of NEWBIS the Newton phase commences immediately; however, when we apply NEWBIS with the initial interval $[0, 3.1]$,
there results the output

```
METHOD          VALUE

BISECTION    0.1550000E 01
NEWTON       0.1499137E 01
NEWTON       0.1499999E 01
NEWTON       0.1499999E 01
LEFT END=  1.4999990  RIGHT END=1.5500000

TERMINAL ITERATE=  1.4999990
```

EXERCISES

2.29 Fill in the details of Example 2.9.

2.30 Apply inequality (2.14) to Example 2.3 and obtain the error bound

$$|x^* - 0.7034671| \le 1.63 \times 10^{-6}$$

Hint: Show that $|f'(x)| \ge e^{-1}$ on $[0, 1]$.

2.31 Apply the programs in the text for modified false position, secant, and Newton's
method to $f(x) = e^{2x} - 1 - 2x - 2x^2$ on the interval $[-0.5, 0.5]$. Observe the speed of
convergence to the root $x^* = 0$. Compare the output with that of the similar problem
given by $g(x) = e^{2x} - 1 - 3x - 2x^2$ on $[-0.5, 0.5]$. Interpret the results.

2.32 On the basis of the output for exercise 2.31 which of the two problems is better
conditioned? Is $x^* = 0$ a multiple root of f? of g?

*2.33 Newton's method can be written as

$$x_{n+1} = F(x_n)$$

where $F(x) = x - f(x)/f'(x)$. For the problem of Example 2.7 we have $F(x) =
x - (x^2 - e^{-x})/(2x + e^{-x})$. A simple calculation gives $F'(x) = (x^2 - e^{-x})(2 - e^{-x})/
(2x + e^{-x})^2$. Show that $|F'(x)| \le 0.06 = M$ for x in $[0.7, 0.8]$. Use this M, (2.15), and
the output in Example 2.7 to get

$$|x^* - 0.7034674| \le 0.000043 + 1.07|\delta_4| \simeq 0.000043$$

2.34 Newton's method for finding the square root of $A > 0$ is

$$x_{n+1} = \tfrac{1}{2}\left(x_n + \frac{A}{x_n}\right)$$

(a) Let $F(x) = (x + A/x)/2$ and show that a bound for F' on $[\sqrt{A}, L]$ is given by
$(1 - A/L^2)/2$.

(b) For $A = 7$, use the program of Example 2.7 to compute $\sqrt{7}$ approximately. Use the
initial guess X0 = 7.

(c) Use the results of parts (a) and (b) and (2.15) to estimate the accuracy of the terminal iterate in part (b).

2.35 The order of convergence of the secant method is $p \simeq 1.61803$. In the proximity of a root we have by (2.16) that $|e_{n+1}| \simeq C_p |e_n|^p$ and hence $\log|e_{n+1}| - p \log|e_n| \simeq$ constant. Use the program of Example 2.4 applied to the function of exercise 2.22 to print several values of this logarithm expression. These printed values should be nearly constant as x_n approaches $x^* = 0.7034674$.

2.36 Use the program of Example 2.4 and the function of exercise 2.10 to find the root x^* in $[1, 5]$. Take X0 = 5. Print several values of the expression $\log|e_{n+1}| - 2 \log|e_n|$, where $e_n = x^* - x_n$ and $x^* = 1.672981$.

2.37 Apply NEWBIS to the function $f(x) = e^{2x} - 1 - 2x - 2x^2$ on $[-0.5, 0.5]$. Compare the output with that of exercise 2.31.

2.38 Find the root of $f(x) = x + \tan x$ that is closest to 10 by using NEWBIS. Caution: Select the initial interval so that it does not contain any numbers of the form $(2n + 1)\pi/2, n = 0, \pm 1, \ldots$.

2.39 Suppose you are in charge of library subroutines at company XYZ. In choosing a program for root finding, there are many considerations. Which of the following attributes of a program do you think are most important and which do you consider essential:

(1) Is efficient
(2) Is guaranteed to converge for all continuous functions (given suitable starting values)
(3) Is easy to read
(4) Is well-documented
(5) Is short
(6) Is guaranteed to converge for all continuously differentiable functions (given suitable starting values)
(7) Uses structured programming
(8) Is restricted to IBM machines
(9) Is written in ANSI standard FORTRAN
(10) Has been thoroughly tested and debugged with a guarantee in writing
(11) Is limited to double-precision computation
(12) Furnishes an absolute error bound
(13) Furnishes a relative error bound
(14) Requires the user to supply the maximum number of iterations as a parameter
(15) Signals the reason for termination of iteration
(16) Produces a small interval that brackets the root

2.40 Consider the following modification of NEWBIS. Change the definition of TOLX to

$$\text{TOLX} = \text{TOL*ABS(XTEST)} + \text{TOL}$$

If TOL = 1.E−06, estimate the value of TOLX for a root $\simeq 1$, a root $\simeq 100$. Discuss the relative merits of each definition of TOLX.

2.41 In NEWBIS the Newton iterate XN is used as the test point provided it falls within the most recent bracketing interval. Suppose we also require that $|F(XN)| < |F(X)|$, where X is the previous test point. Modify NEWBIS to incorporate this additional requirement. Apply the modified NEWBIS to the problems of exercises 2.37 and 2.38. Compare the results and discuss the effect, if any, of this modification.

2.42 Modify NEWBIS to select the test point based on the following criterion. Assuming the Newton iterate XN falls within the most recent bracketing interval, choose XTEST = XN only if $|F(XN)| < |F(XM)|$, where XM[$= (A + B)/2$.] is the midpoint of the latest bracketing interval. Compare this version of NEWBIS with the original by application to the problems in exercise 2.37 and 2.38. Discuss your results.

Project 2

The object of this project is to modify NEWBIS so that it is more like a library subroutine.

(a) It is good practice to avoid repeated function evaluation at the same point; the value should be saved for future use in the program. Does NEWBIS incorporate this strategy? If not, make appropriate changes.

(b) Does NEWBIS provide protection against an unreasonably small TOL specification? If not, modify the subroutine. Hint: See BISECT.

(c) The termination criterion used in NEWBIS is return if $|XTEST - X| \leq TOL|XTEST|$. However, it also makes sense to return if $A*B > 0$ and $|A - B| \leq TOL|A + B|/2$. (See Theorem 2.1.) Modify NEWBIS to include both criteria. Also use IFLG to signal which termination criterion causes the RETURN, say IFLG = 1 for the former and IFLG = 2 for the latter criterion.

(d) The Newton iterate is used as the test point whenever it falls within the most recent bracketing interval [A, B]. It is more effective to require, in addition, that $\min\{|A - XN|, |B - XN|\} \geq \alpha|B - A|$, where $0 < \alpha \leq 0.5$ is a given constant. The effect of this requirement is to prevent the (Newton) test point from being too close to either endpoint. Add this new requirement to NEWBIS with $\alpha = 0.25$.

(e) Change NEWBIS so that it returns when FX = 0, in which case set IFLG = 3.

(f) Change the computation of the midpoint to XTEST = $A + (B - A)/2$.

Test your subroutine on the problems of Examples 2.1, 2.2, 2.10, and 2.11 and exercise 2.28. Compare the results and discuss the relative merits of your revised subroutine. Mention any additional improvements to NEWBIS that you deem appropriate and give the rationale for their usage.

Project 3

As was mentioned in the text, one of the best root finding algorithms currently available uses a hybrid secant-bisection method. In this project you are required to write a subroutine SECBIS (A,B,F,TOL,X,IFLG) that accepts

A,B: endpoints of interval that contains a root of F [that is, F(A)F(B) < 0]

F: external function subprogram

TOL: relative error tolerance

and returns

A,B: endpoints of final interval that contains a root of F

X: approximate root

IFLG: integer signal of the mode of return

The subroutine is to function in the same way as NEWBIS, with the role of Newton's method being replaced by the secant method. If XS denotes the secant iterate, it should be used as XTEST if

(a) $A \leq XS \leq B$

(b) $\min\{|A - XS|, |B - XS|\} \geq 0.25 |B - A|$

where [A, B] is the most recent bracketing interval.

Use the same variable names as in NEWBIS with the exception of XN; change it to XS.

In writing your subroutine, keep in mind the following desirable features:

1   Avoid repeated function evaluation at the same point

2   Protect against an unreasonable TOL

3   Protect against division by zero

4   Use the form (2.6)

Improved value = approximate value $\pm$ correction term

5   Terminate if

$|XTEST - X| \leq TOL|XTEST|$

or $|A - B| \leq TOL|A + B|/2$, $AB > 0$

or $FX = 0$

Set IFLG = 1, 2, or 3, respectively.

Test your subroutine on the problems of Examples 2.1, 2.2, 2.10, 2.11, and exercise 2.28. Compare the results and discuss the relative merits of SECBIS. Mention any improvements you think are appropriate. You may want to consult Rheinboldt (1981).

2.4   REAL ROOTS OF POLYNOMIALS

Because polynomial equations arise frequently in applications, there is a vast literature devoted to polynomial root finding techniques. Most of these methods take advantage of the special form of polynomials, and the mathematical theory required for a good understanding of these methods is quite deep. We give only a brief introduction to this class of problems. We restrict our attention to polynomials with real coefficients and consider only the determination of real roots. Before discussing any approximation methods, we need to develop some of the basic theory about polynomials and their roots.

A **polynomial** of degree n is a function of the form

$$p(x) = a_0 + a_1 x + a_2 x^2 + \cdots + a_n x^n \tag{2.19}$$

where the **coefficients** a_i are real constants and $a_n \neq 0$. The most basic result about polynomials is the fundamental theorem of algebra, which we now state.

THEOREM 2.6

Suppose $p(x)$ is given by (2.19) with $n \geq 1$. Then there exists a number x^* (possibly complex) such that $p(x^*) = 0$.

If $w = u + iv$ ($i = \sqrt{-1}$) is a root of p, then, because the coefficients are real, it is not difficult to show that the complex conjugate of w, $\bar{w} = u - iv$, is also a root of p. For example, if $p(x) = x^2 - 4x + 5$, then the roots of p are found by the quadratic formula to be $2 + i$ and $2 - i$. Thus complex roots always occur in conjugate pairs.

In any root finding method we need to repeatedly evaluate the polynomial at a point. The most efficient means of polynomial evaluation is based on rewriting (2.19) in the form

$$p(x) = a_0 + x(a_1 + x(a_2 + x(a_3 + \cdots + x(a_{n-1} + a_n x) \cdots) \tag{2.20}$$

EXAMPLE 2.13

The polynomial $p(x) = -3 + 2x + 4x^2 - 6x^3 + 5x^4$ is written in the form (2.20) as:

$$p(x) = -3 + x(2 + x(4 + x(-6 + 5x)))$$

For the polynomial of Example 2.13 we see that p can be evaluated at a point, say x_0, by using four additions and four multiplications. Specifically we could use the algorithm

$$b_3 \leftarrow -6 + 5x_0$$
$$b_2 \leftarrow 4 + b_3 x_0$$
$$b_1 \leftarrow 2 + b_2 x_0$$
$$p(x_0) = b_0 \leftarrow -3 + b_1 x_0$$

This procedure works for any polynomial and is commonly called nested multiplication or Horner's rule.

ALGORITHM 2.6

NESTED MULTIPLICATION   Given a polynomial in the form (2.19) and a point x_0,

$$b_n \leftarrow a_n$$

Do $k = n - 1, n - 2, \ldots, 0:$

$$b_k \leftarrow a_k + b_{k+1} x_0$$

then $p(x_0) = b_0$.

VALUE := c{n};
FOR i! = n-1 DOWNTO 0 DO
*VALUE := VALUE * X + c{i}*

EXAMPLE 2.14

We evaluate the polynomial of Example 2.13 at $x_0 = -2$ by the nested multiplication algorithm. We have

$$b_3 \leftarrow -6 + 5(-2) = -16$$

$$b_2 \leftarrow 4 + (-16)(-2) = 36$$

$$b_1 \leftarrow 2 + 36(-2) = -70$$

$$b_0 \leftarrow -3 + (-70)(-2) = 137$$

Next we show how this algorithm can be used for division of a polynomial by a linear term. Let $q(x)$ be the polynomial of degree $n - 1$ given by

$$q(x) = b_1 + b_2 x + b_3 x^2 + \cdots + b_n x^{n-1}$$

and consider the expression $b_0 + (x - x_0)q(x)$. We have   *why/where*

$$b_0 + (x - x_0)q(x) = b_0 + (x - x_0)(b_1 + b_2 x^1 + \cdots + b_n x^{n-1})$$

$$= (b_0 - b_1 x_0) + (b_1 - b_2 x_0)x + \cdots$$

$$+ (b_{n-1} - b_n x_0)x^{n-1} + b_n x^n$$

and we recall from Algorithm 2.6 that $b_n = a_n$ and that, for $k = 0, 1, \ldots, n-1$, $b_k - b_{k+1} x_0 = a_k$. Therefore we find

$$p(x) = b_0 + (x - x_0)q(x) \tag{2.21}$$

Equation (2.21) says that $q(x)$ is the **quotient** if $p(x)$ is divided by $(x - x_0)$ and b_0 is the **remainder**.

EXAMPLE 2.15

For the polynomial of Example 2.13, division by the linear term $x + 2$ gives

$$\frac{p(x)}{x + 2} = -70 + 36x - 16x^2 + 5x^3 + \frac{137}{x + 2}$$

EXAMPLE 2.16

The following is a subroutine for Algorithm 2.6.

```
      SUBROUTINE HORNER(A,B,M,K,X,VALUE)
C  A:ARRAY FOR COEFFICIENTS OF POLY. P
C  B:ARRAY FOR COEFFICIENTS OF POLY. Q
C  M,K:DIMENSIONS OF A,B RESPECTIVELY
C  X:POINT AT WHICH VALUE OF P IS FOUND
C  VALUE: VALUE OF P(X)
C  K SHOULD EQUAL M-1
      DIMENSION A(M),B(K)
      MEND=M-2
      B(M-1)=A(M)
      IF(M.EQ.2) GO TO 2
      DO 1 J=1,MEND
        L=M-J-1
        LL=L+1
        B(L)=A(LL)+B(LL)*X
    1 CONTINUE END DO
    2 VALUE=A(1)+B(1)*X
      RETURN
      END
```

In this subroutine b_0 is stored in VALUE, b_1 in B(1), b_2 in B(2), and so forth. The coefficients of p in form (2.19) are stored in A with a_0 in A(1), a_1 in A(2), and so on. The degree of p is $M - 1$, and M must be at least 2.

If x_0 is a root of p, then (2.21) gives

$$p(x) = (x - x_0)q(x) \tag{2.22}$$

which says that x_0 is a root of p if and only if $x - x_0$ is a **factor** of p. In this case we see that every root of q is also a root of p. Hence we can find other roots of p by finding roots of a polynomial of reduced degree, namely q. This process is known as **deflation**.

Another consequence of (2.21) is that it provides a convenient means of evaluating the derivative of p. By differentiation of (2.21) we find

$$p'(x) = (x - x_0)q'(x) + q(x)$$

and hence $p'(x_0) = q(x_0)$. Since the coefficients of q are known, we can easily evaluate $q(x_0)$ by application of Algorithm 2.6 to q. We make use of this observation in the following version of Newton's method for polynomials.

ALGORITHM 2.7

NEWTON'S METHOD FOR POLYNOMIALS Given a polynomial of the form (2.19) and an initial guess x_0,

$$b_n \leftarrow a_n, \, c_n \leftarrow a_n$$

Do $N = 0, 1, \ldots, $ NMAX:

  Do $k = n - 1, n - 2, \ldots, 0$:

  $\qquad b_k \leftarrow a_k + b_{k+1} x_N$

  Do $k = n - 1, n - 2, \ldots, 1$:

  $\qquad c_k \leftarrow b_k + c_{k+1} x_N$

  $\qquad x_{N+1} \leftarrow x_N - \dfrac{b_0}{c_1}$

In this algorithm the two inner loops correspond to Algorithm 2.6 applied to p and $q(p')$.

EXAMPLE 2.17

In the following program we apply Algorithm 2.7 to the polynomial $p(x) = x^5 + 3x^4 - 5x^3 - 15x^2 + 4x + 12$, which has roots -3, -2, -1, 1, and 2.

```
      DIMENSION A(10),B(9),C(8)
      DATA (A(J),J=1,6)/12.,4.,-15.,-5.,3.,1./
      DATA M/6/
      X=7.5
      WRITE(6,105)
105   FORMAT(' ',3X,'N',7X,'X(N)',10X,'X(N)-X(N-1)'/)
      TOL=1.E-6
      UNIT=1.
1     UNIT=.5*UNIT
      DELTA=UNIT+1.
      IF(DELTA.GT.1.) GO TO 1
      TOL=TOL+UNIT
      NTOL=25
      MM=M-1
      M2=M-2
      DO 2 J=1,NTOL
         CALL HORNER(A,B,M,MM,X,PX)
         CALL HORNER(B,C,MM,M2,X,DPX)
         RATIO=PX/DPX
         X=X-RATIO
         DIFFX=ABS(RATIO)
         WRITE(6,100)J,X,DIFFX
100      FORMAT(' ',2X,I3,2(2X,E14.7))
         TOLX=TOL*ABS(X)
         IF(DIFFX.LE.TOLX.AND.ABS(PX).LE.TOL) STOP
2     CONTINUE  END DO
      WRITE(6,101)NTOL
```

```
101 FORMAT(' ','TOL TOO SMALL FOR  ',I2,2X,'ITERATIONS')
    STOP
    END
```

N	X(N)	X(N)-X(N-1)
1	0.5970510E 01	0.1529490E 01
2	0.4770670E 01	0.1199841E 01
3	0.3841132E 01	0.9295378E 00
4	0.3136437E 01	0.7046946E 00
5	0.2622935E 01	0.5135012E 00
6	0.2277111E 01	0.3458241E 00
7	0.2081802E 01	0.1953078E 00
8	0.2009938E 01	0.7186371E-01
9	0.2000172E 01	0.9766217E-02
10	0.2000000E 01	0.1716201E-03
11	0.2000000E 01	0.0000000E 00

EXAMPLE 2.18

In the previous example we saw that the Newton sequence converges to the root that is closest to the initial guess ($7.5 = x_0$). However, for the initial guess $x_0 = 1.5$ we get the following output:

N	X(N)	X(N)-X(N-1)
1	0.3986015E 00	0.1101398E 01
2	0.1559913E 01	0.1161311E 01
3	-0.9118776E 00	0.2471790E 01
4	-0.9960576E 00	0.8417994E-01
5	-0.9999896E 00	0.3932014E-02
6	-0.9999999E 00	0.1041079E-04
7	-0.9999999E 00	0.0000000E 00

Thus the Newton sequence converges to the root -1 even though x_0 is closer to the root $+1$.

These two examples illustrate Newton's method for finding a real root. Suppose it is required to find all real roots. As previously mentioned we can use deflation after each approximate root has been determined and thereby reduce the degree of the polynomial at each stage of the computation. Specifically we consider the polynomial

$$p(x) = x^5 - 12x^4 - 293x^3 + 3444x^2 + 20{,}884x - 240{,}240 \qquad (2.23)$$

which has roots -11, -13, 10, 12, and 14. Using the method of Example 2.17 applied to (2.23) with initial guess $X = 11$, we find the approximate root $\hat{x}_1 = 13.99990$. Then we deflate (2.23) to get

$$p(x) = (x - \hat{x}_1)q(x) + b_0$$

where $q(x) = x^4 + 1.999895x^3 - 265.0015x^2 - 265.9927x + 17,160.13$. Since $\hat{x}_1$ is not an exact root of p, we must have $b_0 \neq 0$ and we are not guaranteed that roots of q are also roots of p. Nonetheless we apply Newton's method to q with initial guess $X = 11$ to get the next approximate root, $\hat{x}_2 = 12.00016$. Then we deflate q to get

$$q(x) = (x - \hat{x}_2)r(x) + b_0'$$

where $r(x) = x^3 + 14.00005x^2 - 96.99867x - 1429.992$. Again we are not assured that roots of r are also roots of q or p, but we persist by applying Newton's method to r with initial guess $X = 9$ to find $\hat{x}_3 = 9.999949$. Deflation of r gives

$$r(x) = (x - \hat{x}_3)(x^2 + 23.99998x + 142.9999) + b_0''$$

The two remaining roots are found by the quadratic formula:

$$\hat{x}_4, \hat{x}_5 = \tfrac{1}{2}\{-23.99998 \pm [(-23.99998)^2 - 4(142.9999)]^{1/2}\}$$
$$= -12.99991, -11.00006$$

EXAMPLE 2.19

Below is the output using Newton's method and deflation applied to (2.23). In addition to the iterates we have also printed the final contents of array B, which contain the coefficients of the deflated polynomial for the computation of the next root.

N	X(N)	X(N)-X(N-1)
1	0.1506154E 02	0.4061538E 01
2	0.1441228E 02	0.6492534E 00
3	0.1409404E 02	0.3182408E 00
4	0.1400651E 02	0.8753669E-01
5	0.1399999E 02	0.6515343E-02
6	0.1399992E 02	0.6944429E-04
7	0.1399999E 02	0.6945318E-04
8	0.1399992E 02	0.6944433E-04
9	0.1399998E 02	0.5787761E-04
10	0.1399994E 02	0.3472304E-04
11	0.1399990E 02	0.4629967E-04
12	0.1399990E 02	0.0000000E 00

THE FINAL CONTENTS OF B ARE:

```
0.1716013E 05  -0.2659927E 03  -0.2650015E 03   0.1999895E 01
0.1000000E 01
     X=11.0 IN THIS INITIAL RUN
```

N	X(N)	X(N)-X(N-1)
1	-0.4638090E 00	0.1146381E 02
2	0.8929688E 03	0.8934326E 03
3	0.6696387E 03	0.2233299E 03

```
 4     0.5021536E 03     0.1674849E 03
 5     0.3765564E 03     0.1255972E 03
 6     0.2823806E 03     0.9417569E 02
 7     0.2117783E 03     0.7060229E 02
 8     0.1588657E 03     0.5291266E 02
 9     0.1192333E 03     0.3963232E 02
10     0.8957832E 02     0.2965501E 02
11     0.6742876E 02     0.2214957E 02
12     0.5093744E 02     0.1649132E 02
13     0.3872675E 02     0.1221069E 02
14     0.2977170E 02     0.8955047E 01
15     0.2330981E 02     0.6461881E 01
16     0.1876930E 02     0.4540511E 01
17     0.1570962E 02     0.3059681E 01
18     0.1377536E 02     0.1934257E 01
19     0.1267148E 02     0.1103880E 01
20     0.1215737E 02     0.5141132E 00
21     0.1201259E 02     0.1447769E 00
22     0.1200025E 02     0.1234631E-01
23     0.1200016E 02     0.9169152E-04
24     0.1200016E 02     0.3396350E-05
25     0.1200016E 02     0.0000000E 00
```

THE FINAL CONTENTS OF B ARE:

-0.1429992E 04   -0.9699867E 02   0.1400005E 02   0.1000000E 01
        X=11. IN THIS SECOND RUN

```
N        X(N)              X(N)-X(N-1)

1     0.1010546E 02     0.1105462E 01
2     0.1000095E 02     0.1045139E 00
3     0.9999947E 01     0.1000648E-02
4     0.9999948E 01     0.1516410E-05
5     0.9999949E 01     0.1010940E-05
6     0.9999949E 01     0.0000000E 00
```

THE FINAL CONTENTS OF B ARE:

0.1429999E 03   0.2399998E 02   0.1000000E 01
        X=9. IN THIS THIRD RUN

USING THE REMAINING DEFLATED POLY GIVEN BY
142.9999+23.99998*X+X**2 WE FIND, BY THE
QUADRATIC FORMULA, THE REMAINING ROOTS TO BE
-12.99991 AND -11.00006

In applying the deflation process, we are using a polynomial whose coefficients are perturbed because the previous approximate roots are not exact. For instance, we used q in the second stage of Example 2.19 whereas the correct polynomial is $x^4 + 2x^3 - 265x^2 - 266x + 17{,}160$. How does the perturbation of the coefficients of a polynomial affect the roots of the polynomial?

Suppose x^* is a simple root of $p(x) = \sum_{k=0}^{n} a_k x^k$ and let $\tilde{p}(x) = \sum_{k=0}^{n} \tilde{a}_k x^k$, where $\tilde{a}_k = (1 - \rho_k)a_k$ and $|\rho_k| \ll 1$. Then it can be shown that there is a root $\tilde{x}$ of $\tilde{p}$ that satisfies

$$\frac{x^* - \tilde{x}}{x^*} \simeq \sum_{k=0}^{n} \frac{\rho_k a_k (x^*)^{k-1}}{p'(x^*)} \qquad (2.24)$$

Thus the sensitivity of a root to changes in the coefficients depends on which root we are considering and on the changes. Suppose only one coefficient is changed, say $\rho_j \neq 0$ and $\rho_k = 0$ if $k \neq j$. Then (2.24) becomes

$$\frac{x^* - \tilde{x}}{x^*} \simeq \frac{\rho_j a_j (x^*)^{j-1}}{p'(x^*)} \qquad (2.25)$$

EXAMPLE 2.20

Suppose we want to compute the roots of the polynomial $p(x) = x^4 - 16.1x^3 + 66.6x^2 - 56.5x + 5$ by using an appropriate modification of the program in Example 2.17. Unfortunately we use the incorrect DATA statement,

DATA (A(J),J = 1,5)/5., $-$ 56.5,66.6, $-$ 16.5,1./

and so we find the roots of the polynomial $\tilde{p}(x) = x^4 - 16.5x^3 + 66.6x^2 - 56.5x + 5$. Let us use (2.25) to estimate the relative error in the roots. We have $\rho_3 = (a_3 - \tilde{a}_3)/a_3 \simeq -0.0248$ and $\rho_3 a_3 = -0.4$. The roots of p are 0.1, 1, 5, and 10, and some simple calculations using (2.25) give the following estimated relative errors:

Root	Estimated Relative Error
$x^* = 0.1$	0.9×10^{-4}
$x^* = 1$	-0.12×10^{-1}
$x^* = 5$	0.45×10^{-1}
$x^* = 10$	-0.72×10^{-1}

We expect the root $x^* = 0.1$ to be least affected by the error in coefficient a_3. We have used deflation and Newton's method to compute the roots of $\tilde{p}$ in increasing order of magnitude. The results are

x^*	$\tilde{x}$	$(x^* - \tilde{x})/x^*$
0.1	0.9999084E $-$ 01	0.92×10^{-4}
1	0.1012702E $-$ 01	-0.13×10^{-1}
5	0.4560791E $-$ 01	0.88×10^{-1}
10	0.1082649E $-$ 02	-0.83×10^{-1}

A comparison of the two tables shows considerable agreement between the estimated and computed relative errors.

If $r_1, \ldots, r_k$ are roots of p, then by induction we may write p as

$$p(x) = (x - r_1)(x - r_2)(x - r_2) \cdots (x - r_k)r(x)$$

where the degree of $r(x)$ is $n - k$. It follows that a polynomial of degree n has exactly n roots (possibly complex), counting their multiplicities. For example, the polynomial $x^3 + 2x^2 - x - 2$ has three distinct roots, namely 1, -1, and -2, whereas the polynomial $x^3 + 3x^2 - 4$ has roots 1, 2, and 2. In the latter case 2 is a root of multiplicity 2. If the roots of (2.19) are $r_1, \ldots, r_n$, then p may be factored as

$$p(x) = a_n(x - r_1)(x - r_2) \cdots (x - r_n) = a_n \prod_{j=1}^{n} (x - r_j) \qquad (2.26)$$

For example, the polynomial $p(x) = x^3 + 2x^2 - x - 2$ is written in factored form as $p(x) = (x - 1)(x + 1)(x + 2)$.

In attempting to use Newton's method, it is usually essential that the initial guess be close to the desired root. For other methods knowledge of the locale of the roots may not be necessary but is certainly advantageous. We give some techniques that are useful in determining the locale of some or all of the roots. The first result in this direction gives a useful method for determining the number of positive (real) roots of a polynomial.

DESCARTES' RULE OF SIGNS

Write down the signs (in order) of the nonzero coefficients of the polynomial. Count the number of sign changes in the list, call it N. Let k be the number of positive roots. Then (a) $k \le N$ and (b) $N - k = $ an even integer.

EXAMPLE 2.21

Suppose it is required to determine the number of positive roots of $p(x) = x^4 + 2x^3 + x^2 + 8x - 12$. The sign pattern is $+, +, +, +, -$, and so $N = 1$. Hence $1 - k$ must be a nonnegative even integer. The only possibility is $k = 1$; that is, p has exactly one positive root.

Actually Descartes' rule of signs allows us to find the number of negative roots as well. The basis for this is the observation that x^* is a root of $p(x)$ if and only if $-x^*$ is a root of $p(-x)$. For the polynomial of Example 2.21 we have $p(-x) = x^4 - 2x^3 + x^2 - 8x - 12$. The sign pattern is $+, -, +, -, -$, and hence $N = 3$. Therefore, the number of negative roots of p is either one or three.

In order to get more information about the roots of p, we can make a change of variables, say $x = t + \alpha$ for some constant α. If $q(t) = p(t + \alpha)$, then positive roots of q correspond to roots of p that are greater than α and negative roots of q correspond to roots of p that are less than α. As an illustration of this technique we write the polynomial of Example 2.21 in its

Taylor expansion about 2. By elementary calculus we find $p(2) = 40$, $p'(2) = 68$, $p''(2) = 74$, $p'''(2) = 60$, $p^{(4)}(2) = 24$, and $p^{(k)}(2) = 0$ for all $k \geq 5$. Thus by Taylor's theorem

$$p(x) = 40 + 68(x - 2) + \frac{74(x - 2)^2}{2!} + \frac{60(x - 2)^3}{3!} + \frac{24(x - 2)^4}{4!}$$

Now let $q(t) = p(t + 2) = 40 + 68t + 37t^2 + 10t^3 + t^4$. There is no sign change in the sign pattern for q, and it follows that p has no roots that are greater than 2. On the other hand we have $q(-t) = 40 - 68t + 37t^2 - 10t^3 + t^4$, which has sign pattern $+, -, +, -, +$, with four sign changes. Hence q has either zero or two negative roots, which implies that p has either zero or two roots that are less than 2. Let us summarize what we know about the real roots of p. First we found that p has one root in $(0, \infty)$ and no roots in $(2, \infty)$. Thus $(0, 2)$ must contain exactly one root of p. Also we found that $(-\infty, 2)$ contains zero or two roots and $(-\infty, 0)$ contains one or three roots of p. It follows that p has exactly one negative root. Therefore p has only two real roots. By considering other points α, one can use this technique to determine smaller intervals, each of which contains a (real) root of p.

Some other results that are useful in **root localization** are given in the following.

THEOREM 2.7

Suppose p is given by (2.19), and let $r_1, r_2, \ldots, r_n$ denote its roots. Then

(i) $-a_{n-1}/a_n = \displaystyle\sum_{k=1}^{n} r_k$

(ii) $(-1)^n a_0/a_n = r_1 r_2 \cdots r_n = \displaystyle\prod_{k=1}^{n} r_k$

(iii)   If $\rho = 1 + \displaystyle\max_{0 \leq k \leq n-1} |a_k|/|a_n|$, then $|r_k| < \rho$ for all k

(iv)   Let $R_1 = n|a_0|/|a_n|$ and $R_2 = [|a_0|/|a_n|]^{1/n}$ and set $R = \min\{R_1, R_2\}$; then for some k, $|r_k| \leq R$.

For the polynomial of Example 2.13 we have $\sum_{k=1}^{4} r_k = \frac{6}{5}$ and $\prod_{k=1}^{4} r_k = -\frac{3}{5}$. By (iii) of Theorem 2.7 every root satisifies $|r_k| \leq \frac{11}{5}$, and by (iv) there is a root that satisfies $|r_k| \leq (\frac{3}{5})^{1/4} = 0.8801 \ldots$.

We use Theorem 2.7 to derive a bound that can be used as a termination criterion for Newton's method.

THEOREM 2.8

Let p be given by (2.19) and suppose α is such that $p(\alpha)$, $p'(\alpha) \neq 0$. Then there is a root x^* of p such that

$$|x^* - \alpha| \leq \frac{n|p(\alpha)|}{|p'(\alpha)|}$$

Proof

Let $r_1, \ldots, r_n$ be the roots of p and suppose x^* is that root of p that is closest to α. By Taylor's theorem we have

$$p(x) = p(\alpha) + p'(\alpha)(x - \alpha) + \cdots + \frac{p^{(n)}(\alpha)(x - \alpha)^n}{n!}$$

Make the change of variables $x = t + \alpha$ and define

$$P(t) = p(t + \alpha) = p(\alpha) + p'(\alpha)t + \cdots + \frac{p^{(n)}(\alpha)t^n}{n!}$$

The roots of P are $r_1 - \alpha, \ldots, r_n - \alpha$. Next we define

$$Q(t) = t^n P\left(\frac{1}{t}\right) = p(\alpha)t^n + p'(\alpha)t^{n-1} + \cdots + \frac{p^{(n)}(\alpha)}{n!}$$

The roots of Q are the reciprocals of the roots of P. By (ii) of Theorem 2.7 applied to Q we have

$$\frac{-p'(\alpha)}{p(\alpha)} = \sum_{k=1}^{n} \frac{1}{r_k - \alpha}$$

from which it follows that

$$\left|\frac{p'(\alpha)}{p(\alpha)}\right| \leq \sum_{k=1}^{n} \frac{1}{|r_k - \alpha|} \leq \frac{n}{|x^* - \alpha|}$$

and the proof is complete.

If Algorithm 2.7 is used for root finding, it is a simple matter to perform the test

$$\text{Is } \frac{n|p(x_N)|}{|p'(x_N)|} \leq \text{TOL}|x_N| ? \tag{2.27}$$

If x_N passes the test, then there is a root x^* of p such that $|x^* - x_N|/|x_N| \leq \text{TOL}$.

We complete the analysis by deriving an a posteriori error bound that can be used for any polynomial root finding method.

THEOREM 2.9

Let p be given by (2.19) and suppose $\hat{x}$ is an approximate root of p in the sense that $|p(\hat{x})| \leq \varepsilon$; then there is a root x^* of p that satisfies

$$\left|\frac{x^* - \hat{x}}{x^*}\right| \leq \left(\frac{\varepsilon}{|a_0|}\right)^{1/n}$$

Proof

By virtue of (2.26) we have

$$|p(\hat{x})| = |a_n| \prod_{k=1}^{n} |\hat{x} - r_k|$$

and hence by (ii) of Theorem 2.7 we find

$$|p(\hat{x})| = |a_0| \prod_{k=1}^{n} \left| \frac{\hat{x} - r_k}{r_k} \right|$$

Let x^* be that root of p that satisfies $|\hat{x} - x^*|/|x^*| \leq |\hat{x} - r_k|/|r_k|$ for all k. Then we have

$$|p(\hat{x})| \geq |a_0| \left| \frac{\hat{x} - x^*}{x^*} \right|^n \tag{2.28}$$

from which the result immediately follows.

If x_N denotes the Nth Newton iterate obtained by applying Algorithm 2.7 to p, then by (2.28) there is a root x^* of p such that

$$\left| \frac{x^* - x_N}{x^*} \right| \leq \left| \frac{p(x_N)}{a_0} \right|^{1/n} \tag{2.29}$$

This inequality furnishes an a posteriori relative error bound for Newton's method.

In this section we have provided some tools for polynomial root finding. These tools should enable the reader to apply Newton's method to a polynomial with the expectation of producing a reasonably accurate approximation to a real root. However, we should warn the reader that polynomial root finding is fraught with computational pitfalls when all the roots are required. If a root has been found, then we can use deflation and work with a polynomial of reduced degree. In so doing, we expect a loss in accuracy because the reduced, or deflated, polynomial has errors in its coefficients due to the error in the initial roots. In using deflation, the best strategy is to determine the roots in ascending order of magnitude. (This requires prior knowledge of the locale of the roots.) This is clearly illustrated in the following example.

EXAMPLE 2.22

We use the program in Example 2.17 (with appropriate modifications) and deflation to find the roots of $p(x) = x^5 - 15.8x^4 + 63.8x^3 - 10.6x^2 - 112.8x - 21.6$. The exact roots of p are -0.2, -1, 2, 6, and 9. As starting values we use 0.9 times the exact root; for example, in seeking the root -0.2, we use the initial guess -0.18. In the table below we give the roots and the relative error for increasing and decreasing order of magnitude.

Increasing Order		Decreasing Order	
Root	Relative Error	Root	Relative Error
-0.2000001	-0.5×10^6	8.999998	0.2×10^{-6}
-1.000000	0.0	6.000018	0.3×10^{-5}
2.000000	0.0	1.999953	0.2×10^{-4}
6.000000	0.0	-1.000051	-0.5×10^{-4}
9.000000	0.0	-0.1999144	0.4×10^{-3}

From the table we see that the relative error increases for each successive root when the roots are found in decreasing order of magnitude.

In the deflation process we determine the real roots one after the other until the deflated polynomial is a quadratic and then use the quadratic formula to find the remaining roots. A better approach to use is as follows. Compute the first root $\hat{x}_1$ as before. Deflate and compute the second root $\hat{x}_2$. Instead of using $\hat{x}_2$ to deflate again, use $\hat{x}_2$ as the initial guess and apply Newton's method to the original polynomial to generate an improved approximation $\tilde{x}_2$ to the second root. Then use $\tilde{x}_2$ in place of $\hat{x}_2$ to perform the next deflation. This process should be used for all the other approximate roots as well.

We have limited our discussion to real root finding techniques. In order to find complex roots, one can employ the methods of Müller or Bairstow, each of which is reasonably effective. The method of Müller is related to the secant method and applies to functions other than polynomials as well. Bairstow's method is related to Newton's method (in two variables) and, as such, requires an accurate initial guess. Both of these methods find roots one at a time.

There are methods that give all the approximate roots of a polynomial simultaneously. We mention the root squaring method of Graeffe and the quotient difference algorithm due to Rutishauser. If one of these methods is employed, then Newton's method can be used to obtain more accurate approximations to the real roots.

EXERCISES

2.43 For a polynomial of degree n how many multiplications and additions are required for evaluation at a point by (a) Algorithm 2.6 and (b) "brute force" using the form (2.19)? Compare (a) and (b) for $n = 10$.

2.44 Use subroutine HORNER to print a table of values of $p(x) = x^4 - 3x^3 + 3x^2 - 3x + 2$ for $x = -1, -0.9, \ldots, 0.9, 1$.

2.45 Modify HORNER to accept a polynomial of degree ≥ 1 in the form (2.19) where the coefficients are stored as $a_{M-1} = A(1)$, $a_{M-2} = A(2)$, $\ldots$, $a_0 = A(M)$. Test your version of HORNER on the polynomial in exercise 2.44.

2.46 Use the program of Example 2.17 to find the real roots of the polynomial $-x^3 + 3x^2 + 1 = p(x)$.

2.47 The polynomial $p(x) = x^5 - 9x^4 - 43x^3 + 327x^2 + 120x - 900$ has zeros $\pm\sqrt{3}$, 5, -6, and 10. Use Newton's method and deflation to find these zeros on a computer in increasing order of magnitude. Use initial guesses that are 0.8 times the exact zeros.

2.48 Repeat exercise 2.47 by finding the zeros in decreasing order of magnitude. Compare the results.

2.49 Change the coefficient of x^2 to 328 in exercise 2.47 and solve for the zeros again. Discuss the results.

2.50 Use (2.25) to estimate the relative change in the roots as a result of the coefficient change in exercise 2.49. Compare these estimates with the relative change in the computed roots obtained in exercise 2.49 (see Example 2.19).

2.51 How many real roots does $p(x) = x^4 - 3x^3 - 7x^2 + 15x + 18$ have? Determine, as best you can, the locale of the real roots of p.

2.52 Apply Theorem 2.7 to $p(x) = 18x^4 + 15x^3 - 7x^2 - 3x + 1$.

2.53 Find a bound for the root of largest magnitude of the polynomial of exercise 2.44. Show that at least one root satisfies $|x| \le 1$.

2.54 How are the nonzero roots of $p(x)$ related to those of $q(x) = x^n p(1/x)$, where p is given by (2.19)?

2.55 Use the results of exercises 2.52 and 2.54 to estimate one or more of the roots of the polynomial of exercise 2.51.

2.56 Use the result of exercise 2.54 to show that every root of the polynomial of exercise 2.44 satisfies $|x| \ge 0.5$. Does this polynomial have any real roots?

2.57 The polynomial of exercise 2.51 has a root of multiplicity 2 at $x = 3$. Use the program of Example 2.17 with initial guess of 2.7. Print the iterates and the values of p and p' at these iterates. Is the convergence quadratic?

2.58 Use (2.29) on the output of exercise 2.57 to bound the relative error.

2.59 Use (2.27) and (2.29) on the output of exercise 2.47.

2.60 Observe that the left-hand side of (2.27) is equal to $n|x_{N+1} - x_N|$. Modify the program of Example 2.17 to use (2.27) as the termination criterion.

2.61 Use the program in Example 2.17 as the basis for a subroutine called POLY with parameter list (A, M, X, NTOL, TOL) where

A:	array of coefficients of the polynomial
M:	dimension of A
X:	initial guess and approximate root
NTOL:	upper limit on number of iterations
TOL:	relative error tolerance

Do not call HORNER but rather include its function as part of POLY. Use (2.27) for the termination criterion. Test your subroutine on several of these exercises.

2.62 Use the approximate roots of Example 2.22, which were found in decreasing order of magnitude, as initial guesses for Newton's method applied to the original polynomial. Compute and print the improved approximate roots by this method.

NOTES AND COMMENTS

Section 2.1

The bisection method is presented in virtually every text on numerical analysis. Our presentation of false position and its modifications is similar to that given in Conte and de Boor (1980) and Ralston and Rabinowitz (1978). The latter book gives other modifications of the false position method with a detailed analysis of order of convergence and efficiency.

Section 2.2

The methods of this section are also a standard part of most texts on the subject. For a rigorous analysis of the error in the secant method and Newton's method in one or more variables, see Conte and de Boor (1980) or Johnson and Riess (1982).

Section 2.3

Our presentation of NEWBIS is intended to provide the reader with an introductory development of several features present in state-of-the-art software for root finding. Subroutine NEWBIS, when modified via Project 2, provides a safe, simple, and reasonably effective program for applying Newton's method to smooth functions. The reader is invited to gain additional insight by consulting Rheinboldt (1981), Brent (1971), and Forsythe et al. (1977). The first reference is particularly accessible to undergraduate students and should be consulted prior to writing subroutine SECBIS. The last reference gives a subroutine ZEROIN, written in FORTRAN, for the algorithm of Brent.

Section 2.4

The method of Müller is carefully explained and illustrated in Conte and de Boor (1980), and Bairstow's method is nicely presented in Johnson and Riess (1982). We suggest the book by Henrici (1964) for an explanation of the quotient difference algorithm; for Graeffe's method see Hildebrand (1956). The book by Ralston and Rabinowitz (1978) gives more advanced techniques for determining the locale of polynomial roots; in particular the use of Sturm sequences in this regard is discussed in section 8.9-1.

Some texts of general interest on root finding are Traub (1964), Householder (1970), and Ortega and Rheinboldt (1970). All three require a substantial mathematical background of the reader.

LINEAR SYSTEMS OF EQUATIONS

3

One mathematical problem that arises quite frequently in applications is the solution of a system of simultaneous linear equations. A general system of such equations is of the form

$$a_{11} x_1 + a_{12} x_2 + \quad \cdots \quad + a_{1n} x_n = b_1$$
$$a_{21} x_1 + a_{22} x_2 + \quad \cdots \quad + a_{2n} x_n = b_2$$
$$\cdots\cdots\cdots\cdots\cdots\cdots\cdots\cdots\cdots\cdots\cdots$$
$$a_{m1} x_1 + a_{m2} x_2 + \quad \cdots \quad + a_{mn} x_n = b_m$$

(3.1)

where the coefficients a_{ij}, $1 \le i \le m$ and $1 \le j \le n$, and the b_i, $1 \le i \le m$, are given real constants. The problem is to find $x_1, \ldots, x_n$ so that the m equations in (3.1) are satisfied simultaneously.

3.1 MATRIX TERMINOLOGY

In discussing the theoretical aspects of (3.1) as well as its numerical solution, it is natural and convenient to use matrix notation and terminology. A **matrix** is a rectangular array of real numbers. The coefficients of (3.1) determine a matrix, say **A**, which we write as

$$\mathbf{A} = \begin{bmatrix} a_{11} & a_{12} & \cdots & a_{1n} \\ a_{21} & a_{22} & \cdots & a_{2n} \\ \cdots\cdots\cdots\cdots\cdots\cdots \\ a_{m1} & a_{m2} & \cdots & a_{mn} \end{bmatrix}$$

(3.2)

The matrix of (3.2) has m rows and n columns, or, more briefly, we say that **A** is an $m \times n$ matrix. A more compact notation for **A** is $\mathbf{A} = (a_{ij})_{m, n}$, where a_{ij} is the (i, j) entry of **A**.

If a matrix has only one row, we call it a **row vector**, and one that has only one column is called a **column vector**. We denote vectors by lowercase letters in order to distinguish them from other matrices. Moreover, we are concerned primarily with column vectors and refer to them simply as vectors. The right-hand side of (3.1) forms an m-vector:

$$\mathbf{b} = \begin{bmatrix} b_1 \\ b_2 \\ \vdots \\ b_m \end{bmatrix} \tag{3.3}$$

and the unknowns of (3.1) form an n-vector:

$$\mathbf{x} = \begin{bmatrix} x_1 \\ x_2 \\ \vdots \\ x_n \end{bmatrix} \tag{3.4}$$

The concept of matrix multiplication enables us to write (3.1) in terms of **A**, **b**, and **x**. If $\mathbf{A} = (a_{ij})_{m, n}$ and $\mathbf{B} = (b_{ij})_{n, p}$, then the matrix product **AB** is defined to be the $m \times p$ matrix **C** whose (i, j) entry is given by

$$c_{ij} = \sum_{k=1}^{n} a_{ik} b_{kj}$$

In words, the (i, j) entry of **C** is found by summing the products of corresponding entries of row i of **A** and column j of **B**.

EXAMPLE 3.1

If $\mathbf{A} = \begin{bmatrix} 2 & 3 & -1 \\ 0 & 4 & 1 \end{bmatrix}$ and $\mathbf{B} = \begin{bmatrix} 0 & 1 \\ -2 & 4 \\ 3 & -2 \end{bmatrix}$, then $\mathbf{AB} = \begin{bmatrix} -9 & 16 \\ -5 & 14 \end{bmatrix}$.

For instance, the $(2, 1)$ entry is obtained by summing the products of the corresponding entries of row 2 of **A** and column 1 of **B**: $0 \cdot 0 + 4(-2) + 1 \cdot 3 = -5$. For these matrices it is also possible to calculate the matrix product **BA**:

$$\mathbf{BA} = \begin{bmatrix} 0 & 1 \\ -2 & 4 \\ 3 & -2 \end{bmatrix} \begin{bmatrix} 2 & 3 & -1 \\ 0 & 4 & 1 \end{bmatrix} = \begin{bmatrix} 0 & 4 & 1 \\ -4 & 10 & 6 \\ 6 & 1 & -5 \end{bmatrix}$$

EXAMPLE 3.2

Another example of a matrix product is

$$\begin{bmatrix} 2 & -1 & 0 \\ 1 & 4 & -3 \\ 5 & -2 & 1 \end{bmatrix} \begin{bmatrix} x_1 \\ x_2 \\ x_3 \end{bmatrix} = \begin{bmatrix} 2x_1 - x_2 \\ x_1 + 4x_2 - 3x_3 \\ 5x_1 - 2x_2 + x_3 \end{bmatrix}$$

Using summation notation, the system (3.1) can be written in more compact form as

$$\sum_{k=1}^{n} a_{ik} x_k = b_i \qquad 1 \le i \le m \tag{3.5}$$

In (3.5) we recognize that the left-hand side is the sum of products of corresponding entries of row i of **A**, given by (3.2), and vector **x**, given by (3.4). It follows that the system (3.1) can be written as

$$\mathbf{Ax} = \mathbf{b} \tag{3.6}$$

where **A**, **b**, and **x** are given in (3.2), (3.3), and (3.4), respectively.

EXAMPLE 3.3

The system

$$7x_1 - x_2 + 4x_3 + x_4 = 12$$
$$-x_1 \qquad - 3x_3 + 2x_4 = 7$$
$$5x_2 + x_3 - x_4 = 23$$

can be written in the form (3.6) with

$$\mathbf{A} = \begin{bmatrix} 7 & -1 & 4 & 1 \\ -1 & 0 & -3 & 2 \\ 0 & 5 & 1 & -1 \end{bmatrix} \qquad \mathbf{b} = \begin{bmatrix} 12 \\ 7 \\ 23 \end{bmatrix} \qquad \mathbf{x} = \begin{bmatrix} x_1 \\ x_2 \\ x_3 \\ x_4 \end{bmatrix}$$

Observe that, in order for the matrix product **AB** to be defined, the number of columns of **A** must be the same as the number of rows of **B**. If **AB** is defined, it does not follow that **BA** is defined. Even if **AB** and **BA** are both defined, they need not be equal. Thus matrix multiplication is not commutative. However, it is associative; this means that $\mathbf{A(BC)} = \mathbf{(AB)C}$, where **A** is $m \times n$, **B** is $n \times r$, and **C** is $r \times p$.

We say that **A** is a **square matrix** if the number of rows equals the number of columns. A square matrix $\mathbf{B}$ is said to be **upper** (lower) **triangular** if $b_{ij} = 0$ for $i > j$ $(i < j)$. If $\mathbf{B}$ is both upper and lower triangular, then it is called a **diagonal** matrix.

EXAMPLE 3.4

$\mathbf{L}$ is lower triangular, $\mathbf{U}$ is upper triangular, and $\mathbf{D}$ is a diagonal matrix:

$$\mathbf{L} = \begin{bmatrix} 1 & 0 & 0 \\ -2 & 4 & 0 \\ 3 & 0 & -5 \end{bmatrix} \quad \mathbf{U} = \begin{bmatrix} 2 & 3 & 6 \\ 0 & -1 & 4 \\ 0 & 0 & 0 \end{bmatrix} \quad \mathbf{D} = \begin{bmatrix} 2 & 0 & 0 \\ 0 & 1 & 0 \\ 0 & 0 & 7 \end{bmatrix}$$

If a diagonal matrix $\mathbf{D}$ has all of its diagonal entries equal to 1, $d_{ii} = 1$ for $1 \le i \le n$, then we call $\mathbf{D}$ the **identity** matrix and denote it by the symbol $\mathbf{I}_n$. This matrix has the special property that

$$\mathbf{I}_n \mathbf{A} = \mathbf{A} \quad \text{for all } n \times m \text{ matrices } \mathbf{A}$$

$$\mathbf{B} \mathbf{I}_n = \mathbf{B} \quad \text{for all } m \times n \text{ matrices } \mathbf{B}$$

We note, if $\mathbf{A}$ is $n \times n$, that $\mathbf{I}_n \mathbf{A} = \mathbf{A} \mathbf{I}_n = \mathbf{A}$. Henceforth we use $\mathbf{I}$ instead of $\mathbf{I}_n$, as its size is given implicitly by its usage.

There is no meaningful definition of matrix division; however, for square matrices we may define a similar process called matrix inversion. We say that a square matrix $\mathbf{A}$ is **invertible** if there exists a square matrix $\mathbf{B}$ such that

$$\mathbf{AB} = \mathbf{BA} = \mathbf{I} \tag{3.7}$$

Clearly the size of $\mathbf{A}$ and $\mathbf{B}$ must be the same in (3.7).

EXAMPLE 3.5

The matrix $\mathbf{A} = \begin{bmatrix} 1 & -2 \\ 0 & 1 \end{bmatrix}$ is invertible because

$$\begin{bmatrix} 1 & -2 \\ 0 & 1 \end{bmatrix}\begin{bmatrix} 1 & 2 \\ 0 & 1 \end{bmatrix} = \begin{bmatrix} 1 & 2 \\ 0 & 1 \end{bmatrix}\begin{bmatrix} 1 & -2 \\ 0 & 1 \end{bmatrix} = \begin{bmatrix} 1 & 0 \\ 0 & 1 \end{bmatrix}$$

However, the matrix

$$\mathbf{C} = \begin{bmatrix} 1 & -2 \\ -0.5 & 1 \end{bmatrix}$$

is not invertible. Indeed, if $\mathbf{C}$ were invertible, then for some 2×2 matrix $\mathbf{B}$ we would have $\mathbf{CB} = \mathbf{I}$ or, equivalently,

$$\begin{bmatrix} b_{11} - 2b_{21} & b_{12} - 2b_{22} \\ -0.5b_{11} + b_{21} & -0.5b_{12} + b_{22} \end{bmatrix} = \begin{bmatrix} 1 & 0 \\ 0 & 1 \end{bmatrix}$$

Therefore we would have $b_{11} - 2b_{21} = 1$ and $-0.5(b_{11} - 2b_{21}) = 0$, which is obviously impossible.

Next we show that there can be at most one matrix $\mathbf{B}$ that satisfies (3.7). Suppose $\mathbf{C}$ also satisfies (3.7); then, since multiplication is associative,

$$\mathbf{B} = \mathbf{BI} = \mathbf{B(AC)} = (\mathbf{BA})\mathbf{C} = \mathbf{IC} = \mathbf{C}$$

Henceforth we denote the unique matrix $\mathbf{B}$ that satisfies (3.7) by $\mathbf{A}^{-1}$, which is called the **inverse** of $\mathbf{A}$. Thus if $\mathbf{A}$ is invertible, then $\mathbf{AA}^{-1} = \mathbf{A}^{-1}\mathbf{A} = \mathbf{I}$.

We can also define the sum of two matrices and the multiplication of a matrix by a scalar. If $\mathbf{A} = (a_{ij})_{m, n}$ and $\mathbf{B} = (b_{ij})_{m, n}$, then we define the sum $\mathbf{A} + \mathbf{B}$ to be the matrix $\mathbf{C} = (c_{ij})_{m, n}$ whose (i, j) entry is $c_{ij} = a_{ij} + b_{ij}$. Thus addition is done entrywise. If α is a scalar (real number), then $\alpha\mathbf{A}$ is the $m \times n$ matrix whose (i, j) entry is αa_{ij}.

EXAMPLE 3.6

Let

$$\mathbf{A} = \begin{bmatrix} 2 & 0 & 1 \\ -1 & 4 & 3 \end{bmatrix} \quad \text{and} \quad \mathbf{B} = \begin{bmatrix} 5 & 1 & 0 \\ -2 & 6 & -3 \end{bmatrix}$$

Then

$$\mathbf{A} + \mathbf{B} = \begin{bmatrix} 7 & 1 & 1 \\ -3 & 10 & 0 \end{bmatrix} \quad \text{and} \quad 6\mathbf{A} = \begin{bmatrix} 12 & 0 & 6 \\ -6 & 24 & 18 \end{bmatrix}$$

For the convenience of the reader we summarize the basic properties of these matrix operations. Let $\mathbf{A}$, $\mathbf{B}$, and $\mathbf{C}$ be matrices and α be a scalar. Assuming the matrix operations are well-defined, we have

(a) $\mathbf{A(BC)} = (\mathbf{AB})\mathbf{C}$

(b) $\mathbf{A(B + C)} = \mathbf{AB} + \mathbf{AC}$

(c) $(\mathbf{B + C})\mathbf{A} = \mathbf{BA} + \mathbf{CA}$

(d) $\alpha(\mathbf{A + B}) = \alpha\mathbf{A} + \alpha\mathbf{B}$

(e) $\alpha(\mathbf{AB}) = (\alpha\mathbf{A})\mathbf{B} = \mathbf{A}(\alpha\mathbf{B})$

(f) $\mathbf{A + B} = \mathbf{B + A}$

(g) $(\mathbf{A + B}) + \mathbf{C} = \mathbf{A} + (\mathbf{B + C})$

(h) $\mathbf{A} + (-1)\mathbf{A} = \mathbf{0}$

where $\mathbf{0}$ denotes the matrix (of the same size as $\mathbf{A}$) whose entries are all zero.

We emphasize that **A** does not have the same meaning in each of the above properties. In (a) **A** must have the proper size so that **AB** is well-defined, whereas in (f) **A** must be such that **A** + **B** is well-defined. The reader who has not previously been exposed to matrices should verify by example that (a) through (h) are valid.

We state in the following theorem some basic results concerning the existence and uniqueness of solutions of equation (3.6).

THEOREM 3.1

   (i)   If **x** is a solution of (3.6), then any other solution **y** can be written as **y** = **x** + **z** with **Az** = **0**, where **0** denotes the vector whose entries are all zero.

   (ii)   Equation (3.6) has at most one solution if and only if the only solution of **Az** = **0** is **z** = **0**.

   (iii)   If $m < n$, then there exists **z** ≠ **0** such that **Az** = **0**.

   (iv)   If (3.6) has a solution for every m-vector **b**, then $m \leq n$.

The proofs of (i) and (ii) are straightforward. For (i) we suppose that **x** and **y** are solutions and set **z** = **x** − **y**. Then **Az** = **A**(**x** − **y**) = **Ax** − **Ay** = **b** − **b** = **0**. For (ii) we first suppose that (3.6) has a unique solution **x**. If there exists **z** ≠ **0** such that **Az** = **0**, then **y** = **z** + **x** satisfies **Ay** = **Az** + **Ax** = **0** + **b** = **b**, so that **y** is also a solution of (3.6). This contradicts our assumption that **x** is unique, and so we must have **z** = **0**. Conversely, suppose that **Az** = **0** has only the solution **z** = **0**. Then by (i) equation (3.6) has at most one solution.

The proofs of (iii) and (iv) are more difficult, and we omit them. Instead let us examine some consequences of these results. Suppose (3.6) has a unique solution for every right-hand-side **b**. By (ii) the only solution of **Az** = **0** is **z** = **0**, and hence by (iii) we must have $m \geq n$. Statement (iv) may be rephrased as, If $m > n$, then for some m-vector **b** (3.6) has no solution. It follows that if (3.6) is uniquely solvable for all m-vectors **b**, then we must have $m = n$. In other words, for **Ax** = **b** to have a unique solution for all **b**, **A** must be a square matrix.

EXAMPLE 3.7

Consider the system in Example 3.3. It is an easy matter to verify that two solutions are given by

$$\mathbf{x} = \begin{bmatrix} 1.5 \\ 5.5 \\ 0.5 \\ 5 \end{bmatrix} \quad \text{and} \quad \mathbf{y} = \begin{bmatrix} 6 \\ 5\frac{1}{3} \\ -5\frac{2}{3} \\ -2 \end{bmatrix}$$

Moreover, if $\mathbf{z} = \mathbf{x} - \mathbf{y}$, then

$$\mathbf{Az} = \begin{bmatrix} 7 & -1 & 4 & 1 \\ -1 & 0 & -3 & 2 \\ 0 & 5 & 1 & -1 \end{bmatrix} \begin{bmatrix} -4\frac{1}{2} \\ \frac{1}{6} \\ 6\frac{1}{6} \\ 7 \end{bmatrix} = \begin{bmatrix} 0 \\ 0 \\ 0 \end{bmatrix}$$

Thus the system of Example 3.3 does not have a unique solution, and the difference between two solutions $\mathbf{z}$ satisfies $\mathbf{Az} = \mathbf{0}$.

EXAMPLE 3.8

Consider the system

$$\begin{bmatrix} 7 & -1 & 4 & 1 \\ -1 & 0 & -3 & 2 \\ 0 & 5 & 1 & -1 \end{bmatrix} \begin{bmatrix} x_1 \\ x_2 \\ x_3 \\ x_4 \end{bmatrix} = \begin{bmatrix} 11 \\ -2 \\ 5 \end{bmatrix}$$

which has more unknowns than equations. Then for any scalar α the vector $\mathbf{w} = \mathbf{u} + \alpha\mathbf{z}$ is a solution, where

$$\mathbf{u} = \begin{bmatrix} 1 \\ 1 \\ 1 \\ 1 \end{bmatrix} \quad \text{and} \quad \mathbf{z} = \begin{bmatrix} -4\frac{1}{2} \\ \frac{1}{6} \\ 6\frac{1}{6} \\ 7 \end{bmatrix}$$

Since α is completely arbitrary, this system has infinitely many solutions. In particular, if $\alpha = 3$, then a solution is

$$\mathbf{w} = \begin{bmatrix} -12.5 \\ 1.5 \\ 19.5 \\ 22 \end{bmatrix}$$

EXAMPLE 3.9

Consider the system $\mathbf{Ax} = \mathbf{b}$ given by

$$\begin{bmatrix} 1 & -1 \\ 3 & 2 \\ 1 & 1 \end{bmatrix} \begin{bmatrix} x_1 \\ x_2 \end{bmatrix} = \begin{bmatrix} 0 \\ 5 \\ 2 \end{bmatrix}$$

Here there are more equations than unknowns. It is easy to verify that a solution is given by $x_1 = x_2 = 1$. In fact this is the only solution of the system.

By (ii) of Theorem 3.1 this means that the only solution of $\mathbf{Az} = \mathbf{0}$ is $\mathbf{z} = \mathbf{0}$. By (iv) of Theorem 3.1 there must be a right-hand-side vector, call it $\mathbf{c}$, such that $\mathbf{Ax} = \mathbf{c}$ has no solution. An example of such a vector is

$$\mathbf{c} = \begin{bmatrix} 2 \\ 1 \\ 1 \end{bmatrix}$$

In the previous two examples we saw systems of linear equations that have no solution, precisely one solution, and infinitely many solutions. In fact it can be shown, using standard results from linear alegbra, that these are the only possibilities. Thus the number of solutions of a system of linear equations is either zero, one, or infinite.

The following theorem gives sufficient conditions for existence and uniqueness of solutions of $\mathbf{Ax} = \mathbf{b}$ for square matrices.

THEOREM 3.2

If $\mathbf{A}$ is a square matrix, then the following are equivalent:

(i)   The only solution of $\mathbf{Az} = \mathbf{0}$ is $\mathbf{z} = \mathbf{0}$.

(ii) $\mathbf{Ax} = \mathbf{b}$ has a solution for every $\mathbf{b}$.

(iii) $\mathbf{A}$ is invertible.

This theorem says that uniqueness of solutions is equivalent to existence of solutions for every right-hand-side vector. In addition the theorem provides a means of finding $\mathbf{A}^{-1}$. Let $\mathbf{e}^j$ denote the jth column of the $n \times n$ identity matrix. Then by (ii) of Theorem 3.2 the problem $\mathbf{Ax} = \mathbf{e}^j$ has a solution, say $\mathbf{x}^j$. Let $\mathbf{B}$ be the matrix whose jth column is $\mathbf{x}^j$ and consider the product $\mathbf{AB}$. The jth column of the product is determined by multiplying the jth column of $\mathbf{B}$ by $\mathbf{A}$:

$$j\text{th column of } \mathbf{AB} = \mathbf{Ax}^j = \mathbf{e}^j$$

Hence $\mathbf{AB} = \mathbf{I}$ or $\mathbf{B} = \mathbf{A}^{-1}$. Thus the jth column of $\mathbf{A}^{-1}$ is the solution of the problem $\mathbf{Ax} = \mathbf{e}^j$. Once $\mathbf{A}^{-1}$ has been determined, we can solve $\mathbf{Ax} = \mathbf{b}$ for any $\mathbf{b}$ as follows. We multiply both sides by $\mathbf{A}^{-1}$ to yield

$$\mathbf{x} = \mathbf{Ix} = \mathbf{A}^{-1}\mathbf{Ax} = \mathbf{A}^{-1}\mathbf{b} \tag{3.8}$$

From a theoretical point of view the problem $\mathbf{Ax} = \mathbf{b}$ may be solved by finding $\mathbf{A}^{-1}$ and calculating the product $\mathbf{A}^{-1}\mathbf{b}$ as in (3.8). As we shall see, this approach is not so efficient as Gaussian elimination.

The reader has no doubt heard of determinants in connection with matrices or linear systems of equations and may even have been led to believe that they are useful in finding the inverse of a matrix or in solving linear systems. In fact determinants are only rarely needed in practical matrix computations. Thus we refer to them only occasionally in passing but not in any essential way.

In this section we have discussed the basic operations of matrix algebra and established some essential facts regarding existence and uniqueness of solutions of $\mathbf{Ax} = \mathbf{b}$. As yet we have not given any techniques for solving a system of linear equations—this is the subject of the next section.

EXERCISES

3.1 Write the system

$$4x_1 + 6x_2 + 3x_3 = 5$$
$$2x_1 + \ x_2 \qquad = 4$$
$$x_1 - \ x_2 + \ x_3 = 7$$
$$x_2 - \ x_3 = 9$$

in the form $\mathbf{Ax} = \mathbf{b}$.

3.2 Let $\mathbf{A} = \begin{bmatrix} 1 & 0 & 4 \\ -1 & 2 & 1 \end{bmatrix}$ and $\mathbf{B} = \begin{bmatrix} -1 & 6 \\ 2 & 3 \\ 0 & 4 \end{bmatrix}$. Calculate $\mathbf{AB}$ and $\mathbf{BA}$.

3.3 If $\mathbf{A}$ and $\mathbf{B}$ are $n \times n$ upper triangular matrices, show that $\mathbf{AB}$ is also upper triangular.

3.4 Let $\mathbf{A} = \begin{bmatrix} 2 & -1 & 3 \\ 0 & 4 & 6 \\ 5 & 1 & -2 \end{bmatrix}$ and $\mathbf{D} = \begin{bmatrix} d_1 & 0 & 0 \\ 0 & d_2 & 0 \\ 0 & 0 & d_3 \end{bmatrix}$. Compute $\mathbf{DA}$ and $\mathbf{AD}$.

3.5 Can you find, by inspection, the inverse of $\mathbf{A} = \begin{bmatrix} 2 & 0 & 0 \\ 0 & 4 & 0 \\ 0 & 0 & -1 \end{bmatrix}$?

What is the inverse of a diagonal matrix whose diagonal entries are nonzero?

3.6 If $\mathbf{A}$ is upper triangular, is it easy to solve $\mathbf{Ax} = \mathbf{b}$? Try a 3×3 example.

3.7 Does $\mathbf{Ax} = \mathbf{0}$ have a solution other than $\mathbf{x} = \mathbf{0}$, where $\mathbf{A} = \begin{bmatrix} 2 & -3 & 1 \\ 1 & 4 & -1 \end{bmatrix}$?
Hint: Choose $x_3 = 1$; then solve for x_1 and x_2.

3.8 Show that the problem $\mathbf{Ax} = \mathbf{b}$ has no solution where

$$\mathbf{A} = \begin{bmatrix} 2 & 1 \\ -3 & 4 \\ 1 & -1 \end{bmatrix} \quad \text{and} \quad \mathbf{b} = \begin{bmatrix} 3 \\ 1 \\ 2 \end{bmatrix}$$

Hint: Solve for x_1 in the first equation and substitute the result in the second and third equations.

3.9 How many multiplications are required to perform the product $\mathbf{Ax}$ where $\mathbf{A}$ is an

$n \times n$ matrix and $\mathbf{x}$ is an n-vector? How many multiplications are required to find the product $\mathbf{AB}$ where $\mathbf{A}$ and $\mathbf{B}$ are $n \times n$ matrices?

3.10 If $\mathbf{A}$ is an $m \times n$ matrix, then the **transpose** of $\mathbf{A}$, denoted $\mathbf{A}^T$, is the $n \times m$ matrix whose (i, j) entry is a_{ji}. Thus the kth row of $\mathbf{A}^T$ is the kth column of $\mathbf{A}$. Find $\mathbf{A}^T$ and $\mathbf{B}^T$ for the matrices of exercise 3.2. Compute $\mathbf{B}^T\mathbf{A}^T$ and $(\mathbf{AB})^T$. Show that in general $(\mathbf{AB})^T = \mathbf{B}^T\mathbf{A}^T$.

3.11 We say that square matrix $\mathbf{A}$ is **symmetric** if $\mathbf{A}^T = \mathbf{A}$. Let $\mathbf{B}$ be an $m \times n$ matrix and set $\mathbf{A} = \mathbf{B}^T\mathbf{B}$. Show that $\mathbf{A}$ is symmetric. Hint: $(\mathbf{B}^T)^T = \mathbf{B}$.

3.12 Let matrices $\mathbf{A}$, $\mathbf{B}$, and $\mathbf{C}$ be given by

$$\mathbf{A} = \begin{bmatrix} 1 & 2 & 3 \\ -1 & 0 & 1 \\ 1 & 1 & 1 \end{bmatrix} \qquad \mathbf{B} = \begin{bmatrix} 2 & 1 \\ 1 & 0 \\ 1 & -4 \end{bmatrix} \qquad \mathbf{C} = \begin{bmatrix} 3 & 2 \\ 1 & 0 \end{bmatrix}$$

(a) Compute $\mathbf{A}(\mathbf{BC})$ and $(\mathbf{AB})\mathbf{C}$.
(b) Find $\mathbf{AB} + \mathbf{BC}$.
(c) Find $\mathbf{CB}^T$ and $(\mathbf{BC})^T$.
(d) Find $\mathbf{B}^T\mathbf{B}$.
(e) Is $\mathbf{B}^T\mathbf{B} = \mathbf{BB}^T$?
(f) Verify that $\mathbf{A}$ times (second column of $\mathbf{B}$) = second column of $\mathbf{AB}$.

3.13 Find $\mathbf{x}^T\mathbf{x}$ and $\mathbf{xx}^T$ if $\mathbf{x} = \begin{bmatrix} 2 & 1 & -1 \end{bmatrix}^T$.

3.14 We define powers of square matrices inductively by $\mathbf{A}^2 = \mathbf{AA}$, $\mathbf{A}^3 = \mathbf{AA}^2$, and so forth. Find $\mathbf{A}^3$ if

$$\mathbf{A} = \begin{bmatrix} 0 & 1 & 0 \\ 0 & 0 & 1 \\ 0 & 0 & 0 \end{bmatrix}$$

3.15 Suppose $\mathbf{A}$ and $\mathbf{B}$ are $n \times n$ matrices with $\mathbf{AB} = \mathbf{BA}$. Show that $(\mathbf{A} + \mathbf{B})^2 = \mathbf{A}^2 + 2\mathbf{AB} + \mathbf{B}^2$.

3.16 Does the conclusion of exercise 3.15 hold if $\mathbf{AB} \neq \mathbf{BA}$? Hint: Consider a 2×2 example.

3.17 If $\mathbf{A}$ and $\mathbf{B}$ are $n \times n$ matrices and $\mathbf{AB} = \mathbf{0}$, does it follow that $\mathbf{A} = \mathbf{0}$ or $\mathbf{B} = \mathbf{0}$? Hint: Consider

$$\mathbf{A} = \begin{bmatrix} 1 & -1 \\ -2 & 2 \end{bmatrix} \qquad \mathbf{B} = \begin{bmatrix} 0 & 1 \\ 0 & 1 \end{bmatrix}$$

3.18 The brute-force method of calculating the determinant of an $n \times n$ matrix requires $n!$ multiplications. Assume that each multiplication requires 10^{-6} seconds on a computer. Determine the amount of time required to calculate the determinant for $n = 20$ by brute force.

3.19 Write a simple FORTRAN subroutine SUMM(A,B,C,M,N) that accepts two $M \times N$ matrices $\mathbf{A}$ and $\mathbf{B}$ and returns their sum in matrix $\mathbf{C}$. Test your program on several examples.

3.20 Write a program that uses the code

```
      DO 2 I=1,M
        DO 2 J=1,L
          SUM=0.0
          DO 1 K=1,N
            SUM=SUM+A(I,K)*B(K,J)
1         CONTINUE
        C(I,J)=SUM
2     CONTINUE
```

to compute the matrix product $\mathbf{AB}$ where $\mathbf{A}$ is M $\times$ N and $\mathbf{B}$ is N $\times$ L. Verify the code on problems where you know the exact product.

3.2 NAIVE GAUSSIAN ELIMINATION

In this section we present an elimination method for the solution of $\mathbf{Ax} = \mathbf{b}$ where $\mathbf{A}$ is a square matrix. We show how to solve the problem using exact arithmetic and postpone the consideration of roundoff error to the next section.

We begin by considering the special case where the coefficient matrix is upper triangular. Consider the system

$$
\begin{aligned}
b_{11}x_1 + b_{12}x_2 + \quad \cdots \quad + b_{1n}x_n \quad &= c_1 \\
b_{22}x_2 + \quad \cdots \quad + b_{2n}x_n \quad &= c_2 \\
\cdots\cdots\cdots\cdots\cdots\cdots\cdots\cdots\cdots\cdots\cdots\cdots\cdots \\
b_{n-1,n-1}x_{n-1} + b_{n-1,n}x_n &= c_{n-1} \\
b_{nn}x_n &= c_n
\end{aligned}
\tag{3.9}
$$

The last equation involves only x_n, and hence if $b_{nn} \neq 0$, we obtain $x_n = c_n/b_{nn}$. The penultimate equation involves only x_n and x_{n-1}, and, since x_n is known, we can solve for x_{n-1} to get $x_{n-1} = (c_{n-1} - b_{n-1,n}x_n)/b_{n-1,n-1}$ provided $b_{n-1,n-1} \neq 0$. Continuing in this manner, we can solve for the unknowns x_n, $\ldots$, x_1 (in this order) using equations $n, n-1, \ldots, 1$. This process is known as solution by backsubstitution.

ALGORITHM 3.1

BACKSUBSTITUTION   Given an $n \times n$ upper triangular matrix $\mathbf{B}$ with $b_{ii} \neq 0$ for all i and an n-vector $\mathbf{c}$, then the solution of $\mathbf{Bx} = \mathbf{c}$ is obtained by

$$
x_n \leftarrow \frac{c_n}{b_{nn}}
$$

Do $k = n-1, n-2, \ldots, 1$:

$$
x_k \leftarrow \frac{c_k - \displaystyle\sum_{j=k+1}^{n} b_{kj}x_j}{b_{kk}}
$$

EXAMPLE 3.10

We use backsubstitution to solve the system

$$4x_1 - 3x_2 + x_3 = 8$$

$$x_2 + 2x_3 = 1$$

$$-2x_3 = 4$$

From the last equation we get $x_3 = -2$. Then the penultimate equation gives $x_2 = 1 - 2x_3 = 5$, and finally from the first equation we find $x_1 = (8 + 3x_2 - x_3)/4 = 6.25$.

An immediate consequence of the backsubstitution algorithm is the following. For an $n \times n$ upper triangular matrix $\mathbf{U}$, all of whose diagonal entries are nonzero, the system $\mathbf{Ux} = \mathbf{c}$ has a solution for every n-vector $\mathbf{c}$. By Theorem 3.2 this is equivalent to the invertibility of $\mathbf{U}$. It follows that $\mathbf{U}$ is invertible if and only if all of its diagonal entries are nonzero.

The method of Gaussian elimination is a procedure that reduces the problem $\mathbf{Ax} = \mathbf{b}$ to a problem $\mathbf{Ux} = \mathbf{c}$, which is in the form (3.9). The justification for the elimination process involves elementary row operations and the concept of equivalent problems. We say that the problems $\mathbf{Ax} = \mathbf{b}$ and $\mathbf{Ux} = \mathbf{c}$ are **equivalent** if every solution of the former is a solution of the latter and vice versa.

Given the problem $\mathbf{Ax} = \mathbf{b}$ with $\mathbf{A} = (a_{ij})_{n,n}$, form the $n \times (n + 1)$ matrix $\mathbf{W}$ given by

$$\mathbf{W} = [\mathbf{A} \mid \mathbf{b}] \tag{3.10a}$$

$\mathbf{W}$ is called the **augmented** matrix for the system $\mathbf{Ax} = \mathbf{b}$. Suppose $\mathbf{W}$ is subjected to a finite sequence of operations of the following type (elementary row operations):

(i)   Interchange two rows

(ii)   Multiply a row by a nonzero scalar

(iii)   Add a multiple of one row to another

to produce a new matrix

$$\tilde{\mathbf{W}} = [\mathbf{U} \mid \mathbf{c}] \tag{3.10b}$$

Then $\mathbf{Ax} = \mathbf{b}$ and $\mathbf{Ux} = \mathbf{c}$ are equivalent.

EXAMPLE 3.11

Consider the system $\mathbf{A}\mathbf{x} = \mathbf{b}$ given by

$$x_1 + 2x_2 + 3x_3 = 2$$
$$3x_1 + x_2 + 5x_3 = 7 \qquad\qquad (3.11)$$
$$-2x_1 - 4x_2 + x_3 = 4$$

The augmented matrix is

$$\mathbf{W}_1 = \begin{bmatrix} 1 & 2 & 3 & | & 2 \\ 3 & 1 & 5 & | & 7 \\ -2 & -4 & 1 & | & 4 \end{bmatrix}$$

Suppose we add -3 times row 1 to row 2, thereby giving the new matrix

$$\mathbf{W}_2 = \begin{bmatrix} 1 & 2 & 3 & | & 2 \\ 0 & -5 & -4 & | & 1 \\ -2 & -4 & 1 & | & 4 \end{bmatrix}$$

The augmented matrix $\mathbf{W}_2$ corresponds to a system of linear equations found by adding -3 times the first equation of (3.11) to the second equation of (3.11):

$$x_1 + 2x_2 + 3x_3 = 2$$
$$-5x_2 - 4x_3 = 1$$
$$-2x_1 - 4x_2 + x_3 = 4$$

Next we add 2 times row 1 of $\mathbf{W}_2$ to row 3 to yield

$$\mathbf{W}_3 = \begin{bmatrix} 1 & 2 & 3 & | & 2 \\ 0 & -5 & -4 & | & 1 \\ 0 & 0 & 7 & | & 8 \end{bmatrix}$$

which corresponds to an upper triangular system of equations. The solution of the system corresponding to $\mathbf{W}_3$ is easily found, by backsubstitution, to be $\mathbf{x} = [\frac{4}{5}, -\frac{39}{35}, \frac{8}{7}]^T$. The reader should verify that $\mathbf{x}$ solves (3.11) and the system corresponding to $\mathbf{W}_2$. Thus the systems corresponding to augmented matrices $\mathbf{W}_1$, $\mathbf{W}_2$, and $\mathbf{W}_3$ are all equivalent—they have the same solution.

The process involved in transforming $\mathbf{W}_1$ into $\mathbf{W}_3$ is indicative of the elimination method for a general system of linear equations. The remainder of this section is devoted to the development of an algorithm for the elimination method. The main idea of Gaussian elimination is to use elementary row operations on **W** in (3.10a) to produce $\tilde{\mathbf{W}}$ in (3.10b) such that **U** is upper triangular. Then backsubstitution may be used to find the solution **x**.

The first step of the process consists in eliminating x_1 from every equation except the first. In terms of **W** this corresponds to using elementary row operations in order to produce a new augmented matrix whose (1, 1) entry is nonzero and whose (i, 1) entry is zero for $i = 2, \ldots, n$. If the (1, 1) entry of **W** is nonzero, then by adding $-(w_{i1}/w_{11})$ times row 1 to row i we produce a zero in the (i, 1) entry. If the (1, 1) entry of **W** is zero, we search the first column of **W** for the first nonzero entry, say it occurs in row r; then we interchange rows 1 and r and proceed as before.

Suppose we have completed the $(k-1)$st step and **W** has been reduced to the form

$$
\begin{bmatrix}
w_{11} & w_{12} & \cdots & & w_{1k} & \cdots & w_{1n} & w_{1,\,n+1} \\
0 & w_{22} & \cdots & & w_{2k} & \cdots & w_{2n} & w_{2,\,n+1} \\
\hline
& & & 0 & w_{kk} & \cdots & w_{kn} & w_{k,\,n+1} \\
& & & 0 & w_{k+1,\,k} & \cdots & w_{k+1,\,n} & w_{k+1,\,n+1} \\
\hline
0 & \cdots & & 0 & w_{nk} & \cdots & w_{nn} & w_{n,\,n+1}
\end{bmatrix}
$$

The kth step consists in eliminating x_k from equations $k+1, \ldots, n$. This is accomplished as follows.

(a)   Search column k, rows k through n, for the first nonzero entry, say it occurs in row r.

(b)   If $k \neq r$, then interchange rows k and r.

(c)   Compute multipliers

$$
m_{ik} = \frac{-w_{ik}}{w_{kk}} \qquad i = k+1, \ldots, n
$$

where w_{kk} is the "new" (k, k) entry.

(d)   Compute for $i = k+1, \ldots, n$

$$
w_{ij}^{\text{new}} = w_{ij}^{\text{old}} + m_{ik} w_{kj}^{\text{old}} \qquad j = k+1, \ldots, n+1
$$

EXAMPLE 3.12

We use Gaussian elimination for $\mathbf{Ax} = \mathbf{b}$ where

$$
\mathbf{A} = \begin{bmatrix} 1 & 2 & 1 \\ 3 & 6 & 0 \\ 2 & 8 & 4 \end{bmatrix} \qquad \text{and} \qquad \mathbf{b} = \begin{bmatrix} 2 \\ 9 \\ 6 \end{bmatrix}
$$

We have

$$\mathbf{W} = \begin{bmatrix} 1 & 2 & 1 & 2 \\ 3 & 6 & 0 & 9 \\ 2 & 8 & 4 & 6 \end{bmatrix}$$

In step 1 we find $m_{21} = -3$ and $m_{31} = -2$, and so $\mathbf{W}$ is transformed to

$$\begin{bmatrix} 1 & 2 & 1 & 2 \\ 0 & 0 & -3 & 3 \\ 0 & 4 & 2 & 2 \end{bmatrix}$$

and in step 2, since the (2, 2) entry is 0, we must interchange rows 2 and 3 to give

$$\begin{bmatrix} 1 & 2 & 1 & 2 \\ 0 & 4 & 2 & 2 \\ 0 & 0 & -3 & 3 \end{bmatrix}$$

Since $m_{32} = 0$, that is, the (3, 2) entry is 0, we are finished. Backsubstitution gives the solution $x_3 = -1$, $x_2 = 1$, and $x_1 = 1$.

ALGORITHM 3.2

NAIVE GAUSSIAN ELIMINATION   Given an $n \times n$ matrix $\mathbf{A}$ and an n-vector $\mathbf{b}$, form the augmented matrix $\mathbf{W} = [\mathbf{A} \mid \mathbf{b}]$.

Do $k = 1, 2, \ldots, n - 1$:

Search column k, rows k through n, for the first nonzero entry, say w_{rk}

If no such r exists then stop

If $r \neq k$ then Do $j = k, \ldots, n + 1$:

$$\text{temp} \leftarrow w_{kj}$$

$$w_{kj} \leftarrow w_{rj}$$

$$w_{rj} \leftarrow \text{temp}$$

Do $i = k + 1, \ldots, n$:

$$m_{ik} \leftarrow -w_{ik}/w_{kk}$$

$$w_{ik} \leftarrow 0$$

Do $j = k + 1, \ldots, n + 1$:

$$w_{ij} \leftarrow w_{ij} + m_{ik} w_{kj}$$

If in step k we find that rows k through n of column k each contain zero, then the algorithm requires us to stop the computation. The reason for this termination is that A is not invertible in this case. If $w_{kk} = w_{k+1, k} = \cdots = w_{nk} = 0$ at step k, then the last $n - k + 1$ equations

$$0 \cdot x_k + w_{k, k+1} x_{k+1} + \quad \cdots \quad + w_{kn} x_n = w_{k, n+1}$$
$$\dotfill$$
$$0 \cdot x_k + w_{n, k+1} x_{k+1} + \quad \cdots \quad + w_{nn} x_n = w_{n, n+1}$$

$$(3.12)$$

involve only the $n - k$ unknowns $x_{k+1}, \ldots, x_n$. There are more equations than unknowns in (3.12), and hence, by Theorem 3.1, part (iv), this subsystem does not have a solution for some right-hand side. Therefore the entire original system does not have a solution for some right-hand side. It follows from Theorem 3.2 that A is not invertible. The following example illustrates this point.

EXAMPLE 3.13

Consider the system $Ax = b$ whose augmented matrix is

$$\left[\begin{array}{cccc|c}
12 & 6 & 1 & -1 & 3 \\
4 & 2 & 0 & 2 & -1 \\
2 & 1 & 5 & 4 & 2 \\
6 & 3 & -2 & 0 & 6
\end{array}\right]$$

After the first step of elimination we find the augmented matrix is transformed to

$$\left[\begin{array}{cccc|c}
12 & 6 & 1 & -1 & 3 \\
0 & 0 & -\frac{1}{3} & \frac{7}{3} & -2 \\
0 & 0 & \frac{29}{6} & \frac{25}{6} & \frac{3}{2} \\
0 & 0 & -\frac{5}{2} & \frac{1}{2} & \frac{9}{2}
\end{array}\right]$$

and the subsystem

$$0 \cdot x_2 + (-\tfrac{1}{3})x_3 + \tfrac{7}{3}x_4 = -2$$
$$0 \cdot x_2 + \tfrac{29}{6}x_3 + \tfrac{25}{6}x_4 = \tfrac{3}{2}$$
$$0 \cdot x_2 + (-\tfrac{5}{2})x_3 + \tfrac{1}{2}x_4 = \tfrac{9}{2}$$

has no solution. The reader should verify this as follows. Solve the last two equations (for x_3 and x_4) and substitute x_3 and x_4 into the first equation of the subsystem. Hence $Ax = b$ has no solution and A is not invertible.

The method of Gaussian elimination provides a systematic means of solving a linear system of equations that is conceptually simple and ideally suited for implementation on a digital computer. A by-product of the elimination process is a computational test for the invertibility of A: the matrix A is invertible if and only if the final augmented matrix of Gaussian elimination applied to $\mathbf{A}x = \mathbf{b}$ is of the form (3.10b), where $\mathbf{U}$ is upper triangular with nonzero diagonal entries.

We complete this section by giving a subroutine that implements Algorithm 3.2.

```
      SUBROUTINE ELIM(W,N,M,IFLG)
C   W:AUGMENTED MATRIX
C   N:NUMBER OF ROWS IN W
C   M:NUMBER OF COLUMNS IN W
C   IFLG:SIGNALS MODE OF RETURN;1 NORMAL;-1 ABNORMAL
      DIMENSION W(N,M)
      IFLG=1
      NN=N-1
C   MAIN OUTER LOOP
      DO 7 K=1,NN
         TEST=0.
C   DETERMINE PIVOTAL ROW
         DO 2 J=K,N
          L=J
          IF(TEST.NE.W(J,K)) GO TO 3
2        CONTINUE
C   SIGNAL THAT MATRIX IS SINGULAR
         IFLG=-IFLG
         RETURN
3        CONTINUE
         IF(L.EQ.K) GO TO 5
C   PERFORM ROW INTERCHANGES
         DO 4 JJ=K,M
          TEMP=W(K,JJ)
          W(K,JJ)=W(L,JJ)
          W(L,JJ)=TEMP
4        CONTINUE
5        KK=K+1
C   UPDATE THE WORKING ARRAY
         DO 6 I=KK,N
          WMULT=-W(I,K)/W(K,K)
          W(I,K)=0.
          DO 6 J=KK,M
           W(I,J)=W(I,J)+WMULT*W(K,J)
6        CONTINUE
7     CONTINUE
      IF(W(N,N).EQ.0.) IFLG=-IFLG
      RETURN
      END
```

The parameter N is the row dimension of the coefficient matrix $\mathbf{A}$, and M should normally equal $N + 1$. These parameters allow the subroutine to be called several times within the same calling (main) program for arrays of different size. The flag IFLG is used to signal the mode of return. If IFLG =

−1, then the calling program should print a message that the given system does not possess an invertible coefficient matrix.

This subroutine performs the elimination phase of the process and returns (if IFLG = 1) the final working array whose first N columns are in upper triangular form and (if M = N + 1) whose last column contains the updated **b** vector. The solution process is completed by use of the following implementation of Algorithm 3.1.

```
      SUBROUTINE BAKSUB(W,N,M,X)
C  W:AUGMENTED MATRIX
C  N:NUMBER OF ROWS IN W
C  M:NUMBER OF COLUMNS IN W
C  X:ON RETURN CONTAINS SOLUTION VECTOR
      DIMENSION W(N,M),X(N)
      X(N)=W(N,M)/W(N,N)
      NN=N-1
      DO 2 I=1,NN
        II=N-I
        SUM=W(II,M)
        JJ=M-I
        DO 1 J=JJ,N
          SUM=SUM-W(II,J)*X(J)
1       CONTINUE
        X(II)=SUM/W(II,II)
2     CONTINUE
      RETURN
      END
```

These codes are by no means optimal in any sense of the word. We give these subroutines simply so that the reader has a reasonable model that may be easily modified in order to implement the more sophisticated versions of Gaussian elimination in the next section. For the system of Example 3.12, a simple main program and output are as follows:

Main program:

```
      DIMENSION W(3,4),X(3)
      DATA W/1.,3.,2.,2.,6.,8.,1.,0.,4.,2.,9.,6./
      CALL ELIM(W,3,4,IFLG)
      IF(IFLG.LE.0) GO TO 4
      CALL BAKSUB(W,3,4,X)
      WRITE(6,100) (X(J),J=1,3)
100   FORMAT(' ',3(2X,E14.7))
      STOP
4     WRITE(6,200)
200   FORMAT(' ','TROUBLE:MATRIX NOT INVERTIBLE')
      STOP
      END
```

Output:

0.1000000E 01   0.1000000E 01   -0.1000000E 01

EXERCISES

3.21 Show that the number of multiplication/divisions in Algorithm 3.1 is $n(n + 1)/2$.
Hint: $\sum_{k=1}^{n} k = n(n + 1)/2$.

3.22 Calculate $M_1 W_1$ and $M_2 M_1 W_1$, where W_1 is given in Example 3.11 and

$$
M_1 = \begin{bmatrix} 1 & 0 & 0 \\ 3 & 1 & 0 \\ 0 & 0 & 1 \end{bmatrix} \qquad M_2 = \begin{bmatrix} 1 & 0 & 0 \\ 0 & 1 & 0 \\ 2 & 0 & 1 \end{bmatrix}
$$

Compare the resulting matrices with W_2 and W_3.

3.23 Use naive Gaussian elimination to solve the system

$$x_1 - x_2 + x_3 = 2$$
$$3x_1 - x_2 + 2x_3 = -6$$
$$3x_1 + x_2 - x_3 = 12$$

3.24 Use Gaussian elimination to show that the system

$$x_1 - x_2 + x_3 = 2$$
$$3x_1 - x_2 + 2x_3 = -6$$
$$3x_1 + x_2 + x_3 = -18$$

has infinitely many solutions. If the last equation is changed to

$$3x_1 + x_2 + x_3 = 6$$

how many solutions are there? Is the coefficient matrix for the system invertible?

3.25 Let A and M be given as

$$
A = \begin{bmatrix} 2 & -3 & 7 \\ 1 & 5 & -1 \\ 2 & 0 & 3 \end{bmatrix} \qquad M = \begin{bmatrix} 1 & 0 & 0 \\ 0 & 1 & 0 \\ 0 & 5 & 1 \end{bmatrix}
$$

Compute the product MA.

3.26 Let M be the $n \times n$ matrix that is obtained by changing the (i, j) entry, $i > j$, of I from 0 to $m \neq 0$. Describe the result of the matrix product MA in terms of elementary row operations. Hint: Try a 3×3 example.

3.27 Let P denote the 4×4 matrix obtained by interchanging rows 2 and 4 of I_4. By considering the product $P^2 = PP$, determine if P is invertible. Can your conclusion be generalized to $n \times n$ matrices obtained by row interchanges of I_n?

3.28 Let $\mathbf{P}$ be the $n \times n$ matrix that is obtained by interchanging rows i and k of $\mathbf{I}_n$. Describe the result of the matrix product $\mathbf{PA}$ in terms of elementary row operations. Hint: Try a 4×4 example first.

3.29 In view of exercises 3.26 and 3.28, describe the method of Gaussian elimination in terms of premultiplication by appropriate $\mathbf{P}$ and $\mathbf{M}$ matrices.

3.30 Is the matrix $\mathbf{M}$ of exercise 3.25 invertible? Describe the inverse of the matrix $\mathbf{M}$ of exericse 3.26.

3.31 Show that the number of multiplication/divisions in Algorithm 3.2 is $n^3/3 + n^2/2 - 5n/6$. Hint: $\sum_{k=1}^{m} k^2 = m(m+1)(2m+1)/6$.

3.32 For n large, say $n \geq 15$, show that the number of multiplication/divisions in Algorithm 3.2 is approximately $n^3/3$. Hint: Use exercise 3.31.

3.33 In subroutine ELIM show that the statement $W(I,K) = 0.$ is unnecessary.

3.34 Why is it necessary to check that $W(N,N) \neq 0$ in subroutine ELIM?

3.35 In the next section we see that it is advantageous to save the multipliers. Can you think of an appropriate place in existing memory to store the multiplier corresponding to row i during step k of the elimination process? Hint: See exercise 3.33.

3.36 What happens in ELIM if WMULT = 0 for some I? Modify ELIM to skip the array updating for such values of I.

3.37 Write a main program that calls subroutines ELIM and BAKSUB. Use your program to solve the system

$$\begin{bmatrix} 3.146 & 5.011 & -0.010 & 1.001 \\ 0.976 & 2.010 & 0.989 & 0.002 \\ -0.002 & 1.107 & 1.965 & 2.989 \\ -0.003 & 0.002 & 1.001 & 1.978 \end{bmatrix} \begin{bmatrix} x_1 \\ x_2 \\ x_3 \\ x_4 \end{bmatrix} = \begin{bmatrix} 9.168 \\ 1.999 \\ 2.129 \\ 0.976 \end{bmatrix}$$

Compare your computed solution with the exact solution, given by $x_1 = x_2 = -x_3 = x_4 = 1$.

3.38 Suppose it is required to solve two systems with the same coefficient matrix, say $\mathbf{Ax} = \mathbf{b}^1$ and $\mathbf{Ax} = \mathbf{b}^2$. This could be accomplished by calling ELIM and BAKSUB twice, once for each system. However, this is not very efficient. A better approach is to form an augmented matrix with N rows and N + 2 columns that handles both systems simultaneously in ELIM. Thus $\mathbf{b}^1$ and $\mathbf{b}^2$ would be placed in columns N + 1 and N + 2, respectively, of $\mathbf{W}$. This requires no modifications of ELIM. How would BAKSUB need to be changed in order to solve each of the two systems with a single call?

3.39 Modify BAKSUB so that backsubstitution is carried out for columns N + 1, N + 2, ..., M of $\mathbf{W}$. Hint: Make $\mathbf{X}$ an array of dimension N $\times$ (M − N) and return the result of backsubstitution on column N + 1 of $\mathbf{W}$ in column 1 of $\mathbf{X}$, on column N + 2 of $\mathbf{W}$ in column 2 of $\mathbf{X}$, and so on.

3.40 You can check the modified version of BAKSUB in exercise 3.39 as follows. Call ELIM with $\mathbf{W}$ given by the 4×8 matrix

$$\begin{bmatrix} 5 & -2 & 4 & 1 & 1 & 0 & 0 & 0 \\ -2 & 1 & 1 & -1 & 0 & 1 & 0 & 0 \\ 4 & 1 & 0 & 0 & 0 & 0 & 1 & 0 \\ 1 & -1 & 0 & 1 & 0 & 0 & 0 & 1 \end{bmatrix}$$

Then call BAKSUB and print the 4×4 array $\mathbf{X}$. Array $\mathbf{X}$ should contain the inverse of the matrix $\mathbf{A}$, given by the first four columns of $\mathbf{W}$. Compare your output with

$$\mathbf{A}^{-1} = \tfrac{1}{12} \begin{bmatrix} 1 & -4 & 1 & -5 \\ -4 & 16 & 8 & 20 \\ 1 & 8 & 1 & 7 \\ -5 & 20 & 7 & 37 \end{bmatrix}$$

3.41 Use ELIM and BAKSUB to solve $\mathbf{Ax} = \mathbf{b}$, where

$$\mathbf{A} = \begin{bmatrix} 5 & 4 & 7 & 5 & 6 & 7 & 5 \\ 4 & 12 & 8 & 7 & 8 & 8 & 6 \\ 7 & 8 & 10 & 9 & 8 & 7 & 7 \\ 5 & 7 & 9 & 11 & 9 & 7 & 5 \\ 6 & 8 & 8 & 9 & 10 & 8 & 9 \\ 7 & 8 & 7 & 7 & 8 & 10 & 10 \\ 5 & 6 & 7 & 5 & 9 & 10 & 10 \end{bmatrix} \qquad \mathbf{b} = \begin{bmatrix} 39 \\ 53 \\ 56 \\ 53 \\ 58 \\ 57 \\ 52 \end{bmatrix}$$

Compare with the exact solution, given by $x_i = 1$ for $i = 1, 2, \ldots, 7$.

3.3 PIVOTING AND ROUNDOFF ERROR

In order to illustrate some of the effects of roundoff error in the elimination process, we apply Algorithm 3.2 to the system

$$\begin{bmatrix} 16 & -9 & 1 \\ -2 & 1.127 & 8 \\ 4 & 3 & 1 \end{bmatrix} \begin{bmatrix} x_1 \\ x_2 \\ x_3 \end{bmatrix} = \begin{bmatrix} 38 \\ -12.873 \\ 14 \end{bmatrix} \tag{3.13}$$

on a hypothetical machine that uses five decimal digit floating point arithmetic with chopping. In the first step we find $m_{21} = 0.125$ and $m_{31} = -0.25$. The augmented matrix after the first step is

$$\begin{bmatrix} 16 & -9 & 1 & 38 \\ 0 & 0.002 & 8.125 & -8.123 \\ 0 & 5.25 & 0.75 & 4.5 \end{bmatrix}$$

For the second (and final) step we get $m_{32} = -5.25/0.002 = -2625$. The new $(3, 3)$ entry is

$$w_{33}^{\text{old}} + m_{32} w_{23}^{\text{old}} = 0.75 + (-2625)(8.125)$$
$$= 0.75 - 21{,}328.125$$

but $m_{32} w_{23}^{\text{old}}$ must be chopped to $-21{,}328$, which when added to 0.75 gives $w_{33}^{\text{new}} = -21{,}327$. Similarly we find that $w_{34}^{\text{new}} = 21{,}326$. The final augmented matrix is

$$\begin{bmatrix} 16 & -9 & 1 & 38 \\ 0 & 0.002 & 8.125 & -8.123 \\ 0 & 0 & -21{,}327 & 21{,}326 \end{bmatrix}$$

By backsubstitution we find $x_3 = -0.99995$, $x_2 = 0.75000$, and $x_1 = 2.8593$. The exact solution is $x_3 = -1$, $x_2 = 1$, and $x_1 = 3$, and we see that the floating point solution has a good relative error in x_3, namely 5×10^{-5}, but the relative error in x_2 and x_1 is 0.25 and 0.0469, respectively.

In order to determine the source of this large relative error, let us reconsider the final step (the first step resulted in no roundoff error) of the algorithm. The term $m_{32} w_{23}^{old}$ is large in magnitude relative to w_{33}^{old}, and the information in w_{33}^{old} is essentially lost because of the floating point arithmetic. Note that if $w_{33}^{old} = 0.25$, then w_{33}^{new} would be unchanged; that is, a 50 percent change in w_{33}^{old} does not change w_{33}^{new}. It would appear that the reason for this unacceptable state of affairs is the size of the multiplier m_{32}.

Let us consider the effect of interchanging rows 2 and 3 at the initiation of step 2. Thus we have the augmented matrix

$$\left[\begin{array}{ccc|c} 16 & -9 & 1 & 38 \\ 0 & 5.25 & 0.75 & 4.5 \\ 0 & 0.002 & 8.125 & -8.123 \end{array}\right]$$

and now the multiplier is $m_{32} = -0.002/5.25 = -0.38095 \times 10^{-3}$. The final matrix is

$$\left[\begin{array}{ccc|c} 16 & -9 & 1 & 38 \\ 0 & 5.25 & 0.75 & 4.5 \\ 0 & 0 & 8.1247 & -8.1247 \end{array}\right]$$

and by backsubstitution we find that the floating point solution is the exact solution. We should point out that the final (3, 3) and (3, 4) entries are not exact; we were fortunate that the roundoff errors in these entries happened to result in the exact computation of x_3.

In this second approach to our example we observe that all the multipliers are less than unity in magnitude, thereby avoiding the difficulty we previously encountered. There are several strategies that can be used to ensure that the multipliers are no larger than unity in absolute value. Perhaps the simplest strategy is called **partial pivoting**. At the kth step of Gaussian elimination we search column k, rows k through n, for the entry of largest magnitude, say w_{rk}. Row r is called the **pivotal row** for step k. Then rows k and r of the augmented matrix are interchanged so that the pivotal row becomes the kth row.

Let us return to our example, in modified form, to illustrate why partial pivoting is not entirely satisfactory. The system

$$\left[\begin{array}{ccc} 16 \times 10^4 & -9 \times 10^4 & 10^4 \\ -2 \times 10^4 & 1.127 \times 10^4 & 8 \times 10^4 \\ 4 & 3 & 1 \end{array}\right] \left[\begin{array}{c} x_1 \\ x_2 \\ x_3 \end{array}\right] = \left[\begin{array}{c} 38 \times 10^4 \\ -12.873 \times 10^4 \\ 14 \end{array}\right] \quad (3.14)$$

is equivalent to (3.13). The augmented matrix after the first step, using partial pivoting, is

$$\left[\begin{array}{ccc|c} 16 \times 10^4 & -9 \times 10^4 & 10^4 & 38 \times 10^4 \\ 0 & 20 & 81{,}250 & -81{,}230 \\ 0 & 5.25 & 0.75 & 4.5 \end{array}\right]$$

and the pivotal row for step 2, via partial pivoting, is row 2. Upon completion of elimination we find by backsubstitution that the solution is the same as we obtained in our initial attempt. Verify this! The reader has no doubt noticed that (3.14) is obtained by multiplying the first two rows of (3.13) by 10^4. We see that partial pivoting leads to the same difficulty that occurred in our initial usage of naive Gaussian elimination for (3.13). As the choice of pivotal row, using partial pivoting, is directly influenced by the "size" of the rows, it would seem appropriate to use a pivoting strategy that at the kth step chooses as the pivotal row that one whose kth column entry is largest relative to the size of its row.

The strategy of **scaled partial pivoting** works as follows. Prior to the first step compute the size of row i by

$$s_i = \max_{1 \le j \le n} |w_{ij}|$$

for each $i = 1, 2, \ldots, n$. In the kth step choose the pivotal row by selecting the smallest index $r \ge k$ such that

$$\frac{|w_{rk}|}{s_r} \ge \frac{|w_{ik}|}{s_i} \qquad k \le i \le n$$

Then interchange rows k and r.

With this pivoting strategy the proper choice of the pivotal row is made in (3.13) and is not affected by rescaling the equations as we did to get (3.14). Scaled partial pivoting is a widely used and accepted pivoting strategy. It is designed so as to minimize, in some sense, the growth of roundoff error in the computed solution. In addition to this important consideration the user of Gaussian elimination should be concerned with the efficiency of implementation and execution.

Frequently one is faced with the problem of solving several linear systems that have the same coefficient matrix. Specifically, suppose we want to solve $\mathbf{Ax} = \mathbf{b}^j$ for $j = 1, 2, \ldots, m$. If we solve $\mathbf{Ax} = \mathbf{b}^1$ by elimination, we obtain an equivalent problem, say $\tilde{\mathbf{A}}\mathbf{x} = \hat{\mathbf{b}}^1$, where $\tilde{\mathbf{A}}$ is upper triangular. The same upper triangular matrix results when elimination is applied to any of these m problems. Rather than repeat this process m times, we require a means of transforming $\mathbf{b}^j$ into $\hat{\mathbf{b}}^j$ such that $\tilde{\mathbf{A}}\mathbf{x} = \hat{\mathbf{b}}^j$ is equivalent to $\mathbf{Ax} = \mathbf{b}^j$. The transformation of $\mathbf{b}^j$ into $\hat{\mathbf{b}}^j$ is accomplished by using the same set of multipliers that are used to transform $\mathbf{A}$ into $\tilde{\mathbf{A}}$. These multipliers should be stored in memory while $\tilde{\mathbf{A}}$ is being computed. The procedure should be clear with the aid of the following example.

EXAMPLE 3.14

Suppose we want to solve $\mathbf{Ax} = \mathbf{b}^1$ and $\mathbf{Ax} = \mathbf{b}^2$ where

$$\mathbf{A} = \begin{bmatrix} 4 & 1 & -2 \\ -1 & 2 & 5 \\ 2 & 3 & 6 \end{bmatrix} \qquad \mathbf{b}^1 = \begin{bmatrix} 6 \\ -6 \\ 4 \end{bmatrix} \qquad \mathbf{b}^2 = \begin{bmatrix} 3 \\ 6 \\ 11 \end{bmatrix}$$

In solving $\mathbf{Ax} = \mathbf{b}^1$ by naive Gaussian elimination, the augmented matrices are

$$\begin{bmatrix} 4 & 1 & -2 & 6 \\ 0 & \frac{9}{4} & \frac{9}{2} & -\frac{9}{2} \\ 0 & \frac{5}{2} & 7 & 1 \end{bmatrix} \qquad \text{with } m_{21} = 0.25, \, m_{31} = -0.5$$

and

$$\begin{bmatrix} 4 & 1 & -2 & 6 \\ 0 & \frac{9}{4} & \frac{9}{2} & -\frac{9}{2} \\ 0 & 0 & 2 & 6 \end{bmatrix} \qquad \text{with } m_{32} = -\frac{10}{9}$$

In step 1 there are two multipliers, and since the two zeros in column 1 are never explicitly used, we may store m_{21} in the $(2, 1)$ entry and m_{31} in the $(3, 1)$ entry. Similarly m_{32} may be stored in the $(3, 2)$ entry. With this storage our final augmented matrix is

$$\begin{bmatrix} 4 & 1 & -2 & 6 \\ \frac{1}{4} & \frac{9}{4} & \frac{9}{2} & -\frac{9}{2} \\ -\frac{1}{2} & -\frac{10}{9} & 2 & 6 \end{bmatrix}$$

By backsubstitution we find $x_3 = 3$, $x_2 = -8$, and $x_1 = 5$. In order to solve $\mathbf{Ax} = \mathbf{b}^2$, we need to determine $\tilde{\mathbf{b}}^2$. The first step transforms $\mathbf{b}^2$ into

$$\begin{bmatrix} 3 \\ \frac{27}{4} \\ \frac{19}{2} \end{bmatrix} = \begin{bmatrix} 3 \\ 6 + 3m_{21} \\ 11 + 3m_{31} \end{bmatrix} = \mathbf{c}$$

and the second step transforms $\mathbf{c}$ into

$$\begin{bmatrix} 3 \\ \frac{27}{4} \\ 2 \end{bmatrix} = \begin{bmatrix} c_1 \\ c_2 \\ c_3 + m_{32} c_2 \end{bmatrix} = \tilde{\mathbf{b}}^2$$

Then $\mathbf{A}\mathbf{x} = \mathbf{b}^2$ is solved by backsubstitution of the equivalent system:

$$\begin{bmatrix} 4 & 1 & -2 \\ 0 & \frac{9}{4} & \frac{9}{2} \\ 0 & 0 & 2 \end{bmatrix} \begin{bmatrix} x_1 \\ x_2 \\ x_3 \end{bmatrix} = \begin{bmatrix} 3 \\ \frac{27}{4} \\ 2 \end{bmatrix}$$

In Example 3.14 no row interchanges were required. In general, row interchanges must be taken into account when $\mathbf{b}$ is transformed into $\tilde{\mathbf{b}}$. The simplest method of keeping track of row interchanges is as follows. For solving $\mathbf{A}\mathbf{x} = \mathbf{b}$, where $\mathbf{A}$ is $n \times n$, introduce an n-vector $\mathbf{p}$ whose entries are initialized by $p_i = i$ for $i = 1, 2, \ldots, n$. If, at step k, rows k and r of $\mathbf{W}$ are to be interchanged, then interchange rows k and r of $\mathbf{p}$. At the end of the elimination process, the vector $\mathbf{p}$ furnishes a set of pointers to the pivotal rows that were used in elimination. Thus the final p_1 points to the row of the original $\mathbf{W}$ that was the pivotal row for step 1, p_2 gives the row of the original $\mathbf{W}$ used as the pivotal row for step 2, and so on. A moment's reflection reveals that, with the proper use of $\mathbf{p}$, no row interchanges in $\mathbf{W}$ are necessary. The only reason we used row interchanges in $\mathbf{W}$ was to make $\tilde{\mathbf{A}}$ upper triangular. This makes it apparent by inspection which equation involves only x_n, which involves only x_n and x_{n-1}, and so on, but the final entries of $\mathbf{p}$ give this information as well.

Therefore we suggest the following scheme for keeping track of the pivotal rows. At step k we use a pivoting strategy that involves searching rows $p_k, p_{k+1}, \ldots, p_n$ of $\mathbf{W}$. If the strategy selects row p_r as the pivotal row for step k, we exchange rows k and r of $\mathbf{p}$. Then p_k is a pointer to the row of $\mathbf{W}$ that is the pivotal row for step k of elimination. For the convenience of the reader we modify Algorithm 3.2 to include scaled partial pivoting, storage of the multipliers, and the pointer vector. In addition, we split the elimination process on $\mathbf{W}$ into two parts: one for $\mathbf{A}$ and one for $\mathbf{b}$.

ALGORITHM 3.3

GAUSSIAN ELIMINATION WITH SCALED PARTIAL PIVOTING Given an $n \times n$ matrix $\mathbf{A}$,

Do $i = 1, 2, \ldots, n$:

$s_i \leftarrow \max\limits_{1 \le j \le n} |a_{ij}|$

If $s_i = 0$ then stop

$p_i \leftarrow i$

Do $k = 1, 2, \ldots, n - 1$:

    Select the least index r $(\geq k)$ such that

$$\frac{|a_{p_r, k}|}{s_{p_r}} \geq \frac{|a_{p_i, k}|}{s_{p_i}} \qquad i = k, \ldots, n$$

    If $a_{p_r, k} = 0$ then stop

    If $k \neq r$ then temp $\leftarrow p_k$

                    $p_k \leftarrow p_r$

                    $p_r \leftarrow$ temp

    Do $i = p_{k+1}, \ldots, p_n$:

        $a_{ik} \leftarrow -a_{ik}/a_{p_k, k}$

        Do $j = k + 1, \ldots, n$:

            $a_{ij} \leftarrow a_{ij} + a_{ik} a_{p_k, j}$

ALGORITHM 3.4

Given the output from Algorithm 3.3 and an n-vector $\mathbf{b}$,

$$\tilde{b}_{p_1} \leftarrow b_{p_1}$$

Do $i = 2, 3, \ldots, n$:

$$\tilde{b}_{p_i} \leftarrow b_{p_i} + \sum_{j=1}^{i-1} a_{p_i, j} \tilde{b}_{p_j}$$

then $\tilde{\mathbf{b}}$ is ready for backsubstitution.

ALGORITHM 3.5

BACKSUBSTITUTION   Given the output from Algorithms 3.3 and 3.4,

$$x_n \leftarrow \frac{\tilde{b}_{p_n}}{a_{p_n, n}}$$

Do $i = n - 1, n - 2, \ldots, 1$:

$$x_i \leftarrow \frac{\tilde{b}_{p_i} - \sum_{j=i+1}^{n} a_{p_i, j} x_j}{a_{p_i, i}}$$

then $\mathbf{x}$ is the solution of $\mathbf{Ax} = \mathbf{b}$.

At first inspection these algorithms may appear to unduly complicate the elimination process. For use in hand calculations this is certainly a valid criticism; however, these algorithms are more efficient on a digital computer than analogous procedures that use row interchanges. The following example illustrates these algorithms, and its study should facilitate the implementation of the algorithms in a high level language such as FORTRAN.

EXAMPLE 3.15

Consider the system

$$\begin{bmatrix} 1 & 2 & 1 \\ 3 & 4 & 0 \\ 2 & 10 & 4 \end{bmatrix} \begin{bmatrix} x_1 \\ x_2 \\ x_3 \end{bmatrix} = \begin{bmatrix} 3 \\ 3 \\ 10 \end{bmatrix}$$

which we shall solve, with exact arithmetic, using Algorithms 3.3 to 3.5. Initially the vector $\mathbf{p}$ is $\mathbf{p} = [1, 2, 3]^T$, and we calculate $s_1 = 2$, $s_2 = 4$, and $s_3 = 10$. Then we find $|a_{p_1, 1}|/s_{p_1} = \frac{1}{2}$, $|a_{p_2, 1}|/s_{p_2} = \frac{3}{4}$, and $|a_{p_3, 1}|/s_{p_3} = \frac{1}{5}$ so that $p_2 = 2$ is the pivotal row. We update $\mathbf{p}$ to $\mathbf{p} = [2, 1, 3]^T$, and the multipliers are

$$a_{11} = a_{p_2, 1} \leftarrow -\frac{a_{p_2, 1}}{a_{p_1, 1}} = -\frac{1}{3}$$

$$a_{31} = a_{p_3, 1} \leftarrow -\frac{a_{p_3, 1}}{a_{p_1, 1}} = -\frac{2}{3}$$

The updated matrix after step 1 is

$$\begin{bmatrix} -\frac{1}{3} & \frac{2}{3} & 1 \\ 3 & 4 & 0 \\ -\frac{2}{3} & \frac{22}{3} & 4 \end{bmatrix}$$

For step 2 we find that $|a_{p_2, 2}|/s_{p_2} = \frac{1}{3}$ and $|a_{p_3, 2}|/s_{p_3} = \frac{22}{30}$ so that $p_3 = 3$ is the pivotal row. We update $\mathbf{p}$ to $\mathbf{p} = [2, 3, 1]^T$ and compute the multiplier $a_{12} = a_{p_3, 2} \leftarrow -a_{p_3, 2}/a_{p_2, 2} = -\frac{1}{11}$. The final matrix is

$$\begin{bmatrix} -\frac{1}{3} & -\frac{1}{11} & \frac{7}{11} \\ 3 & 4 & 0 \\ -\frac{2}{3} & \frac{22}{3} & 4 \end{bmatrix}$$

Next we use Algorithm 3.4 to compute the updated vector $\tilde{\mathbf{b}}$ so that the system may be solved by backsubstitution. We have

$$\tilde{b}_{p_1} = b_2 = 3$$

$$\tilde{b}_{p_2} = \tilde{b}_3 \leftarrow b_{p_2} + a_{p_2, 1} \tilde{b}_{p_1}$$

$$= 10 + (-\tfrac{2}{3})3 = 8$$

$$\tilde{b}_{p_3} = \tilde{b}_1 \leftarrow b_{p_3} + a_{p_3, 1} \tilde{b}_{p_1} + a_{p_3, 2} \tilde{b}_{p_2}$$

$$= 3 + (-\tfrac{1}{3})3 + (-\tfrac{1}{11})8 = \tfrac{14}{11}$$

Finally by backsubstitution we find the solution:

$$x_3 \leftarrow \frac{\tilde{b}_{p3}}{a_{p3,\,3}} = \frac{\tilde{b}_1}{a_{13}} = \frac{\frac{14}{11}}{\frac{7}{11}} = 2$$

$$x_2 \leftarrow \frac{\tilde{b}_{p2} - a_{p2,\,3}\,x_3}{a_{p2,\,2}} = \frac{8 - 4 \cdot 2}{\frac{22}{3}} = 0$$

$$x_1 \leftarrow \frac{\tilde{b}_{p1} - a_{p1,\,3}\,x_3 - a_{p1,\,2}\,x_2}{a_{p1,\,1}} = \frac{3 - 0 \cdot 2 - 4 \cdot 0}{3} = 1$$

An example of a situation where one may want to solve several linear systems with the same coefficient matrix is that of determining the inverse of a matrix. Given an $n \times n$ invertible matrix $\mathbf{A}$, recall that $\mathbf{x}^j$, the jth column of $\mathbf{A}^{-1}$, satisfies $\mathbf{A}\mathbf{x}^j = \mathbf{e}^j$, where $\mathbf{e}^j$ is the jth column of $\mathbf{I}$. Let us estimate the amount of work required for the computation of $\mathbf{A}^{-1}$ by performing an operation count for the algorithms. Traditionally one counts only the number of multiplication/divisions since these operations are more expensive than addition or subtraction. Of course the work required by a given algorithm depends on the computer used and the operating system. We perform only a traditional work estimate. By exercise 3.32 it requires about $n^3/3$ operations to apply Algorithm 3.3 to $\mathbf{A}$. The updating of $\mathbf{e}^j$ to $\tilde{\mathbf{e}}^j$ requires $n(n-1)/2$ operations. Each backsubstitution requires, by exercise 3.21, $n(n+1)/2$ operations. Therefore the computation of $\mathbf{A}^{-1}$, for n large, requires about

$$\frac{n^3}{3} + \frac{n^2(n-1)}{2} + \frac{n^2(n+1)}{2} \simeq \tfrac{4}{3}n^3$$

operations. The solution of $\mathbf{A}\mathbf{x} = \mathbf{b}$ requires, for n large, approximately

$$\frac{n^3}{3} + \frac{n(n-1)}{2} + \frac{n(n+1)}{2} \simeq \frac{n^3}{3}$$

operations. The matrix product $\mathbf{A}^{-1}\mathbf{b}$ requires n^2 operations and hence the solution of $\mathbf{A}\mathbf{x} = \mathbf{b}$ by matrix inversion requires roughly four times the work in Gaussian elimination. In some situations the entries of $\mathbf{A}^{-1}$ may in themselves be of interest, but in general the computation of $\mathbf{A}^{-1}$ should be avoided.

EXERCISES

3.42 Solve the system (3.14) by using partial pivoting with five decimal digit, chopped floating point arithmetic. Remember to chop the result of each arithmetic operation to a normalized five decimal digit floating point number.

3.43 Suppose each multiplication/division requires 10^{-6} seconds of computation time. Estimate the cost of solving $\mathbf{A}\mathbf{x} = \mathbf{b}$ by Algorithms 3.3 to 3.5 assuming that $\mathbf{A}$ is a 200×200 matrix and computer time costs \$600 per hour. How much would it cost to compute $\mathbf{A}^{-1}$ and $\mathbf{A}^{-1}\mathbf{b}$?

3.44 Suppose $\varepsilon \neq 0$ is close to 0. Find the exact solution of the system

$$\varepsilon x_1 + 2x_2 = 2$$

$$x_1 + x_2 = 2$$

Estimate the size of x_1 and x_2 for $|\varepsilon| \ll 1$.

3.45 Use $\varepsilon = 10^{-4}$ and chopped, decimal digit floating point arithmetic with precision 3 to compute the solution of the system of exercise 3.44 without using row interchanges. Repeat the computation using partial pivoting.

3.46 Solve, as in Example 3.15, the system

$$4x_1 - 3x_2 + x_3 = 5$$

$$-x_1 + 2x_2 - 2x_3 = -3$$

$$2x_1 + x_2 - x_3 = 1$$

with chopped, four decimal digit floating point arithmetic. Compare with the exact solution, $x_1 = x_3 = 1$, $x_2 = 0$.

3.47 Modify subroutine ELIM to implement Algorithm 3.3. Call the new subroutine GAUSS with parameter list A, N, IPVT, IFLG, AA. These parameters are used as follows:

   A: coefficient matrix

   N: row and column dimension of A

IPVT: integer pivot vector

  AA: matrix of dimension N × N that results after the elimination process is applied to A

IFLG: integer signal of the mode of return

The subroutine should begin by copying A onto AA. All elimination takes place on AA, and hence A is not destroyed. Do not perform any row interchanges on AA; rather this should be done in IPVT. The DO 6 loop in ELIM can be modified for use in GAUSS as follows:

```
    DO 6 I=KK,N
      II=IPVT(I)
      AA(II,K)=-A(II,K)/AA(IPVT(K),K)
      DO 6 J=KK,N
        A(II,J)=AA(II,J)+AA(II,K)*AA(IPVT(K),J)
6   CONTINUE
```

Test your subroutine on the matrix of Example 3.15 by printing the final contents of A, AA, and IPVT.

3.48 Write a subroutine called SOLVE that implements Algorithms 3.4 and 3.5. Subroutine SOLVE should have parameter list AA, N, IPVT, B, X, where AA, N, and IPVT are output from GAUSS. The vectors B and X are of dimension N, with the right-hand-side vector stored in B and the solution returned in X. Do not destroy the contents of B; instead use an array BB for the computation in Algorithm 3.4. Test SOLVE on the system in Example 3.15.

3.49 Apply GAUSS and SOLVE to the system $\mathbf{Ax} = \mathbf{b}$ where

$$
\mathbf{A} = \begin{bmatrix}
3 & -5 & 6 & 4 & -2 & -3 & 8 \\
1 & 1 & -9 & 15 & 1 & -9 & 2 \\
2 & -1 & 7 & 5 & -1 & 6 & 11 \\
-1 & 1 & 3 & 2 & 7 & -1 & -2 \\
4 & 3 & 1 & -7 & 2 & 1 & 1 \\
2 & 9 & -8 & 11 & -1 & -4 & -1 \\
7 & 2 & -1 & 2 & 7 & -1 & 9
\end{bmatrix}
\qquad
\mathbf{b} = \begin{bmatrix}
11 \\
2 \\
29 \\
9 \\
5 \\
8 \\
25
\end{bmatrix}
$$

Compare with the exact solution, $x_i = 1$, $1 \le i \le 7$.

3.50 Compute the inverse of matrix $\mathbf{A}$ in exercise 3.49 by calling GAUSS once and SOLVE seven times. Check the computed inverse by forming the matrix products $\mathbf{AC}$ and $\mathbf{CA}$, where $\mathbf{C}$ is the computed inverse of $\mathbf{A}$. Discuss the results.

Project 4

The following subroutine is a modification of ELIM that uses scaled partial pivoting:

```
      SUBROUTINE ELIMIN(W,N,M,IFLG,S)
C  GAUSSIAN ELIMINATION WITH SCALED PARTIAL PIVOTING
C  W:AUGMENTED MATRIX
C  N:NUMBER OF ROWS IN W
C  M:NUMBER OF COLUMNS IN W
C  IFLG:SIGNALS MODE OF RETURN;1 NORMAL;-1 ABNORMAL
C  S:ARRAY OF ROW 'SIZES' USED IN PIVOTING STRATEGY
      DIMENSION W(N,M),S(N)
      IFLG=1
      NN=N-1
      DO 11 I=1,N
        S(I)=0.0
        DO 10 J=1,N
          S(I)=S(I)+ABS(W(I,J))
10      CONTINUE
        IF(S(I).NE.0.) GO TO 11
        IFLG=-IFLG
        RETURN
11    CONTINUE
C  MAIN OUTER LOOP
      DO 7 K=1,NN
        L=K
        TEST=ABS(W(K,K))/S(K)
C  DETERMINE PIVOTAL ROW USING
C  SCALED PARTIAL PIVOTING
        DO 2 J=K,N
          RATIO=ABS(W(J,K))/S(J)
          IF(RATIO.LE.TEST) GO TO 2
          TEST=RATIO
          L=J
```

```
      2     CONTINUE
            IF(TEST.NE.0.) GO TO 3
C   SIGNAL THAT MATRIX IS SINGULAR
            IFLG=-IFLG
            RETURN
      3     CONTINUE
            IF(L.EQ.K) GO TO 5
C   PERFORM ROW INTERCHANGES
            DO 4 JJ=K,M
               TEMP=W(K,JJ)
               W(K,JJ)=W(L,JJ)
               W(L,JJ)=TEMP
      4     CONTINUE
      5     KK=K+1
C   UPDATE THE WORKING ARRAY
            DO 6 I=KK,N
               W(I,K)=-W(I,K)/W(K,K)
               DO 6 J=KK,M
                  W(I,J)=W(I,J)+W(I,K)*W(K,J)
      6     CONTINUE
      7   CONTINUE
          IF(W(N,N).EQ.0.) IFLG=-IFLG
          RETURN
          END
```

In FORTRAN matrices are stored by columns. This fact has important consequences in the design of efficient linear system solvers because the basic operations in Gaussian elimination are performed on rows. In ELIMIN this means that the inner DO 6 loop varies the array column index, thereby causing a nonsequential access to memory. For many operating systems it is more efficient to use sequential memory accessing. Modify the two DO 6 loops so that all array updating is done column by column.

Test your program on the system of exercise 3.41. Use your program and the modified version of BAKSUB suggested in exercise 3.39 to compute the inverse of

$$
\mathbf{A} = \begin{bmatrix}
3 & -5 & 6 & 4 & -2 & -3 & 8 \\
1 & 1 & -9 & 15 & 1 & -9 & 2 \\
2 & -1 & 7 & 5 & -1 & 6 & 11 \\
1 & 1 & 3 & 2 & 7 & -1 & -2 \\
4 & 3 & 1 & -7 & 2 & 1 & 1 \\
2 & 9 & -8 & 11 & -1 & -4 & -1 \\
7 & 2 & -1 & 2 & 7 & -1 & 9
\end{bmatrix}
$$

If **C** denotes the computed inverse of **A**, then a check on the accuracy of **C** is to compute **AC** and **CA** and compare each to **I**. Perform this check and report your findings.

Ask your instructor if it is possible to find out the execution time of a program in milliseconds. If possible, determine the execution time for computing $\mathbf{A}^{-1}$ using ELIM and using ELIMIN. Compare the results. On the basis of your results can you estimate the comparative execution times for a 200×200 matrix? Give the rationale for your estimates.

3.4 CONDITIONING AND ERROR BOUNDS

In the first two chapters we examined the concept of ill-conditioned problems for function evaluation and root finding. In essence a problem is ill-conditioned if small perturbations of the data result in a large change in the solution. We begin by considering this idea for the square matrix problem $\mathbf{Ax} = \mathbf{b}$. For purposes of illustration we examine the following simple example:

$$\begin{bmatrix} 2.0001 & -1 \\ -2 & 1 \end{bmatrix} \begin{bmatrix} x_1 \\ x_2 \end{bmatrix} = \begin{bmatrix} 1 \\ -1 \end{bmatrix} \qquad (3.15)$$

which has the unique solution $x_1 = 0$, $x_2 = -1$. Suppose we change the right-hand side by a small amount to $[1.0002, -1]^T$; then the solution is $x_1 = 2$, $x_2 = 3$. Thus a 0.02 percent change in b_1 results in a 400 percent change in x_2.

Next suppose we change the (1, 1) entry of the matrix from 2.0001 to 2; then the problem $\mathbf{Ax} = [1, -1]^T$ has a solution given by $x_1 = x_2 = 1$. Note also that this new $\mathbf{A}$ is not invertible.

It is clear that (3.15) is very sensitive to perturbations in the right-hand side or in the coefficient matrix; the problem is ill-conditioned. Later in this section we quantify the notion of ill-conditioning by defining the condition number of a matrix.

If $\mathbf{y}$ is a computed solution of $\mathbf{Ax} = \mathbf{b}$ obtained by Gaussian elimination (or some other algorithm), how can we determine whether $\mathbf{y}$ is close to the exact solution $\mathbf{x}$? Assuming that $\mathbf{A}$ is invertible, we have

$$\mathbf{x} - \mathbf{y} = \mathbf{A}^{-1}(\mathbf{b} - \mathbf{Ay}) = \mathbf{A}^{-1}\mathbf{r} \qquad (3.16)$$

where $\mathbf{r} = \mathbf{b} - \mathbf{Ay}$, called the **residual**, is a computable quantity. Intuitively one might expect that $x - y$ is small if the residual is small.

EXAMPLE 3.16

Consider the system

$$\begin{bmatrix} 2.0001 & -1 \\ -2 & 1 \end{bmatrix} \begin{bmatrix} x_1 \\ x_2 \end{bmatrix} = \begin{bmatrix} 7.0003 \\ -7 \end{bmatrix}$$

and the two approximate solutions $\mathbf{y}$ and $\mathbf{w}$, given by $[2.91, -1.01]^T$ and $[2, -3]^T$, respectively. Which is the more accurate solution? The residuals for $\mathbf{y}$ and $\mathbf{w}$ are $[0.170009, -0.17]^T$ and $[0.0001, 0]^T$, respectively. However, the exact solution is $x_1 = 3$, $x_2 = -1$, and $\mathbf{y}$ is the more accurate solution.

This example illustrates the fact that for ill-conditioned problems a small residual does not imply good accuracy. In order to examine the error in the numerical solution of linear systems as well as to quantify the conditioning, we need a measure of the magnitude of vectors and matrices. The most natural measure of the magnitude (or length) of a vector is the Euclidean length, given by

$$\|\mathbf{x}\|_2 = \left(\sum_{i=1}^{n} |x_i|^2 \right)^{1/2}$$

The reader is no doubt familiar with the geometric interpretation for vectors in space (3-vectors), where $\|\mathbf{x}\|_2$ gives the distance from the origin to the point (x_1, x_2, x_3). For our purposes it is more convenient to define the length, or **norm**, of an n-vector by

$$\|\mathbf{x}\|_1 = \sum_{i=1}^{n} |x_i|$$

This norm shares many of the properties of the Euclidean length, namely

(i) $\|\mathbf{x}\|_1 \geq 0$ for all $\mathbf{x} \neq \mathbf{0}$

(ii) $\|\mathbf{x}\|_1 = 0$ if and only if $\mathbf{x} = \mathbf{0}$

(iii) $\|\mathbf{x} + \mathbf{y}\|_1 \leq \|\mathbf{x}\|_1 + \|\mathbf{y}\|_1$

(iv) $\|\alpha\mathbf{x}\|_1 = |\alpha| \, \|\mathbf{x}\|_1$ for all scalars α

It is a simple matter to verify that properties (i), (ii), and (iv) are valid. For (iii), called the triangle inequality, we recall a property of absolute values: $|a + b| \leq |a| + |b|$ for all real numbers a and b. Then for each i we have $|x_i + y_i| \leq |x_i| + |y_i|$ and (iii) follows by summing over i.

    With this definition of length we have from (3.16) that

$$\|\mathbf{x} - \mathbf{y}\|_1 = \|\mathbf{A}^{-1}\mathbf{r}\|_1$$

and we desire to measure the error $\mathbf{x} - \mathbf{y}$ in terms of the computable quantity $\mathbf{r}$. We saw in Example 3.16 that the norm of $\mathbf{r}$ may be quite different from the norm of $\mathbf{A}^{-1}\mathbf{r}$. Specifically we find that $\|\mathbf{r}\|_1 = 10^{-4}$, where $\mathbf{r} = \mathbf{b} - \mathbf{Aw}$ and $\mathbf{x} - \mathbf{w} = \mathbf{A}^{-1}\mathbf{r}$ so that $\|\mathbf{A}^{-1}\mathbf{r}\|_1 = 3$.

    This change in norm is directly related to the idea of conditioning that we want to quantify. Given a matrix $\mathbf{B}$, by how much can $\|\mathbf{x}\|_1$ differ from $\|\mathbf{Bx}\|_1$? We define

$$\|\mathbf{B}\| = \max_{x \neq 0} \frac{\|\mathbf{Bx}\|_1}{\|\mathbf{x}\|_1}$$

Since the maximum value of this ratio is at least as large as any specific ratio, we have

$$\|\mathbf{B}\| \geq \frac{\|\mathbf{Bx}\|_1}{\|\mathbf{x}\|_1}$$

or, equivalently,

$$\|\mathbf{Bx}\|_1 \leq \|\mathbf{B}\| \, \|\mathbf{x}\|_1$$

The real number $\|\mathbf{B}\|$ is called the norm of the matrix $\mathbf{B}$. This **matrix norm**, in view of the last inequality, gives a measure of the change (in norm) from $\mathbf{x}$ to $\mathbf{Bx}$. How do we actually compute $\|\mathbf{B}\|$? Fortunately there is an easy way to calculate $\|\mathbf{B}\|$, and this is the primary reason we chose to use the vector norm $\|\cdot\|_1$. We claim that the norm is given by

$$\|\mathbf{B}\| = \max_{1 \le j \le n} \sum_{i=1}^{n} |\mathbf{b}_{ij}|$$

Before proving this, we interpret its meaning. If $\mathbf{b}^j$ denotes the jth column of $\mathbf{B}$, then our claim says that $\|\mathbf{B}\| = \max_{1 \le j \le n} \|\mathbf{b}^j\|_1 =$ maximum absolute column sum.

EXAMPLE 3.17

Let

$$\mathbf{B} = \begin{bmatrix} 2 & 16 & 1 \\ -7 & 3 & 0 \\ 11 & -4 & -2 \end{bmatrix}$$

We have $\|\mathbf{b}^1\|_1 = 20$, $\|\mathbf{b}^2\|_1 = 23$, and $\|\mathbf{b}^3\|_1 = 3$ so that $\|\mathbf{B}\| = \|\mathbf{b}^2\|_1 = 23$.

In order to establish the claim, we proceed as follows. First we show that for all $\mathbf{x}$

$$\|\mathbf{Bx}\|_1 \le \left(\max_{1 \le j \le n} \|\mathbf{b}^j\|_1 \right) \|\mathbf{x}\|_1 \tag{3.17}$$

and second we demonstrate, for a particular vector $\mathbf{u}$, that

$$\|\mathbf{Bu}\|_1 = \left(\max_{1 \le j \le n} \|\mathbf{b}^j\|_1 \right) \|\mathbf{u}\|_1 \tag{3.18}$$

The inequality (3.17) implies that $\|\mathbf{B}\| \le \max_{1 \le j \le n} \|\mathbf{b}^j\|_1$, whereas (3.18) says that $\|\mathbf{B}\| \ge \max_{1 \le j \le n} \|\mathbf{b}^j\|_1$. For any $\mathbf{x}$ we have

$$\|\mathbf{Bx}\|_1 = \sum_{i=1}^{n} \left| \sum_{j=1}^{n} b_{ij} x_j \right|$$

$$\le \sum_{i=1}^{n} \sum_{j=1}^{n} |b_{ij}| \, |x_j|$$

and by interchanging the order of summation we get

$$\|\mathbf{Bx}\|_1 \le \sum_{j=1}^{n} |x_j| \left(\sum_{i=1}^{n} |b_{ij}| \right) = \sum_{j=1}^{n} |x_j| \|\mathbf{b}^j\|_1$$

$$\le \left(\max_{1 \le j \le n} \|\mathbf{b}^j\|_1 \right) \|\mathbf{x}\|_1$$

which proves (3.17). To establish (3.18), we suppose that the column of $\mathbf{B}$ that has the largest norm is column k, that is, $\max_{1 \le j \le n} \|\mathbf{b}^j\|_1 = \|\mathbf{b}^k\|_1$. Let $\mathbf{e}^k$ be the kth column of $\mathbf{I}$. Then $\mathbf{Be}^k = \mathbf{b}^k$, $\|\mathbf{Be}^k\|_1 = \|\mathbf{b}^k\|_1$, and we have

$$\|\mathbf{Be}^k\|_1 = \left(\max_{1 \le j \le n} \|\mathbf{b}^j\|_1 \right) \|\mathbf{e}^k\|_1$$

since $\|\mathbf{e}^k\|_1 = 1$.

This discourse has taken us somewhat afield; let us return to the question of error estimation.

THEOREM 3.3

Suppose $\mathbf{A}$ is invertible and $\mathbf{x}$ satisfies $\mathbf{Ax} = \mathbf{b}$; then for any $\mathbf{y}$

$$\frac{\|\mathbf{x} - \mathbf{y}\|_1}{\|\mathbf{x}\|_1} \le \|\mathbf{A}\| \, \|\mathbf{A}^{-1}\| \, \frac{\|\mathbf{r}\|_1}{\|\mathbf{b}\|_1}$$

where $\mathbf{r} = \mathbf{b} - \mathbf{Ay}$.

Proof

By (3.16) and the definition of the matrix norm we have

$$\|\mathbf{x} - \mathbf{y}\|_1 \le \|\mathbf{A}^{-1}\| \, \|\mathbf{b} - \mathbf{Ay}\|_1$$

Moreover we have $\|\mathbf{b}\|_1 = \|\mathbf{Ax}\|_1 \le \|\mathbf{A}\| \, \|\mathbf{x}\|_1$ or, equivalently,

$$\|\mathbf{x}\|_1 \ge \frac{\|\mathbf{b}\|_1}{\|\mathbf{A}\|}$$

and combining these inequalities yields the result.

If we interpret $\mathbf{y}$ as a computed (approximate) solution of $\mathbf{Ax} = \mathbf{b}$, then Theorem 3.3 says that the relative error in $\mathbf{y}$ is bounded by the relative residual $\|\mathbf{r}\|_1/\|\mathbf{b}\|_1$ times the factor $K(\mathbf{A}) = \|\mathbf{A}\| \, \|\mathbf{A}^{-1}\|$. $K(\mathbf{A})$ is a measure of how much the relative residual is magnified to yield the relative error and is called the **condition number** of $\mathbf{A}$. If $K(\mathbf{A})$ is large, then we say that $\mathbf{Ax} = \mathbf{b}$ is ill-conditioned, whereas the problem is well-conditioned for small $K(\mathbf{A})$.

Let us derive some of the properties of the condition number. By exercise 3.55 we have for an invertible matrix $\mathbf{A}$

$$1 = \|\mathbf{I}\| = \|\mathbf{A}^{-1}\mathbf{A}\| \le \|\mathbf{A}\| \, \|\mathbf{A}^{-1}\| = K(\mathbf{A})$$

and so the best possible condition number is 1. If $\mathbf{D}$ is a diagonal matrix, then

$$K(\mathbf{D}) = \frac{\max_{1 \le i \le n} |\mathbf{d}_{ii}|}{\min_{1 \le i \le n} |\mathbf{d}_{ii}|}$$

If $\alpha \ne 0$ is a scalar, then $(\alpha\mathbf{A})^{-1} = (1/\alpha)\mathbf{A}^{-1}$ and so $K(\alpha\mathbf{A}) = K(\mathbf{A})$. The condition number is a reasonable measure of how close a matrix is to being singular (not invertible). The larger the condition number, the closer a matrix is to being singular. If $\mathbf{A}$ is singular, then we define $K(\mathbf{A}) = +\infty$.

The reader who is familiar with determinants knows that a square matrix is singular if and only if its determinant is zero. One frequently hears the statement that a matrix is nearly singular if its determinant is nearly zero. This statement is not without some merit; however, consider the following. Let $\mathbf{A}$ be the matrix of Example 3.16 and $\mathbf{B}$ the 2×2 diagonal matrix whose diagonal entries are 10^{-10}. Then the determinant of $\mathbf{A}$ is 10^{-4} and the determinant of $\mathbf{B}$ is 10^{-20}, but, in solving a linear system $\mathbf{Bx} = \mathbf{c}$, $\mathbf{B}$ essentially acts like the identity matrix and $K(\mathbf{B}) = 1$.

The condition number is also significant in analyzing the propagation of roundoff error during Gaussian elimination. Let $\mathbf{y}$ denote the floating point solution of $\mathbf{Ax} = \mathbf{b}$ obtained by Gaussian elimination with scaled partial pivoting. Then by backward error analysis Wilkinson proved that $\mathbf{y}$ is the exact solution of the perturbed problem

$$(\mathbf{A} + \mathbf{E})\mathbf{y} = \mathbf{b} \tag{3.19}$$

where $\mathbf{E}$ is a matrix whose entries are of the same order of magnitude as the roundoff errors in the entries of $\mathbf{A}$. In other words the entries of $\mathbf{E}$ are comparable to the errors that result when the entries of $\mathbf{A}$ are stored in memory as floating point numbers. We have $\mathrm{fl}(a_{ij}) = (1 - \rho)a_{ij}$ for some ρ satisfying $|\rho| \leq u$, where u is the machine unit. Thus we have

$$|a_{ij} - \mathrm{fl}(a_{ij})| = |\rho a_{ij}| \leq u|a_{ij}| \simeq |e_{ij}| \tag{3.20}$$

What does (3.19) tell us about the relative error in the floating point solution? We have

$$\mathbf{x} - \mathbf{y} = \mathbf{A}^{-1}(\mathbf{b} - \mathbf{Ay}) = \mathbf{A}^{-1}[\mathbf{b} - (\mathbf{b} - \mathbf{Ey})]$$
$$= \mathbf{A}^{-1}\mathbf{Ey}$$

so that

$$\|\mathbf{x} - \mathbf{y}\|_1 \leq \|\mathbf{A}^{-1}\| \, \|\mathbf{E}\| \, \|\mathbf{y}\|_1$$

and hence

$$\frac{\|\mathbf{x} - \mathbf{y}\|_1}{\|\mathbf{y}\|_1} \leq K(\mathbf{A}) \frac{\|\mathbf{E}\|}{\|\mathbf{A}\|}$$

By (3.20) we have

$$\|\mathbf{E}\| = \max_{1 \leq j \leq n} \sum_{i=1}^{n} |e_{ij}|$$

$$\simeq u \max_{1 \leq j \leq n} \sum_{i=1}^{n} |a_{ij}| = u\|\mathbf{A}\|$$

and we expect that

$$\frac{\|\mathbf{x} - \mathbf{y}\|_1}{\|\mathbf{y}\|_1} \simeq uK(\mathbf{A}) \tag{3.21}$$

so that the error relative in $\mathbf{y}$ is expected to be about $K(\mathbf{A})$ machine units.

Unfortunately $K(\mathbf{A})$ requires knowledge of $\mathbf{A}^{-1}$, which is expensive to compute for n large. Normally it is sufficient to obtain a reasonably good estimate of $K(\mathbf{A})$. A recently developed set of programs for the solution of linear systems and related problems, called LINPACK, provides an estimate of $K(\mathbf{A})$ so that the user can judge the quality of the computed solution.

We have made the statement that $\mathbf{A}\mathbf{x} = \mathbf{b}$ is ill-conditioned if $K(\mathbf{A})$ is large. What is a large value of $K(\mathbf{A})$? The answer is given by consideration of (3.21). Certainly if $K(\mathbf{A}) \simeq 1/u$, then the computed solution $\mathbf{y}$ is likely to have no correct digits in each component. Thus the ill-conditioning of $\mathbf{A}\mathbf{x} = \mathbf{b}$ depends on the precision of the floating point arithmetic. For a base β, precision t machine a value of $K(\mathbf{A}) = \beta^k$ implies that the relative error is approximately β^{k-t+1}.

It is possible to estimate $K(\mathbf{A})$ without finding $\mathbf{A}^{-1}$. The basis for this estimation procedure is as follows. If $\mathbf{A}\mathbf{x} = \mathbf{b}$ and $\mathbf{A}$ is invertible, then

$$\mathbf{x} = \mathbf{A}^{-1}\mathbf{b}$$

and

$$\|\mathbf{x}\|_1 \leq \|\mathbf{A}^{-1}\| \, \|\mathbf{b}\|_1$$

or, equivalently,

$$\frac{\|\mathbf{x}\|_1}{\|\mathbf{b}\|_1} \leq \|\mathbf{A}^{-1}\| \tag{3.22}$$

Suppose we select, at random, k vectors $\mathbf{b}^1, \ldots, \mathbf{b}^k$ and solve $\mathbf{A}\mathbf{x}^p = \mathbf{b}^p$ for $1 \leq p \leq k$. Then by (3.22) we have

$$\|\mathbf{A}^{-1}\| \geq \max_{1 \leq p \leq k} \frac{\|\mathbf{x}^p\|_1}{\|\mathbf{b}^p\|_1} \tag{3.23}$$

For a small choice of k the right-hand side of (3.23) should give a ballpark estimate of $\|\mathbf{A}^{-1}\|$. Then we estimate

$$K(\mathbf{A}) \simeq \|\mathbf{A}\| \max_{1 \leq p \leq k} \frac{\|\mathbf{x}^p\|_1}{\|\mathbf{b}^p\|_1} \tag{3.24}$$

This estimate for $K(\mathbf{A})$ requires little additional computation effort. In solving $\mathbf{A}\mathbf{x} = \mathbf{b}$, most of the work is required in Algorithm 3.3. Since the multipliers are stored in memory, the determination of each $\mathbf{x}^p$ by Algorithms 3.4 and 3.5 requires only an additional n^2 multiplication/divisions. For n large this is a small additional price to pay, and for n small all of the computations are inexpensive.

EXAMPLE 3.18

As an illustration of the use of (3.24) to estimate the condition number of a matrix, we consider the matrix $\mathbf{A}$ of exercise 3.40. We have $\|\mathbf{A}\| = 12$ and $\|\mathbf{A}^{-1}\| = \frac{69}{12}$; whence $K(\mathbf{A}) = 69$. For the "randomly" chosen vectors

$$\mathbf{b}^1 = \begin{bmatrix} 100 \\ 2 \\ -17 \\ 26 \end{bmatrix} \qquad \mathbf{b}^2 = \begin{bmatrix} 1 \\ 2 \\ 3 \\ 4 \end{bmatrix} \qquad \mathbf{b}^3 = \begin{bmatrix} -29 \\ 0 \\ 33 \\ -75 \end{bmatrix} \qquad \mathbf{b}^4 = \begin{bmatrix} 86 \\ -43 \\ 21 \\ -10 \end{bmatrix}$$

we apply ELIM and BAKSUB to find the corresponding solution vectors

$$\mathbf{x}^1 = [-4.583301, 1.333617, 23.41670, 31.91692]^T$$

$$\mathbf{x}^2 = [-1.999996, 11.00005, 4.000008, 17.00005]^T$$

$$\mathbf{x}^3 = [31.58328, -93.33386, -43.41675, -199.9172]^T$$

$$\mathbf{x}^4 = [27.41660, -88.66704, -25.58337, -126.0836]^T$$

We find that

$$\frac{\|\mathbf{x}^1\|_1}{\|\mathbf{b}^1\|_1} = 0.422 \qquad \frac{\|\mathbf{x}^2\|_1}{\|\mathbf{b}^2\|_1} = 3.40 \qquad \frac{\|\mathbf{x}^3\|_1}{\|\mathbf{b}^3\|_1} = 2.69 \qquad \frac{\|\mathbf{x}^4\|_1}{\|\mathbf{b}^4\|_1} = 1.67$$

and hence (3.24) gives the estimate $K(\mathbf{A}) \simeq 3.4$. Even though this estimate is not very close to the exact value of $K(\mathbf{A})$, it does suggest that $\mathbf{A}$ is not ill-conditioned. Of course if we chose $\mathbf{b}^i = i$th column of $\mathbf{I}_4$, then we would find $\|\mathbf{A}^{-1}\|$ exactly by use of (3.23) with $k = 4$.

We previously mentioned that LINPACK provides the user with an estimate of $K(\mathbf{A})$. The estimate is based on (3.22) for a particular choice of $\mathbf{b}$. This choice of $\mathbf{b}$ is somewhat technical, and we refer the reader to Dongarra et al. (1978).

The state of the art for linear system solvers is such that virtually any well-conditioned system of form (3.1), with $m = n \leq 200$, can be accurately solved. If the problem is ill-conditioned, the user should be automatically warned by the subroutine.

EXERCISES

3.51 Let $\mathbf{x}$ denote the 6-vector whose ith row entry is $x_i = \sum_{k=1}^i k$. Calculate $\|\mathbf{x}\|_1$ and $\|\mathbf{x}\|_2$.

3.52 Show that $\|\mathbf{x}\|_2^2 = \mathbf{x}^T\mathbf{x}$ for any n-vector $\mathbf{x}$.

3.53 Find n-vectors $\mathbf{u}$ and $\mathbf{v}$ such that $\|\mathbf{u}\|_2 = \|\mathbf{u}\|_1$ and $\|\mathbf{v}\|_1 = \sqrt{n}\|\mathbf{v}\|_2$.

*3.54 Show that for any n-vector $\mathbf{x}$, $\|\mathbf{x}\|_2 \leq \|\mathbf{x}\|_1 \leq \sqrt{n}\|\mathbf{x}\|_2$. Hint: Use the Cauchy-Schwarz inequality

$$\left| \sum_{i=1}^n a_i b_i \right| \leq \left(\sum_{i=1}^n a_i^2 \right)^{1/2} \left(\sum_{i=1}^n b_i^2 \right)^{1/2}$$

which is valid for all real numbers a_i and b_i.

*3.55 If $\mathbf{A}$ and $\mathbf{B}$ are $n \times n$ matrices, show that $\|\mathbf{AB}\| \leq \|\mathbf{A}\| \|\mathbf{B}\|$.

3.56 Determine the condition number of the matrix in Example 3.16.

3.57 Use the result of exercise 3.56 to evaluate the right-hand side of the inequality in Theorem 3.3 for the approximate solution $\mathbf{y}$ of Example 3.16. How does this compare with the left-hand side of the inequality?

3.58 Solve (3.21) for $K(\mathbf{A})$ and use the output of exercise 3.46 to estimate the condition number of the coefficient matrix (here $u = 10^{-3}$).

3.59 Repeat exercise 3.58 for the output of exercise 3.49 (here u is determined by your computer).

3.60 Apply GAUSS and SOLVE to the matrix of exercise 3.41 for three or four right-hand-side vectors. Use (3.24) to estimate the condition number.

*3.61 Using the same notation as in Theorem 3.3, show that

$$\frac{\|\mathbf{x} - \mathbf{y}\|_1}{\|\mathbf{x}\|_1} \geq \frac{1}{K(\mathbf{A})} \frac{\|\mathbf{r}\|_1}{\|\mathbf{b}\|_1}$$

3.62 The $n \times n$ Hilbert matrix is given by $h_{ij} = 1/(i + j - 1)$. Such matrices are highly ill-conditioned. Determine the condition number of the 3×3 Hilbert matrix.

3.63 Use (3.24) to estimate the condition number of the 3×3 Hilbert matrix. Use ELIM and BAKSUB. Repeat for the 5×5 Hilbert matrix.

3.64 In exercise 3.41 change the last entry of $\mathbf{b}$ from 52 to 53. Solve the resulting system using ELIM and BAKSUB. Does the solution change much? Repeat the computations using GAUSS and SOLVE. Discuss your results, keeping in mind the result of exercise 3.60.

3.65 Repeat exercise 3.64 for several other perturbations of the right-hand-side vector of exercise 3.41. Use only GAUSS and SOLVE.

3.66 Write a simple FORTRAN function subprogram NORM(A,N) that accepts an $N \times N$ matrix $\mathbf{A}$ and returns the value of $\|\mathbf{A}\|$. Test your program on several matrices.

3.5 TRIDIAGONAL ALGORITHM

For certain matrix problems $\mathbf{Ax} = \mathbf{b}$, it is possible to significantly reduce the computational effort involved in the solution process by exploiting the structure of $\mathbf{A}$. As an example of this situation we consider the solution of the tridiagonal system

$$
\begin{aligned}
d_1 x_1 + f_1 x_2 &= b_1 \\
c_2 x_1 + d_2 x_2 + f_2 x_3 &= b_2 \\
c_3 x_2 + d_3 x_3 + f_3 x_4 &= b_3 \\
\ddots \qquad \ddots \qquad \ddots \qquad &\;\;\vdots \\
c_{n-1} x_{n-2} + d_{n-1} x_{n-1} + f_{n-1} x_n &= b_{n-1} \\
c_n x_{n-1} + d_n x_n &= b_n
\end{aligned}
\tag{3.25}
$$

If $\mathbf{A}$ denotes the coefficient matrix of (3.25), then the only nonzero entries of $\mathbf{A}$ lie on the diagonal $a_{ii} = d_i$ or on the superdiagonal $a_{i, i+1} = f_i$ or on the subdiagonal $a_{i+1, i} = c_{i+1}$. We see that $a_{ij} = 0$ if $|i - j| > 1$, and $\mathbf{A}$ is called a **tridiagonal matrix**.

Systems of the form (3.25) arise in the determination of a cubic spline interpolant (Section 4.3) and in the numerical solution of boundary value problems for differential equations.

The method of Gaussian elimination may be applied to (3.25); however, let us consider an approach that is designed to take advantage of the structure of tridiagonal matrices. The tridiagonal algorithm consists in determining a lower triangular matrix $\mathbf{L}$ and an upper triangular matrix $\mathbf{U}$ such that

$$\mathbf{A} = \mathbf{LU} \tag{3.26}$$

with $\mathbf{L}$ and $\mathbf{U}$ of the following form

$$\mathbf{L} = \begin{bmatrix} 1 & 0 & 0 \cdots \cdots \cdots 0 \\ l_1 & 1 & 0 \cdots \cdots \cdots \\ 0 & l_2 & 1 \cdots \cdots \cdots 0 \\ \cdots \cdots \cdots \cdots \cdots \cdots \\ 0 & \cdots & \cdots \; 0 \; l_{n-1} \; 1 \end{bmatrix} \tag{3.27}$$

$$\mathbf{U} = \begin{bmatrix} u_1 & v_1 & 0 & \cdots & 0 \\ 0 & u_2 & v_2 & \cdots & 0 \\ 0 & 0 & u_3 & v_3 & \vdots \\ \vdots & \vdots & & \ddots & 0 \\ & & & & v_{n-1} \\ 0 & 0 & \cdots & 0 & u_n \end{bmatrix} \tag{3.28}$$

It is not obvious that such a decomposition of $\mathbf{A}$ is possible. If $\mathbf{L}$ and $\mathbf{U}$ are in the forms given above, then the product $\mathbf{LU}$ is tridiagonal (verify!) so that (3.26) is at least plausible. We assume, for the time being, that the decomposition is possible and proceed to determine how the entries of $\mathbf{L}$ and $\mathbf{U}$ are computed.

We calculate the matrix product $\mathbf{LU}$ and equate the result with the matrix $\mathbf{A}$ of (3.25) to find the following equations for the entries of $\mathbf{A}$:

$$c_{i+1} = l_i u_i \qquad 1 \le i \le n - 1$$

$$d_i = v_{i-1} l_{i-1} + u_i \qquad 2 \le i \le n$$

$$f_i = v_i \qquad 1 \le i \le n - 1$$

and $d_1 = u_1$. Thus the superdiagonal entries of $\mathbf{U}$ must be the same as those of $\mathbf{A}$, and we have for $i = 2, 3, \ldots, n$

$$l_{i-1} = \frac{c_i}{u_{i-1}}$$

$$u_i = d_i - f_{i-1} l_{i-1}$$

These equations may be solved by the following algorithm.

ALGORITHM 3.6

Given the $n \times n$ tridiagonal matrix $\mathbf{A}$ of (3.25) with $d_1 \neq 0$,

$u_1 \leftarrow d_1$

Do $i = 2, 3, \ldots, n$:

$\quad l_{i-1} \leftarrow \dfrac{c_i}{u_{i-1}}$

$\quad u_i \leftarrow d_i - l_{i-1} f_{i-1}$

$\quad$ If $u_i = 0$ then stop

This algorithm uniquely determines $\mathbf{L}$ and $\mathbf{U}$ provided $u_i \neq 0$ for $i = 1$, $\ldots, n$. Once $\mathbf{L}$ and $\mathbf{U}$ have been determined, the original problem $\mathbf{A}\mathbf{x} = \mathbf{b}$ may be split into the two problems

$$\mathbf{L}\mathbf{y} = \mathbf{b} \quad \text{and} \quad \mathbf{U}\mathbf{x} = \mathbf{y} \tag{3.29}$$

We "recover" the original problem by

$$\mathbf{A}\mathbf{x} = \mathbf{L}\mathbf{U}\mathbf{x} = \mathbf{L}\mathbf{y} = \mathbf{b}$$

Each of the problems in (3.29) is easy to solve. $\mathbf{L}\mathbf{y} = \mathbf{b}$ can be solved by forward substitution, that is, solve for y_1 in the first equation, then solve for y_2 in the second, and so on.

ALGORITHM 3.7

FORWARD SUBSTITUTION Given the $n \times n$ matrix $\mathbf{L}$ in form (3.27) from Algorithm 3.6 and an n-vector $\mathbf{b}$,

$y_1 \leftarrow b_1$

Do $i = 2, 3, \ldots, n$:

$\quad y_i \leftarrow b_i - l_{i-1} y_{i-1}$

then $\mathbf{L}\mathbf{y} = \mathbf{b}$.

To find $\mathbf{x}$, we use backsubstitution on $\mathbf{U}\mathbf{x} = \mathbf{y}$.

ALGORITHM 3.8

BACKSUBSTITUTION Given the $n \times n$ matrix $\mathbf{U}$ in form (3.28) from Algorithm 3.6 with $v_i = f_i, 1 \leq i \leq n - 1$, and the vector $\mathbf{y}$ from Algorithm 3.7,

$x_n \leftarrow \dfrac{y_n}{u_n}$

Do $i = 1, 2, \ldots, n - 1$:

$\quad x_{n-i} \leftarrow \dfrac{y_{n-i} - f_{n-i} x_{n-i+1}}{u_{n-i}}$

It is a simple matter to count the number of multiplication/divisions in the tridiagonal algorithm. Algorithm 3.6 requires two such operations per step, for a total of $2(n-1)$. Algorithms 3.7 and 3.8 require $n-1$ and $2n-1$ operations, respectively. Hence the total work in the tridiagonal algorithm is $5n-4 \simeq 5n$ operations. In addition we see that there is no need to store all the entries of $\mathbf{A}$ in memory. One need store only the superdiagonal, diagonal, and subdiagonal of $\mathbf{A}$ in three vectors. The following gives sufficient conditions under which the $\mathbf{LU}$ decomposition of (3.26) is possible.

THEOREM 3.4

Suppose $\mathbf{A}$ is the coefficient matrix for system (3.25) and satisfies the following:

(i) $c_n f_1 \neq 0, c_i f_i \neq 0 \qquad 2 \leq i \leq n-1$

(ii) $|d_1| > |f_1|, |d_i| \geq |c_i| + |f_i|$ for $i = 2, \ldots, n-1$ and $|d_n| > |c_n|$

Then $\mathbf{A}$ is invertible with $\mathbf{A} = \mathbf{LU}$, where $\mathbf{L}$ and $\mathbf{U}$ are uniquely determined by Algorithm 3.6.

The proof of this theorem consists in showing that $u_i \neq 0$ for each i. We leave the details to the reader. Condition (ii) in Theorem 3.4 requires that the diagonal entries dominate the sum of the off-diagonal entries in magnitude. Such matrices are called **diagonally dominant**.

EXAMPLE 3.19

The following matrix satisfies the conditions of the previous theorem:

$$
\mathbf{A} = \begin{bmatrix}
2 & 1 & 0 & 0 & 0 \\
1 & 2 & 1 & 0 & 0 \\
0 & 1 & 2 & 1 & 0 \\
0 & 0 & 1 & 2 & 1 \\
0 & 0 & 0 & 1 & 2
\end{bmatrix}
$$

Moreover, the $\mathbf{LU}$ decomposition (3.26) is valid with

$$
\mathbf{L} = \begin{bmatrix}
1 & 0 & 0 & 0 & 0 \\
\frac{1}{2} & 1 & 0 & 0 & 0 \\
0 & \frac{2}{3} & 1 & 0 & 0 \\
0 & 0 & \frac{3}{4} & 1 & 0 \\
0 & 0 & 0 & \frac{4}{5} & 1
\end{bmatrix}
\qquad
\mathbf{U} = \begin{bmatrix}
2 & 1 & 0 & 0 & 0 \\
0 & \frac{3}{2} & 1 & 0 & 0 \\
0 & 0 & \frac{4}{3} & 1 & 0 \\
0 & 0 & 0 & \frac{5}{4} & 1 \\
0 & 0 & 0 & 0 & \frac{6}{5}
\end{bmatrix}
$$

The following subroutine TRISYS implements Algorithms 3.6, 3.7, and 3.8:

```
      SUBROUTINE TRISYS(LDIAG,DIAG,UDIAG,B,N)
C  LDIAG:LOWER DIAGONAL
C  DIAG:MAIN DIAGONAL
C  UDIAG:UPPER DIAGONAL
C  B:ON ENTRY CONTAINS RIGHT HAND VECTOR
C    ON RETURN CONTAINS SOLUTION VECTOR
C  N:NUMBER OF ROWS AND COLUMNS IN TRIDIAGONAL MATRIX
      REAL LDIAG(N),DIAG(N),UDIAG(N),B(N)
      IF(N.NE.1) GO TO 2
      B(1)=B(1)/DIAG(1)
      RETURN
  2   DO 3 J=2,N
        LDIAG(J)=LDIAG(J)/DIAG(J-1)
        DIAG(J)=DIAG(J)-LDIAG(J)*UDIAG(J-1)
        B(J)=B(J)-LDIAG(J)*B(J-1)
  3   CONTINUE
      B(N)=B(N)/DIAG(N)
      NN=N-1
      DO 4 K=1,NN
        KK=N-K
        B(KK)=(B(KK)-UDIAG(KK)*B(KK+1))/DIAG(KK)
  4   CONTINUE
      RETURN
      END
```

In this subroutine the lower, main, and upper diagonal entries are stored in arrays LDIAG, DIAG, and UDIAG, respectively. Notice that each of these has dimension N even though the upper and lower diagonals of the matrix have only $N - 1$ entries. We pad arrays LDIAG and UDIAG by always requiring that LDIAG(1) = 0 and UDIAG(N) = 0.

EXAMPLE 3.20

We apply TRISYS to the system

$$2x_1 + x_2 = 1$$
$$x_{i-1} + 2x_i + x_{i+1} = 0 \qquad 2 \le i \le n - 1$$
$$x_{n-1} + 2x_n = (-1)^{n+1}$$

for $n = 10$. The solution is given by $x_i = (-1)^{i-1}$ for $1 \le i \le 10$. Use of the main program

```
      REAL C(10),D(10),F(10),B(10)
      DATA C/0.,9*1./,D/10*2./,F/9*1.,0./
      DATA B/1.,8*0.,-1./,N/10/
      CALL TRISYS(C,D,F,B,N)
      WRITE(6,10)
  10  FORMAT(' ','THE SOLUTION IS:'/)
      WRITE(6,11) (B(J),J=1,N)
  11  FORMAT(' ',6X,F9.6)
      STOP
      END
```

results in the output

THE SOLUTION IS:

    1.000001
   -1.000002
    1.000003
   -1.000004
    1.000005
   -1.000005
    1.000005
   -1.000005
    1.000004
   -1.000002

EXERCISES

3.67 For $n = 5$, with $\mathbf{L}$ and $\mathbf{U}$ given by (3.27) and (3.28), respectively, verify that the product $\mathbf{LU}$ is tridiagonal.

3.68 Carry out the details of the derivation of Algorithm 3.6 by equating the entries of $\mathbf{A}$ and $\mathbf{LU}$.

3.69 Perform naive Gaussian elimination on the system $\mathbf{Ax} = \mathbf{b}$, where

$$\mathbf{A} = \begin{bmatrix} 2 & 1 & 0 & 0 & 0 \\ 1 & 4 & 1 & 0 & 0 \\ 0 & 1 & 4 & 1 & 0 \\ 0 & 0 & 1 & 4 & 1 \\ 0 & 0 & 0 & 1 & 2 \end{bmatrix} \quad \text{and} \quad \mathbf{b} = \begin{bmatrix} 1 \\ -2 \\ 2 \\ -2 \\ -1 \end{bmatrix}$$

Was it necessary to perform any row interchanges? Solve the same system using the tridiagonal algorithm.

3.70 Use subroutine TRISYS to solve the tridiagonal system $\mathbf{Ax} = \mathbf{b}$, where $\mathbf{A}$ is the 20×20 analogue of the matrix of exercise 3.69 and $\mathbf{b}$ is the 20-vector whose entries are given by $b_1 = 1$, $b_k = 2(-1)^k$ for $2 \le k \le 19$, and $b_{20} = -1$. Compare your computed solution with the exact solution, whose ith row is $x_i = (-1)^{i+1}$.

3.71 Determine $\mathbf{L}$ and $\mathbf{U}$ for the matrix of exercise 3.69. Verify that the systems (3.29) are equivalent to the system of exercise 3.69.

3.72 Use subroutine TRISYS to solve the system

$$4x_1 + x_2 = 1$$
$$x_{i-1} + 4x_i + x_{i+1} = 1 \quad 2 \le i \le n - 1$$
$$x_{n-1} + 4x_n = 1$$

for $n = 20$.

*3.73 Under conditions (i) and (ii) of Theorem 3.4 show that $u_i \ne 0$ for all i in Algorithm 3.6. Hint: Use induction on i.

3.74 Suppose naive Gaussian elimination is applied to (3.25) with no row interchanges. Are the computations the same as those given in Algorithms 3.6, 3.7, and 3.8? Justify your answer. Hint: See exercise 3.69.

3.75 Solve the following tridiagonal system in two ways: by using ELIM and BAKSUB and by using TRISYS:

$$136.01x_1 + 90.860x_2 \qquad\qquad\qquad = -33.254$$
$$90.860x_1 + 98.810x_2 - 67.590x_3 \qquad\qquad = 49.790$$
$$-67.590x_2 + 132.01x_3 + 46.260x_4 = 28.067$$
$$46.260x_3 + 177.17x_4 = -7.3244$$

Compare the computed solutions with the exact solution, $\mathbf{x} = [-2953.3, 4420.5, 2491.5, -650.59]^T$. Note: This problem is an ill-conditioned example, given on p. 467 of Ralson and Rabinowitz (1978).

NOTES AND COMMENTS

Section 3.1

There is a vast textbook literature on matrix and linear algebra. A readable book that emphasizes methodology is Bronson (1970). The more advanced books by Strang (1976), Noble and Daniel (1977), and Stewart (1973) are written with applications (especially to numerical analysis) in mind.

Sections 3.3 and 3.4

For additional material on norms, error bounds, and conditioning of linear systems see Conte and de Boor (1980) or Johnson and Riess (1982). The former book also gives a careful discussion of roundoff error in Gaussian elimination via backward error analysis. The recent book by Rice (1981) discusses all of the essential questions regarding linear system solvers and describes the primary library programs available commerically. These include the LINPACK and IMSL program packages. A complete guide to LINPACK by Dongarra et al. (1978) is available from the Society for Industrial and Applied Mathematics.

Section 3.5

Tridiagonal systems are the simplest (nontrivial) example of banded sparse problems. Such problems arise in many areas, most notably in numerical methods for differential equations. These systems are typically large (several hundred or thousand equations) and are frequently solved by iterative methods. An introduction to iterative methods is given in Chapter 5 of Conte and de Boor (1980). For a more thorough analysis of iterative methods see Varga (1962), and for direct methods see George and Liu (1980).

We have not discussed the matrix eigenvalue problem because its inclusion would have substantially changed the prerequisites for the book. This important problem is discussed in Bronson (1970) at an elementary level and in Noble and Daniel (1977) at a more advanced level. The numerical aspects of eigenvalue problems are introduced in Conte and de Boor (1980). The definitive works on the numerical aspects of eigenvalue problems are the classic by Wilkinson (1965) and the recent book by Parlett (1980).

INTERPOLATION

4

There are several situations where it is necessary or expedient to replace a given function by a simpler function for the purpose of some numerical computation. Suppose, for instance, that f is known only at a discrete set of points $\{x_i\}_{i=1}^n$ in an interval $[a, b]$. It is required to calculate $f(x^*)$, where x^* is an intermediate point, or to compute the definite integral of f on $[a, b]$. In either case a reasonable procedure consists in determining a simple function p that interpolates the data, that is, satisfies $p(x_i) = f(x_i)$ for $i = 1, 2, \ldots, n$. Then $f(x^*)$ is approximated by $p(x^*)$ and the definite integral of f is approximated by the definite integral of p. Even if f is known by formula throughout $[a, b]$, it may be that the evaluation of f is cumbersome or expensive in computer time. Under such circumstances it is reasonable to replace f by a simple function that furnishes an approximation to f. In Chapter 1 we saw several situations where this approach proved useful, for example, in approximating sinh x by a polynomial in Project 1. Depending on the application at hand, the choice of simple functions might be polynomials, rational functions, or trigonometric functions, to name a few. Polynomials are the most common choice of simple functions.

Historically, polynomial interpolation has been an important subject in numerical computation. Indeed, in the sixteenth through the nineteenth century, polynomial interpolation was a fundamental tool used in solving significant problems in navigation and astronomy. Many of the best mathematicians of that era, including Newton, Gauss, Euler, and Lagrange, made important contributions to this subject. Since computations were done by hand in those days, a large number of interpolation formulas were developed for use in various (difference) tables. Most older books on numerical computation devote considerable space to the study of such interpolation formulas as Newton forward and backward, Gauss, Stirling, and Bessel. Today the use of special interpolation formulas in a table is much less frequent since built-in function subprograms can be used to calculate the values of most functions very efficiently and accurately. We should point out that built-in function subprograms may themselves use polynomial approximations.

In this chapter we study polynomial and piecewise polynomial interpolation. The first two sections are devoted to the study of polynomial interpolation and the resultant interpolation error. The remainder of the chapter deals with piecewise polynomials—an increasingly popular approximation tool.

4.1 POLYNOMIAL INTERPOLATION

Existence and uniqueness

Suppose f is a given function whose values are known at n distinct points $\{x_i\}_{i=1}^n$ in some interval $[a, b]$. It is required to find a polynomial p such that p **interpolates** f at the given points. By this we mean that p satisfies

$$p(x_i) = f(x_i) \qquad i = 1, 2, \ldots, n \tag{4.1}$$

Recall that a polynomial of degree $\leq m$ is a function of the form

$$p_m(x) = a_0 + a_1 x + \cdots + a_m x^m \tag{4.2}$$

and is uniquely determined by specifying the $m + 1$ coefficients $a_0, a_1, \ldots, a_m$. If we seek a polynomial of degree $\leq m$ that interpolates f at $x_1, \ldots, x_n$, then (4.1) gives n constraints on p_m that must be satisfied by a suitable choice of the $m + 1$ coefficients in (4.2). This leads to the following linear system of equations:

$$a_0 + a_1 x_1 + \cdots + a_m x_1^m = f(x_1)$$
$$a_0 + a_1 x_2 + \cdots + a_m x_2^m = f(x_2)$$
$$\cdots\cdots\cdots\cdots\cdots\cdots\cdots\cdots\cdots\cdots\cdots$$
$$a_0 + a_1 x_n + \cdots + a_m x_n^m = f(x_n)$$

which can be written in matrix form as

$$\begin{bmatrix} 1 & x_1 & x_1^2 & \cdots & x_1^m \\ 1 & x_2 & x_2^2 & \cdots & x_2^m \\ \cdots\cdots\cdots\cdots\cdots\cdots \\ 1 & x_n & x_n^2 & \cdots & x_n^m \end{bmatrix} \begin{bmatrix} a_0 \\ a_1 \\ \vdots \\ a_m \end{bmatrix} = \begin{bmatrix} f(x_1) \\ f(x_2) \\ \vdots \\ f(x_n) \end{bmatrix} \tag{4.3}$$

In order that system (4.3) be uniquely solvable for any given set of function values, it is necessary that the number of equations equal the number of unknowns (Theorem 3.1), that is, $m = n - 1$. Henceforth we consider only the problem of interpolating f at $x_1, \ldots, x_n$ by a polynomial of degree $\leq n - 1$. We could solve this problem by showing that (4.3), with $m = n - 1$, has an invertible coefficient matrix; however, it is more useful to actually construct the polynomial interpolate.

Consider the polynomial of degree $n - 1$ given by

$$l_k(x) = (x - x_1)(x - x_2) \cdots (x - x_{k-1})(x - x_{k+1}) \cdots (x - x_n)$$

$$= \prod_{\substack{i=1 \\ i \neq k}}^{n} (x - x_i)$$

[multiplication (not summation)]

For each $k = 1, 2, \ldots, n$, we have $l_k(x_j) = 0$ if $j \neq k$ and $l_k(x_k) \neq 0$. If $L_k(x) = l_k(x)/l_k(x_k)$, then for $j = 1, 2, \ldots, n$

$$L_k(x_j) = \begin{cases} 1 & \text{if } j = k \\ 0 & \text{if } j \neq k \end{cases}$$

From this it follows that

$$p(x) = \sum_{k=1}^{n} y_k L_k(x) \tag{4.4}$$

satisfies $p(x_j) = y_j$ for $j = 1, 2, \ldots, n$. Hence the coefficient y_k of p in formula (4.4) gives the value of p at the interpolation point x_k. Consequently the polynomial

$$P_{n-1}(x) = \sum_{k=1}^{n} f(x_k)L_k(x) \tag{4.5}$$

interpolates f at $x_1, \ldots, x_n$ and is of degree $\leq n - 1$. This shows that the polynomial interpolation problem has a solution; namely, P_{n-1} solves the problem.

Next we show that P_{n-1} is the only solution of the interpolation problem. Suppose, to the contrary, that the problem has another solution, say $Q(x)$. Then Q is a polynomial of degree $\leq n - 1$ that satisfies

$$Q(x_j) = f(x_j) = P_{n-1}(x_j) \qquad j = 1, 2, \ldots, n$$

Thus $r = Q - P_{n-1}$ is a polynomial of degree $\leq n - 1$ that satisfies $r(x_j) = 0$ for $j = 1, 2, \ldots, n$. In Section 2.4, however, we found that a polynomial of degree k has exactly k zeros (counting multiplicities). Therefore r can have at most $n - 1$ zeros, and we have arrived at a contradiction. We have established that (4.5) gives the unique solution of the interpolation problem. Our results are summarized in the following theorem.

THEOREM 4.1

Given the values of f at n distinct points $\{x_j\}_{j=1}^{n}$, there exists a unique polynomial of degree $\leq n - 1$ that interpolates f at $x_1, x_2, \ldots, x_n$.

The unique interpolating polynomial of Theorem 4.1 is given by (4.5), which is called the **Lagrange formula** for polynomial interpolation.

EXAMPLE 4.1

The simplest nontrivial application is $n = 2$. If given $f(x_1)$ and $f(x_2)$ with $x_1 \neq x_2$, then

$$L_1(x) = \frac{x - x_2}{x_1 - x_2} \qquad L_2(x) = \frac{x - x_1}{x_2 - x_1}$$

and the linear interpolate is

$$P_1(x) = f(x_1)L_1(x) + f(x_2)L_2(x)$$

$$= f(x_1)\frac{x - x_2}{x_1 - x_2} + f(x_2)\frac{x - x_1}{x_2 - x_1}$$

By rearranging terms, we can also express $P_1(x)$ as

$$P_1(x) = f(x_1) + \frac{f(x_2) - f(x_1)}{x_2 - x_1}(x - x_1)$$

EXAMPLE 4.2

Historically, polynomial interpolation was used extensively in the computation of logarithms. Consider the following table of values for $\log_{10}$:

x	$\log_{10} x$
1.2	0.079181
1.4	0.146128
1.6	0.204120
1.8	0.255273

We approximate $\log_{10} 14.5$ using a third-degree interpolate. First we use the additive property of logarithms:

$$\log_{10} 14.5 = \log_{10} 10 + \log_{10} 1.45 = 1 + \log_{10} 1.45$$

Thus it suffices to find $\log_{10} 1.45$. We have

$$L_1(1.45) = \frac{(1.45 - 1.4)(1.45 - 1.6)(1.45 - 1.8)}{(1.2 - 1.4)(1.2 - 1.6)(1.2 - 1.8)} = -0.0546875$$

$$L_2(1.45) = \frac{(1.45 - 1.2)(1.45 - 1.6)(1.45 - 1.8)}{(1.4 - 1.2)(1.4 - 1.6)(1.4 - 1.8)} = 0.8203125$$

$$L_3(1.45) = \frac{(1.45 - 1.2)(1.45 - 1.4)(1.45 - 1.8)}{(1.6 - 1.2)(1.6 - 1.4)(1.6 - 1.8)} = 0.2734375$$

$$L_4(1.45) = \frac{(1.45 - 1.2)(1.45 - 1.4)(1.45 - 1.6)}{(1.8 - 1.2)(1.8 - 1.4)(1.8 - 1.6)} = -0.0390625$$

Then by (4.5) we find

$$\log_{10} 1.45 \simeq P_3(1.45) = (0.079181)L_1(1.45) + (0.146128)L_2(1.45)$$

$$+ (0.204120)L_3(1.45) + (0.255273)L_4(1.45)$$

and hence

$$\log_{10} 1.45 \simeq P_3(1.45) = 0.161383$$

Finally we get $\log_{10} 14.5 \simeq 1.161383$, which is correct to five decimal digits.

The Newton form and divided differences

The Lagrange formula has some drawbacks as a computational device. When evaluating (4.5) at a point x^*, it is necessary to perform $2n$ multiplication/divisions and $2n - 1$ additions, assuming the denominators of the L_k's have already been evaluated. In comparison we recall that nested multiplication requires only $n - 1$ multiplications and a like number of additions (Section 2.4). A more serious criticism of the Lagrange form is the following. In determining an approximation to $f(x^*)$ for some intermediate point x^*, it may not be clear how to choose the number of interpolation points. A natural procedure is to successively compute $P_1(x^*)$, $P_2(x^*)$, $P_3(x^*)$, and so on, until (presumably) one finds a satisfactory approximation to $f(x^*)$. Here P_k denotes the polynomial of degree $\leq k$ that interpolates f at $x_1, \ldots, x_{k+1}$. In this process it would be advantageous to make use of P_{k-1} when determining P_k, but the Lagrange form is not suited for this purpose. To compute P_k using the Lagrange form, one must essentially start again from the beginning.

We show that both of those objections disappear when the interpolating polynomial is written in a more convenient form. In addition this new form, which we next develop, is better suited for error analysis. Essentially what we require is a means of obtaining P_k by a simple modification of P_{k-1}. Toward this end, consider $h(x) = P_k(x) - P_{k-1}(x)$. Clearly h is a polynomial of degree $\leq k$ and for $j = 1, 2, \ldots, k$

$$h(x_j) = P_k(x_j) - P_{k-1}(x_j)$$
$$= f(x_j) - f(x_j) = 0$$

Thus the zeros of h are $x_1, \ldots, x_k$ and it follows that for some constant α_k

$$h(x) = \alpha_k(x - x_1)(x - x_2) \cdots (x - x_k)$$

or, equivalently,

$$P_k(x) = P_{k-1}(x) + \alpha_k \prod_{j=1}^{k} (x - x_j) \tag{4.6}$$

If the constant α_k can be determined, then (4.6) can be used to find P_k knowing P_{k-1}. From (4.6) we observe that α_k is the coefficient of x^k in $P_k(x)$:

$$P_k(x) = \alpha_k x^k + (\text{a polynomial of degree} < k)$$

Moreover, by examination of the Lagrange form for P_k, we also have

$$P_k(x) = \sum_{j=1}^{k+1} \frac{f(x_j)}{l_j(x_j)} x^k + (\text{a polynomial of degree} < k)$$

Therefore the coefficient α_k is given by

$$\alpha_k = \sum_{j=1}^{k+1} \frac{f(x_j)}{\displaystyle\prod_{i=1,\, i \neq j}^{k+1} (x_j - x_i)} \tag{4.7}$$

Equation (4.7) is not suitable for the computation of α_k. However, it does show that α_k depends only on the values of f at $x_1, \ldots, x_{k+1}$. In order to explicitly denote its dependence on these values, we henceforth denote α_k by

$$f[x_1, \ldots, x_{k+1}]$$

With this definition we write (4.6) as

$$P_k(x) = P_{k-1}(x) + f[x_1, \ldots, x_{k+1}] \prod_{j=1}^{k} (x - x_j)$$

and hence

$$P_0(x) = f(x_1) = f[x_1]$$
$$P_1(x) = P_0(x) + f[x_1, x_2](x - x_1)$$
$$P_2(x) = P_1(x) + f[x_1, x_2, x_3](x - x_1)(x - x_2)$$

and so forth

Finally we have the so-called **Newton formula** for the polynomial of degree $\leq n - 1$, which interpolates f at $x_1, x_2, \ldots, x_n$:

$$P_{n-1}(x) = f[x_1] + f[x_1, x_2](x - x_1) + f[x_1, x_2, x_3](x - x_1)(x - x_2)$$
$$+ \cdots + f[x_1, \ldots, x_n] \prod_{j=1}^{n-1} (x - x_j) \tag{4.8}$$

We emphasize that (4.8) is simply another way of writing (4.5)—one that is more convenient in applications.

In order to use equation (4.8), it is necessary to show how the coefficients $f[x_1, \ldots, x_k]$ can be (efficiently) computed. We claim that for each $k \geq 2$

$$f[x_1, \ldots, x_k] = \frac{f[x_2, \ldots, x_k] - f[x_1, \ldots, x_{k-1}]}{x_k - x_1} \tag{4.9}$$

Before establishing the validity of (4.9), a few comments are in order. The constant $f[x_1, \ldots, x_k]$ is called the $(k-1)$st **divided difference** of f based on the points $x_1, \ldots, x_k$. The relationship (4.9) says that a $(k-1)$st divided difference is the difference quotient of $(k-2)$nd divided differences. Moreover, from (4.8) we see that P_{n-1} is uniquely determined in Newton form by generating the divided differences $f[x_1, \ldots, x_k]$ for $k = 1, 2, \ldots, n$. This is most easily accomplished by using a **divided difference table**, the format of which is as follows:

x_i	$f[\cdot]$	$f[\cdot,\cdot]$	$f[\cdot,\cdot,\cdot]$	$f[\cdot,\cdot,\cdot,\cdot]$
x_1	$f[x_1]$			
x_2	$f[x_2]$	$f[x_1,x_2]$		
x_3	$f[x_3]$	$f[x_2,x_3]$	$f[x_1,x_2,x_3]$	
x_4	$f[x_4]$	$f[x_3,x_4]$	$f[x_2,x_3,x_4]$	$f[x_1,x_2,x_3,x_4]$

We indicate how the entries $f[x_3, x_4]$ and $f[x_2, x_3, x_4]$ are computed by the dotted and dashed lines, respectively. Specifically these entries are

$$f[x_3, x_4] = \frac{f[x_4] - f[x_3]}{x_4 - x_3}$$

$$f[x_2, x_3, x_4] = \frac{f[x_3, x_4] - f[x_2, x_3]}{x_4 - x_2}$$

The table entries are computed column by column, and the diagonal entries are the coefficients required for the Newton form (4.8). The procedure should be clear with the aid of the following example.

EXAMPLE 4.3

We repeat the computation of Example 4.2 using a divided difference table and the Newton form of the interpolating polynomial. The divided difference table for $f(x) = \log_{10} x$ is

x	$f[\cdot]$	$f[\cdot,\cdot]$	$f[\cdot,\cdot,\cdot]$	$f[\cdot,\cdot,\cdot,\cdot]$
1.2	0.079181			
1.4	0.146128	0.334735		
1.6	0.204120	0.28996	-0.1119375	
1.8	0.255273	0.255765	-0.0854875	0.0440833

The table entries are generated column by column as follows:

$$f[x_1, x_2] = \frac{f[x_2] - f[x_1]}{x_2 - x_1} = \frac{0.146128 - 0.079181}{1.4 - 1.2} = 0.334735$$

$$f[x_2, x_3] = \frac{f[x_3] - f[x_2]}{x_3 - x_2} = \frac{0.204120 - 0.146128}{1.6 - 1.4} = 0.28996$$

$$f[x_3, x_4] = \frac{f[x_4] - f[x_3]}{x_4 - x_3} = \frac{0.255273 - 0.204120}{1.8 - 1.6} = 0.255765$$

The second divided differences are

$$f[x_1, x_2, x_3] = \frac{f[x_2, x_3] - f[x_1, x_2]}{x_3 - x_1} = \frac{0.28996 - 0.334735}{1.6 - 1.2} = -0.1119375$$

$$f[x_2, x_3, x_4] = \frac{f[x_3, x_4] - f[x_2, x_3]}{x_4 - x_2} = \frac{0.255765 - 0.28996}{1.8 - 1.4} = -0.0854875$$

and finally the third divided difference is

$$f[x_1, x_2, x_3, x_4] = \frac{f[x_2, x_3, x_4] - f[x_1, x_2, x_3]}{x_4 - x_1}$$

$$= \frac{-0.0854875 - (-0.1119375)}{1.8 - 1.2} = 0.0440833$$

Thus by (4.8) we get the Newton form of the polynomial interpolate:

$$P_3(x) = 0.079181 + 0.334735(x - 1.2) - 0.1119375(x - 1.2)(x - 1.4)$$

$$+ 0.0440833(x - 1.2)(x - 1.4)(x - 1.6)$$

We now demonstrate the validity of (4.9). In order to accomplish this, define Q_{k-2} to be the polynomial of degree $\leq k - 2$ that interpolates f at x_2, $x_3, \ldots, x_k$. Then consider the function

$$p(x) = \frac{x - x_1}{x_k - x_1} Q_{k-2}(x) + \frac{x_k - x}{x_k - x_1} P_{k-2}(x) \qquad (4.10)$$

where, as before, P_{k-2} interpolates f at $x_1, \ldots, x_{k-1}$. It is easy to verify that p interpolates f at $x_1, x_2, \ldots, x_k$. Since the interpolating polynomial is unique, we must have $p = P_{k-1}$. Next we equate the coefficients of x^{k-1} in (4.10). Recall that the coefficients of x^{k-2} in Q_{k-2} and P_{k-2} are $f[x_2, \ldots, x_k]$ and $f[x_1, \ldots, x_{k-1}]$, respectively. Hence from (4.10) we find

$$f[x_1, \ldots, x_k] = \text{coefficient of } x^{k-1} \text{ in } P_{k-1}$$

$$= \frac{f[x_2, \ldots, x_k]}{x_k - x_1} - \frac{f[x_1, \ldots, x_{k-1}]}{x_k - x_1}$$

and the proof of (4.9) is complete.

It should be noted that the use of the relationship in (4.9) is not restricted to the points $x_1, \ldots, x_k$. Indeed, the procedure given in (4.9) can be used for any set of distinct interpolation points. In particular the following is valid for any $i \geq 1$ ($i < k$):

$$f[x_i, \ldots, x_k] = \frac{f[x_{i+1}, \ldots, x_k] - f[x_i, \ldots, x_{k-1}]}{x_k - x_i} \qquad (4.11)$$

In fact we used (4.11) in Example 4.3 when computing the divided differences below the diagonal.

If it is required to find only the diagonal entries in the divided difference table, the following subroutine can be used:

```
      SUBROUTINE DIVDIF(X,FX,M,DD)
C   X:ARRAY OF POINTS
C   FX:VALUES OF FUNCTION AT POINTS IN X
C   M:NUMBER OF TABULAR DATA
C   DD:ON RETURN CONTAINS DIAGONAL ENTRIES
```

who really cares !

```
C      IN DIVIDED DIFFERENCE TABLE
       DIMENSION X(M),FX(M),DD(M)
       M1=M-1
       DO 1 J=1,M
         DD(J)=FX(J)
   1   CONTINUE
       DO 2 J=1,M1
         MJ=M-J
         DO 2 I=1,MJ
           I2=M-I+1
           DD(I2)=(DD(I2)-DD(I2-1))/(X(I2)-X(I2-J))
   2   CONTINUE
       RETURN
       END
```

$$\frac{DD(I2) - DD(I2-1)}{X(I2) - X(I2-J)}$$

In subroutine DIVDIF the function values and interpolation points are stored in arrays FX and X, respectively. The contents of FX are copied into DD, thereby ensuring that the entries of FX are not destroyed. The computation of the divided differences is done in DD, and on return $D(1) = f[x_1]$, $D(2) = f[x_1, x_2]$, and so forth.

We mentioned that the Newton form is more efficient in computation than the Lagrange form. Let us show how (4.8) can be efficiently evaluated using a modification of the nested multiplication algorithm. Consider a polynomial in the form

$$Q(x) = A_1 + A_2(x - x_1) + A_3(x - x_1)(x - x_2) + \cdots$$

$$+ A_n(x - x_1)(x - x_2) \cdots (x - x_{n-1}) \quad (4.12)$$

We write Q in nested form as

$$Q(x) = A_1 + (x - x_1)(A_2 + (x - x_2)(A_3 + \cdots (A_{n-1} + (x - x_{n-1})A_n) \cdots)$$

Then Q can be evaluated by the following algorithm.

ALGORITHM 4.1

Given a polynomial Q in the form (4.12) and a point x^*

$$b_n \leftarrow A_n$$

Do $k = n - 1, n - 2, \ldots, 1$:

$$b_k \leftarrow A_k + b_{k+1}(x^* - x_k)$$

then $Q(x^*) = b_1$.

It is a simple matter to modify subroutine HORNER (Section 2.4) to accommodate polynomials in the form (4.12). Subroutine NEST implements Algorithm 4.1 where arrays A and X contain the coefficients and points, respectively. Note: The last entry in X is never used; the points $x_1, \ldots, x_{n-1}$ are stored in $X(1), \ldots, X(N - 1)$. On return the value of Q at Z is given in VALUE.

```
      SUBROUTINE NEST(A,N,X,Z,VALUE)
C   A: ARRAY OF COEFFICIENTS IN NESTED FORM
C   X: ARRAY OF POINTS IN NESTED FORM
C   Z: POINT AT WHICH POLYNOMIAL IS EVALUATED
C   VALUE: VALUE OF POLYNOMIAL AT Z
C   N: NUMBER OF POLYNOMIAL COEFFICIENTS
      DIMENSION A(N),X(N),B(20)
      NEND=N-2
      B(N-1)=A(N)
      IF(N.EQ.2) GO TO 2
      DO 1 J=1,NEND
        L=N-J-1
        LL=L+1
        B(L)=A(LL)+B(LL)*(Z-X(LL))
1     CONTINUE
2     VALUE=A(1)+B(1)*(Z-X(1))
      RETURN
      END
```

Before considering the error in polynomial interpolation, several remarks are in order regarding the use of (4.8) in conjunction with a divided difference table. First we note that the interpolation points need not be arranged in any particular order. The divided difference $f[x_1, \ldots, x_k]$ is not changed by rearranging the data $(x_1, f(x_1)), \ldots, (x_k, f(x_k))$ in the table (exercise 4.5). It is not necessary to arrange the points in increasing order as we did in Example 4.3. The reader is asked to verify this in exercise 4.11. Our final observation regarding (4.8) is simply that the degree of the polynomial interpolate generally increases as the number of interpolation points does. In an extreme case of 100 interpolation points the degree is 99. A polynomial of such high degree is unwieldy for computational purposes and in fact is unlikely to produce a good approximation. Polynomial interpolates of high degree tend to be oscillatory, thereby resulting in a poor approximation of the underlying function.

EXERCISES

4.1   Determine the polynomial of degree ≤ 3, in Lagrange form, that interpolates the table

x_i	$f(x_i)$
1.05	1.7433
1.2	2.5722
1.3	3.6021
1.45	8.2381

Use the polynomial interpolate to estimate the value of $f(1.25)$. Compare with the exact value, $f(1.25) = 3.0096$.

4.2 Determine the polynomial of degree ≤ 4 $P_4(x)$ that interpolates $f(x) = 3x^2 - 5x + 2$ at the points -2, -1, 0, 1, and 2. What is the degree of P_4? How does P_4 differ from f (if at all)?

*4.3 (a) Show that $\sum_{k=1}^{n} L_k(x) = 1$ for all x. Hint: Use the uniqueness of polynomial interpolation given in Theorem 4.1.

(b) Let $P_n(x)$ denote the polynomial of degree $\leq n$ that interpolates f at distinct points x_1, ..., x_{n+1}. If f if a polynomial of degree m, $m \leq n$, show that $P_n(x) = f(x)$ for all x (see exercise 4.2).

4.4 Determine the Newton form of the polynomial interpolate of exercise 4.1. Convince yourself that the Newton form and Lagrange form polynomials are the same by evaluating each at several points.

4.5 Show that $f[x_1, x_2, x_3] = f[x_2, x_1, x_3] = f[x_3, x_1, x_2]$ for the data of exercise 4.1. Show that, in general, $f[x_1, x_2, x_3]$ is independent of the order in which the data $\{[x_i, f(x_i)]\}_{i=1}^{3}$ is tabulated.

4.6 Fill in the divided difference table:

x	$f[\cdot]$	$f[\cdot,\cdot]$	$f[\cdot,\cdot,\cdot]$	$f[\cdot,\cdot,\cdot,\cdot]$
0.3	0.97741			
0.37	0.96557	———		
0.41	0.95766	———	———	
0.52	0.93157	———	———	-0.0129185

4.7 Use subroutine DIVDIF to compute the diagonal entries for the table in exercise 4.6.

4.8 Write a simple program that calls subroutines DIVDIF and NEST to evaluate the polynomial interpolate of degree 9 (Newton form) for $f(x) = e^{-x}$. Use the interpolation points 0, 0.1, 0.2, ..., 0.9. Print the values of P_9 and f at 0.05, 0.15, ..., 0.85. Discuss your results.

4.9 Write a simple program that calls subroutines DIVDIF and NEST for the evaluation of a polynomial interpolate in Newton form. Use the data

x	$f(x)$
1.1	1.48661
1.44	1.75317
1.53	1.82781
1.61	1.89528
1.68	1.95510

and evaluate the resultant polynomial interpolate at the intermediate points 1.2, 1.3, 1.5, and 1.65. Note: The correct function values (to six decimal digits) are 1.56205, 1.64012, 1.80278, and 1.92938, respectively.

4.10 Let P_3 denote the cubic interpolate of the table in exercise 4.6. Use the program of exercise 4.9 to evaluate $P_3(0.47)$. Compare with $f(0.47) = 0.94423$.

4.11 Rearrange the table of exercise 4.1 so that the interpolation points are not in increasing order. Determine the Newton form polynomial interpolate for the rearranged table, call it $Q_3(x)$. Is Q_3 the same as the interpolate you found in exercise 4.1? Justify your answer.

4.12 Add the data point $f(0.47) = 0.94423$ at the end of the table in exercise 4.6. Determine P_4 by a simple modification of P_3 (from exercise 4.10). Would your calculations be so easy had the new data point been inserted in the middle of the table?

4.13 Suppose P_2 interpolates f at $\{x_i\}_{i=1}^3$, where $x_i = a + (i-1)h$. By differentiating P_2, show that

$$P_2'(x_2) = \frac{f(x_2 + h) - f(x_2 - h)}{2h}$$

$P_2'(x_2)$ is given by a **centered difference quotient** and furnishes an approximation to $f'(x_2)$.

4.14 Use a centered difference quotient to estimate $f'(-0.05)$ and $f'(0)$ given the table of values

x	$f(x)$
-0.15	0.90125
-0.05	0.965936
0.05	1.035265
0.15	1.109569

The correct values (to six decimal digits) are $f'(-0.05) = 0.669536$ and $f'(0) = 0.693147$.

4.15 Suppose a table of values of f is given at equally spaced points, say $x_i = a + (i-1)h$ for $1 \le i \le n+1$ with $h = (b-a)/n$. (Note: $a = x_1$ and $b = x_{n+1}$.) Denote $f(x_i)$ by f_i and define the **forward difference operator** Δ by

$$\Delta f_i = f_{i+1} - f_i$$

We define powers of Δ inductively as follows:

$$\Delta^2 f_i = \Delta(\Delta f_i) = \Delta(f_{i+1} - f_i) = \Delta f_{i+1} - \Delta f_i = f_{i+2} - 2f_{i+1} + f_i$$

$$\Delta^3 f_i = \Delta(\Delta^2 f_i)$$

and so on

*(a) Show that $f[x_i, x_{i+1}] = h^{-1}\Delta f_i$, $f[x_i, x_{i+1}, x_{i+2}] = (h^{-2}/2)\Delta^2 f_i$, and in general

$$f[x_i, x_{i+1}, \ldots, x_{i+k}] = \frac{h^{-k}}{k!} \Delta^k f_i$$

(b) Let $x = x_i + \tau h$ and show that

$$\prod_{j=0}^{k}(x - x_{i+j}) = h^{k+1} \prod_{j=0}^{k}(\tau - j)$$

(c) Use (a) and (b) to show that the Newton form for P_3, with equally spaced points, can be written as

$$P_3(x) = f_1 + \Delta f_1 \tau + \Delta^2 f_1 \frac{\tau(\tau-1)}{2!} + \Delta^3 f_1 \frac{\tau(\tau-1)(\tau-2)}{3!}$$

where $x = x_1 + \tau h$.

4.16 Use the result of exercise 4.15(c) to write P_3 for the table in exercise 4.6. Compute $P_3(0)$. Note: $0 = x = -0.15 + \tau(0.1)$.

4.17 Suppose that, as a result of roundoff or experimental error, the values of f at $\{a + (k - 1)h\}_{k=1}^{N+1}$ are not known exactly, say

$$\tilde{f}_k = f_k + \varepsilon_k$$

where $f_k = f[a + (k - 1)h]$ is the exact value. Let $\varepsilon = \max_{1 \le k \le N+1} |\varepsilon_k|$.

(a) Show that for $k = 1, \ldots, N$

$$\Delta^k \tilde{f}_1 = \Delta^k f_1 + \Delta^k \varepsilon_1$$

(b) Show that for each $k = 1, 2, \ldots, N$

$$|\Delta \varepsilon_k| \le 2\varepsilon$$

(c) Use (b) to show that $|\Delta^2 \varepsilon_1| \le 4\varepsilon$ and in general that

$$|\Delta^k \varepsilon_1| \le 2^k \varepsilon \qquad 1 \le k \le N$$

(d) From (a) and (c) conclude that

$$|\Delta^k \tilde{f}_1 - \Delta^k f_1| \le 2^k \varepsilon \qquad 1 \le k \le N$$

Thus $2^k \varepsilon$ gives a bound for absolute error in the kth forward difference.

4.2 ERRORS IN POLYNOMIAL INTERPOLATION

In using a polynomial interpolate for f, the error is zero at each interpolation point, and it would seem plausible that the interpolate should approximate f well at intermediate points. It is also natural to expect, at least theoretically, that a better approximation is achieved as the number of interpolation points is increased. Unfortunately the latter expectation of polynomial interpolation is generally unfounded.

EXAMPLE 4.4

This is the classic example that illustrates that the error need not decrease as the number of interpolation points is increased. Let $f(x) = (1 + 25x^2)^{-1}$ and let P_n interpolate f at the $n + 1$ equally spaced points $x_i = -1 + 2(i - 1)/n$, $1 \le i \le n + 1$, in the interval $[-1, 1]$. In the following program we determine P_n for $n = 2, 4, \ldots, 20$ and estimate

$$\max_{-1 \le x \le 1} |P_n(x) - f(x)| = E_n(f)$$

by computing

$$\max_{z_i} |P_n(z_i) - f(z_i)|$$

where $z_i = 0.04(i - 1) - 1$, $1 \le i \le 51$.

```
      DIMENSION X(21),FX(21),DD(21),B(20)
      WRITE(6,99)
99    FORMAT(' ',3X,'N',9X,'ERROR',12X,'POINT'//)
      DO 1 K=2,20,2
        H=2./FLOAT(K)
        NK=K+1
        DO 2 I=1,NK
          X(I)=FLOAT(I-1)*H-1.
          FX(I)=F(X(I))
2       CONTINUE
        CALL DIVDIF(X,FX,NK,DD)
        ERR=0.
        DO 3 J=1,51
          Z=FLOAT(J-1)/25.-1.
          CALL NEST(DD,NK,X,Z,VALU)
          DIFF=ABS(F(Z)-VALU)
          IF(DIFF.LE.ERR) GO TO 3
          ERR=DIFF
          BADZ=Z
3       CONTINUE
        WRITE(6,100)K,ERR,BADZ
100     FORMAT(' ',2X,I3,4X,E14.7,3X,F10.6)
1     CONTINUE
      WRITE(6,101)
101   FORMAT(' ',/2X,'N IS THE DEGREE OF INTERPOLATE')
      STOP
      END
C
      REAL FUNCTION F(X)
      F=1./(1.+25.*X**2)
      RETURN
      END
```

(handwritten annotations: `ENDDO` next to line 2 CONTINUE; `END DO` next to line 3 CONTINUE; `ENDDO` next to line 1 CONTINUE; `P.149 → Z;` near Z=FLOAT line; $\dfrac{1}{1+25x^2}$ next to F=1./(1.+25.*X**2))

N	ERROR		POINT
2	0.6461539E 00	⊖	0.400000
4	0.4381333E 00		0.799999
6	0.6166661E 00	⊕	-0.880000
8	0.1045168E 01		-0.920000
10	0.1801089E 01		0.919999
12	0.3605262E 01		-0.960000
14	0.7195719E 01		0.959999
16	0.1406334E 02	⊖	0.959999
18	0.2696259E 02	⊕	-0.960000
20	0.5118352E 02		0.959999

N IS THE DEGREE OF INTERPOLATE

We print the point at which the maximum error occurs. For large values
of N these points are near the endpoints of $[-1, 1]$.

From the output it is clear that the interpolation error does not decrease as n increases. In fact it can be proved that

$$\lim_{n \to \infty} E_n(f) = +\infty$$

and so the maximum interpolation error actually increases without bound. For this particular function one source of the poor approximation is the choice of equally spaced points. It can be shown that the polynomial that interpolates f at the Chebyshev points (exercise 4.18) gives a good approximation even for moderately large n. In fact the Chebyshev points are generally a better choice than equally spaced points; however, in many situations the choice of points is not at our disposal. We shall not examine this choice of interpolation points any further except to point out that it is not a panacea.

Consider a triangular array of interpolation points in $[a, b]$ where the points in each row are distinct:

$$
\begin{array}{cccc}
x_1 & & & \\
x_1 & x_2 & & \\
x_1 & x_2 & x_3 & \\
x_1 & x_2 & x_3 & x_4 \\
\vdots & & & \ddots
\end{array}
$$

Let P_n be the polynomial of degree $\leq n$ that interpolates f at the points in the $(n + 1)$st row of the table. Then we have the following surprising result.

THEOREM 4.2

For any such triangular array of interpolation points there exists a continuous function f on $[a, b]$ such that

$$\lim_{n \to \infty} E_n(f) = +\infty$$

where, as before, $E_n(f) = \max_{a \leq x \leq b} |P_n(x) - f(x)|$.

This theorem says that there is no single choice of interpolation points that, as $n \to \infty$, works for all continuous functions. To summarize our remarks, we contend that polynomial interpolates of high degree are unlikely to be satisfactory approximations.

In what follows, we obtain a general formula for the interpolation error, but the reader should keep in mind that in applications the cases of primary interest usually involve no more than five interpolation points. Given a function f and distinct points $\{x_i\}_{i=1}^{n+1}$ in $[a, b]$, we seek a formula for the error

$$e_n(f, x) = f(x) - P_n(x)$$

If x^* is an intermediate point, that is, $x^* \neq x_i$ for $i = 1, 2, \ldots, n + 1$, let $Q(x)$ be the polynomial of degree $\leq n + 1$ that interpolates f at the given points and at x^*. Then Q is given by

$$Q(x) = P_n(x) + f[x_1, \ldots, x_{n+1}, x^*] \prod_{i=1}^{n+1} (x - x_i)$$

and $Q(x^*) = f(x^*)$. Therefore,

$$e_n(f, x^*) = Q(x^*) - P_n(x^*) = f[x_1, \ldots, x_{n+1}, x^*] \prod_{i=1}^{n+1} (x^* - x_i) \qquad (4.13)$$

This last formula is of little practical use since $f[x_1, \ldots, x_{n+1}, x^*]$ cannot be evaluated without knowledge of $f(x^*)$. However, we shall demonstrate, for smooth f, that $f[x_1, \ldots, x_{n+1}, x^*]$ is related to the $(n + 1)$st derivative of f. Toward this end we note that for each $k \geq 0$

$$e_k(f, x_i) = f(x_i) - P_k(x_i) = 0 \qquad i = 1, 2, \ldots, k + 1$$

Hence by Rolle's theorem the derivative of the error $e_k'(f, x)$ has at least k distinct zeros (one between each adjacent pair of interpolation points). By applying Rolle's theorem to $e_k'(f, x)$, we conclude that $e_k''(f, x)$ has at least $k - 1$ distinct zeros, and, continuing in this manner, we find that the kth derivative of $e_k(f, x)$ must have at least one zero, say z. Thus we have

$$e_k^{(k)}(f, z) = f^{(k)}(z) - P_k^{(k)}(z) = 0$$

In addition we recall that

$$P_k(x) = f[x_1, \ldots, x_{k+1}]x^k + (\text{a polynomial of degree} < k)$$

and hence for all x

$$P_k^{(k)}(x) = k! \, f[x_1, \ldots, x_{k+1}]$$

Therefore we obtain

$$f^{(k)}(z) = P_k^{(k)}(z) = k! \, f[x_1, \ldots, x_{k+1}]$$

or, equivalently,

$$f[x_1, \ldots, x_{k+1}] = \frac{f^{(k)}(z)}{k!} \qquad (4.14)$$

Our derivation of (4.14) requires only that the points $\{x_i\}_{i=1}^{k+1}$ be distinct and that f have k continuous derivatives. Returning to equation (4.13), we note that the points $x_1, \ldots, x_{n+1}, x^*$ are distinct, and hence for smooth f we can apply (4.14) to $f[x_1, \ldots, x_{n+1}, x^*]$. Thus we have established the following result.

THEOREM 4.3

Suppose $\{x_i\}_{i=1}^{n+1}$ are distinct points in $[a, b]$ and f has $n + 1$ continuous derivatives on $[a, b]$. Then for each x^* in $[a, b]$ there is a point z in (a, b) such that

$$e_n(f, x^*) = f(x^*) - P_n(x^*) = \frac{f^{(n+1)}(z)}{(n+1)!} \prod_{i=1}^{n+1} (x^* - x_i)$$

The point z depends on x^*, and the quantity $f^{(n+1)}(z)$ is generally unknown. However, we can use Theorem 4.3 to obtain a bound (possibly crude) for the interpolation error if a bound for $f^{(n+1)}$ is known. Suppose M_{n+1} is such a bound, that is,

$$\max_{a \le x \le b} |f^{(n+1)}(x)| \le M_{n+1}$$

Then $|f^{(n+1)}(z)| \le M_{n+1}$ and we have the error bound

$$|e_n(f, x^*)| \le \frac{M_{n+1}}{(n+1)!} \prod_{i=1}^{n+1} |x^* - x_i| \qquad (4.15)$$

EXAMPLE 4.5

An important function in many branches of engineering and physics is the **error function**, which is defined by

$$\text{erf}(x) = \frac{2}{\sqrt{\pi}} \int_0^x e^{-t^2}\, dt$$

Tabular values of $f(x) = \text{erf}(x)$ can be found in most handbooks of mathematical tables. Let us estimate the error in linear interpolation between $x_1 = 4$ and $x_2 = 5$. We have

$$P_1(x) = f(4) + f[4, 5](x - 4)$$

and by (4.15)

$$|f(x^*) - P_1(x^*)| \le \frac{M_2}{2!} |x^* - 4|\, |x^* - 5|$$

For x^* in $[4, 5]$ the maximum value of $|(x^* - 4)(x^* - 5)|$ occurs at the midpoint $(x_1 + x_2)/2 = 4.5$, and hence

$$|f(x^*) - P_1(x^*)| \le \frac{M_2}{8}$$

To find M_2, we have, by the fundamental theorem of calculus,

$$f'(x) = \frac{2}{\sqrt{\pi}} e^{-x^2}$$

and hence

$$f''(x) = -\frac{4x}{\sqrt{\pi}} e^{-x^2}$$

Since $f'''(x) > 0$ for x in $[4, 5]$, it follows that

$$\max_{4 \le x \le 5} |f''(x)| = |f''(4)| \le 1.016 \times 10^{-6} = M_2$$

EXAMPLE 4.6

Suppose $f(x) = \cos x$ is tabulated at $n + 1$ equally spaced points on $[0, \pi/2]$. Since $\max_{0 \le x \le \pi/2} |f^{(n+1)}(x)| \le 1$, we have by (4.15)

$$|\cos x - P_n(x)| \le \frac{1}{(n+1)!} \prod_{i=1}^{n+1} |x - x_i|$$

for any $x \in [0, \pi/2]$. We determine n so that the error is no larger than 0.5×10^{-4}. In order to accomplish this, we need to estimate the term $\prod_{i=1}^{n+1} |x - x_i|$, where $x_i = (\pi/2n)(i - 1)$. For a given x in $[0, \pi/2]$ we have $x \in [x_j, x_{j+1}]$ for some index j $(1 \le j \le n)$. The maximum value of the quadratic function $|(x - x_j)(x - x_{j+1})|$ occurs at $\frac{1}{2}(x_j + x_{j+1})$, and hence

$$|(x - x_j)(x - x_{j+1})| \le \frac{h^2}{4}$$

where $h = \pi/2n$. For $x \in [x_j, x_{j+1}]$ we have

$$|x - x_i| \le 2h \qquad \text{for } i = j - 1 \text{ or } i = j + 2$$

$$|x - x_i| \le 3h \qquad \text{for } i = j - 2 \text{ or } i = j + 3$$

and so on

By this analysis the worst case occurs for $j = 1$ or $j = n$ (that is, $x \in [x_1, x_2]$ or $x \in [x_n, x_{n+1}]$), in which case we find

$$\max_{0 \le x \le \pi/2} \prod_{i=1}^{n+1} |x - x_i| \le \frac{h^2}{4} (2h)(3h) \cdots (nh) = h^{n+1} \frac{n!}{4}$$

and therefore we have the bound

$$\max_{0 \le x \le \pi/2} |\cos x - P_n(x)| \le \frac{h^{n+1}}{4(n+1)}$$

Using this error bound, it is easy to verify, for $n = 6$ ($h = \pi/12$), that

$$\max_{0 \le x \le \pi/2} |\cos x - P_6(x)| \le 1.15 \times 10^{-5}$$

In the next example we determine an error bound for quadratic interpolation in a table of values.

EXAMPLE 4.7

Suppose f is tabulated at $N + 1$ equally spaced points $\{x_i\}_{i=1}^{N+1}$ with $x_i = a + (i - 1)h$ and $h = (b - a)/N$. Given an intermediate point x^*, we approximate $f(x^*)$ using a quadratic interpolate. We choose i so that $x^* \in [x_i, x_{i+2}]$, and the resultant quadratic interpolate is

$$p_2(x) = f(x_i) + f[x_i, x_{i+1}](x - x_i) + f[x_i, x_{i+1}, x_{i+2}](x - x_i)(x - x_{i+1})$$

Then by (4.15) we have

$$|f(x) - p_2(x)| \leq \frac{M_3}{3!}|x - x_i||x - x_{i+1}||x - x_{i+2}|$$

where M_3 is a bound for f''' on $[a, b]$. We may write x as $x = x_{i+1} + \tau h$ for some $\tau \in [-1, 1]$ so that

$$(x - x_i)(x - x_{i+1})(x - x_{i+2}) = h^3\tau(1 - \tau^2)$$

and therefore

$$\max_{x_i \leq x \leq x_{i+2}} |f(x) - p_2(x)| \leq \frac{h^3}{3!}M_3 \max_{-1 \leq \tau \leq 1} |\tau(1 - \tau^2)|$$

The critical points of $w(\tau) = \tau(1 - \tau^2)$ are $\pm\sqrt{3}/3$, and hence

$$\max_{-1 \leq \tau \leq 1} |w(\tau)| = \frac{2\sqrt{3}}{9}$$

It follows that

$$\max_{x_i \leq x \leq x_{i+2}} |f(x) - p_2(x)| \leq \frac{\sqrt{3}\,M_3}{27}h^3 \qquad (4.16)$$

In these three examples we obtained rigorous a priori error bounds for interpolation. It was essential to have a formula for derivatives of f in order to calculate M_{n+1}. The next example illustrates a technique for estimating the error without explicit knowledge of the derivatives. The basis for this estimation procedure is equation (4.13).

EXAMPLE 4.8

Consider the following divided difference table, in which the function values are correct to six decimal digits:

x_i	$f[\cdot]$	$f[\cdot, \cdot]$	$f[\cdot, \cdot, \cdot]$	$f[\cdot, \cdot, \cdot, \cdot]$
1.88	1.511693			
1.91	1.536459	0.8255333		
1.97	1.585330	0.8145167	−0.1224067	
2.02	1.625390	0.8012000	−0.1210609	0.9612857×10^{-2}
2.11	1.696001	0.7845555	−0.1188893	1.085800×10^{-2}
2.18	1.749617	0.7659571	−0.1162400	1.261571×10^{-2}

Suppose we decide to use quadratic interpolation. Thus for an intermediate point x^* we choose three interpolation points close to x^* for the determination of the quadratic interpolate. For instance if $x^* = 2$, then choose the points 1.91, 1.97, and 2.02. The resultant quadratic is

$$p_2(x) = 1.536459 + 0.8145167(x - 1.91) - 0.1210609(x - 1.91)(x - 1.97)$$

By (4.13) we have for $x \in [1.91, 2.02]$

$$|f(x) - p_2(x)| = |f[x_2, x_3, x_4, x]| \prod_{i=2}^{4} |(x - x_i)|$$

Note that in the column of third divided differences the entries are all approximately equal to 10^{-2}. This suggests that $f[x_2, x_3, x_4, x] \simeq 10^{-2}$ for $x \in [1.88, 2.18]$, and therefore we make the estimate

$$|f(2) - p_2(2)| \simeq 10^{-2} \prod_{i=2}^{4} |2 - x_i|$$

$$\simeq 10^{-2}(0.09)(0.03)(0.02) = 0.54 \times 10^{-6}$$

Direct computation gives $p_2(2) = 1.609439$, whereas the correct value is $f(2) = \log 5 = 1.6094379$ [$f(x) = \log(1 + x^2)$ in this example].

We conclude this section with an analysis of the effect of perturbations in the function values on the value of the polynomial interpolate. For simplicity we consider only equally spaced points, say $x_i = a + (i - 1)h$, $1 \leq i \leq N + 1$, where $h = (b - a)/N$. Suppose we are given the values $\{\tilde{f}_i\}_{i=1}^{N+1}$, where

$$\tilde{f}_i = f(x_i) + \varepsilon_i$$

and ε_i represents roundoff or experimental error. The polynomial interpolate for f, based on the points $\{x_i\}_{i=1}^{N+1}$, is

$$P_N(x) = \sum_{k=1}^{N+1} f[x_1, \ldots, x_k] \prod_{i=1}^{k-1} (x - x_i)$$

where, by convention, $\prod_{i=1}^{0} (x - x_i) = 1$. However, we can compute only the polynomial

$$\tilde{P}_N(x) = \sum_{k=1}^{N+1} \tilde{f}[x_1, \ldots, x_k] \prod_{i=1}^{k-1} (x - x_i)$$

where $\tilde{f}[x_1] = \tilde{f}_1$, $\tilde{f}[x_1, x_2] = (\tilde{f}_2 - \tilde{f}_1)/(x_2 - x_1)$, and so forth. By the triangle inequality it follows that

$$|P_N(x) - \tilde{P}_N(x)| \leq \sum_{k=1}^{N+1} |f[x_1, \ldots, x_k] - \tilde{f}[x_1, \ldots, x_k]| \prod_{i=1}^{k-1} |x - x_i|$$

Moreover, by exercise 4.21, we have

$$\max_{a \leq x \leq b} \prod_{i=1}^{k-1} |x - x_i| \leq \frac{h^{k-1}}{4} (k - 2)! \qquad 2 \leq k \leq N + 1$$

so that

$$|P_N(x) - \tilde{P}_N(x)| \leq |\varepsilon_1| + \frac{1}{4} \sum_{k=2}^{N+1} h^{k-1}(k - 2)! |f[x_1, \ldots, x_k] - \tilde{f}[x_1, \ldots, x_k]| \tag{4.17}$$

By exercise 4.15 we have

$$f[x_1, \ldots, x_k] = \frac{h^{-k+1}}{(k-1)!} \Delta^{k-1} f_1 \qquad 2 \leq k \leq N+1$$

and similarly for $\tilde{f}[x_1, \ldots, x_k]$. Therefore

$$f[x_1, \ldots, x_k] - \tilde{f}[x_1, \ldots, x_k] = \frac{h^{-k+1}}{(k-1)!} (\Delta^{k-1} f_1 - \Delta^{k-1} \tilde{f}_1)$$

$$= -\frac{h^{-k+1}}{(k-1)!} \Delta^{k-1} \varepsilon_1 \qquad (4.18)$$

Combining (4.17) and (4.18) gives

$$|P_N(x) - \tilde{P}_N(x)| \leq |\varepsilon_1| + \frac{1}{4} \sum_{k=2}^{N+1} |\Delta^{k-1} \varepsilon_1|$$

If we let $\varepsilon = \max_{1 \leq k \leq N+1} |\varepsilon_k|$, then, by part (c) of exercise 4.17, $|\Delta^{k-1} \varepsilon_1| \leq 2^{k-1} \varepsilon$. Thus we have

$$\max_{a \leq x \leq b} |P_N(x) - \tilde{P}_N(x)| \leq \varepsilon \left(1 + \sum_{k=2}^{N+1} 2^{k-3}\right) = \frac{\varepsilon}{2} (2^N + 1) \qquad (4.19)$$

The bound (4.19) shows that a perturbation of magnitude ε in the function values results in a change in the value of the polynomial interpolate that is no larger than $\varepsilon(2^N + 1)/2$. This is actually a statement about the **conditioning** of the Newton formula for polynomial interpolation. The quantity $(2^N + 1)/2$ gives a bound for the amplification factor (or condition) by which the data error is magnified to give the change in the polynomial interpolate.

We can combine (4.15) and (4.19) to obtain an overall error bound. For $x \in [a, b]$ the triangle inequality gives

$$|f(x) - \tilde{P}_N(x)| \leq |f(x) - P_N(x)| + |P_N(x) - \tilde{P}_N(x)|$$

and hence by (4.15) and (4.19)

$$|f(x) - \tilde{P}_N(x)| \leq \frac{M_{N+1}}{(N+1)!} \max_{a \leq x \leq b} \prod_{i=1}^{N+1} |x - x_i| + \frac{\varepsilon}{2} (2^N + 1)$$

By exercise 4.21 we have

$$\max_{a \leq x \leq b} \prod_{i=1}^{N+1} |x - x_i| \leq h^{N+1} \frac{N!}{4}$$

and therefore

$$|f(x) - \tilde{P}_N(x)| \leq \frac{h^{N+1} M_{N+1}}{4(N+1)} + \frac{\varepsilon(2^N + 1)}{2} \qquad (4.20)$$

We summarize this in the following theorem.

THEOREM 4.4

Suppose $\tilde{P}_N$ denotes the polynomial interpolate of the equally spaced data

$\{(x_i, \tilde{f}_i)\}_{i=1}^{N+1}$, where $\tilde{f}_i = f(x_i) + \varepsilon_i$. If f has $N + 1$ continuous derivatives on $[a, b]$, then for any $x \in [a, b]$ the error $f(x) - \tilde{P}_N(x)$ satisfies (4.20).

In (4.20) the term $h^{N+1} M_{N+1}/4(N + 1)$ corresponds to the **discretization error** for polynomial interpolation whereas the term $\varepsilon(2^N + 1)/2$ bounds the roundoff or experimental error. The overall error in the process depends on which of these two terms is dominant. If the roundoff term is dominant, then the addition of more interpolation points cannot improve the overall error.

EXAMPLE 4.9

An experimenter plans to measure the value of some function f at equally spaced points with spacing h. Quadratic interpolation will then be used to find function values at intermediate points. The values of f are measured with a device whose relative error is no more than 10^{-4}; that is, the measured value $\tilde{f}$ satisfies

$$\tilde{f}(x) = f(x)(1 - \rho) \qquad |\rho| \le 10^{-4}$$

Also it is known that $|f(x)|$ and $|f'''(x)|$ are bounded by 10 for all points x of interest. How should h be chosen? We have

$$|f(x) - \tilde{f}(x)| = |\rho||f(x)| \le 10^{-4}(10) = 10^{-3}$$

and hence by (4.20)

$$|f(x) - \tilde{P}_2(x)| \le \frac{h^3 M_3}{12} + \frac{10^{-3}(9)}{2} \quad \overset{5}{\underset{(2^2 + 1)}{}}$$

On the basis of this bound it is reasonable to choose h so as to balance the interpolation error and experimental error:

$$\frac{h^3 M_3}{12} = 4.5 \times 10^{-3}$$
$$\underset{2.5}{} \qquad h = 1.442$$

Since $M_3 = 10$, we find $h = 0.175$.

EXERCISES

4.18 The Chebyshev points for $[a, b]$ are defined by

$$\hat{x}_i = \frac{a + b}{2} + \frac{b - a}{2} \cos \frac{(2i - 1)\pi}{2(n + 1)} \qquad i = 1, \ldots, n + 1$$

Repeat the computation of Example 4.4 using $\{\hat{x}_i\}_{i=1}^{N+1}$ instead of equally spaced points. Compare the output and note the marked improvement in the interpolation error.

4.19 In the application of (4.14) to $f[x_1, \ldots, x_{n+1}, x^*]$ we required $x^* \ne x_i$, $1 \le i \le n + 1$. However, there is no restriction on $x^* \in [a, b]$ in Theorem 4.3. Explain why the conclusion of Theorem 4.3 is valid if $x^* = x_i$ for some i.

4.20 Use the ideas in Example 4.5 to obtain the error bound for linear interpolation at x_i, x_{i+1}:

$$\max_{x_i \le x \le x_{i+1}} |f(x) - P_1(x)| \le \frac{M_2}{8}(x_{i+1} - x_i)^2 = \frac{M_2}{8}h_i^2$$

where $M_2 \ge \max_{x_i \le x \le x_{i+1}} |f''(x)|$.

4.21 Let $x_i \in [a, b]$ be given by $x_i = a + (i-1)h$, where $h = (b-a)/n$. Define $w(x) = \prod_{i=1}^{n+1}(x - x_i)$ and show that

$$\max_{a \le x \le b} |w(x)| \le \frac{h^{n+1}}{4}n! = \frac{h^n}{4}(b-a)(n-1)!$$

Hint: Fill in the details of the analysis given in Example 4.6.

4.22 If f is a polynomial of degree $\le n$, what does Theorem 4.3 tell you about the error $e_n(f, x) = f(x) - P_n(x)$? Hint: Calculate $f^{(n+1)}$.

4.23 If f is a polynomial of degree $\le n - 1$ and $\{x_i\}_{i=1}^{n+1}$ are distinct points, show that

$$f[x_1, \ldots, x_{n+1}] = 0$$

Hint: Use (4.14).

4.24 Let $f(x) = x^{1/3}$ on $[2, 3]$. Use the result of Example 4.7 to determine the number of equally spaced interpolation points that are necessary for quadratic interpolation to have an absolute error of no more than 0.5×10^{-6}.

4.25 Calculate the quadratic interpolate $P_2(x)$ for $f(x) = \sqrt{|x|}$ based on the interpolation points -0.1, 0.15, 0.2. Compare $0 = f(0)$ and $P_2(0)$. Does Theorem 4.3 apply to this problem?

4.26 For $f(x) = \log x$ on $[0.5, 1]$, show that

$$\max_{0.5 \le x \le 1} |f^{(n+1)}(x)| = |f^{(n+1)}(0.5)| = 2^{n+1}(n!)$$

Would you expect high degree polynomial interpolation of $\log x$ on $[0.5, 1]$ to provide an accurate approximation? Justify your answer.

4.27 Does $M_1 = \max_{0 \le x \le 1} |f'(x)|$ exist for $f(x) = \sqrt{x}$?

4.28 Develop a divided difference table for the data

x	$f(x)$
0.3	1.04403
0.42	1.08462
0.5	1.11803
0.58	1.15603
0.66	1.19817
0.72	1.23223

Use the technique of Example 4.8 to estimate the error at $x^* = 0.55$ for linear interpolation, at $x^* = 0.60$ for quadratic interpolation, and at $x^* = 0.46$ for cubic interpolation. The correct values of f at these points are $f(0.46) = 1.10073$, $f(0.55) = 1.14127$, and $f(0.6) = 1.16619$, to six decimal digits. How do your estimates compare with the actual errors?

4.29 For the following divided difference table would it be reasonable to use the technique of Example 4.8 for error estimation? Note: The divided differences are rounded to six digits.

x	$f[\cdot]$	$f[\cdot,\cdot]$	$f[\cdot,\cdot,\cdot]$	$f[\cdot,\cdot,\cdot,\cdot]$	$f[\cdot,\cdot,\cdot,\cdot,\cdot]$
1.32	1.6322				
1.44	−2.8105	37.0225			
1.50	4.8609	12.7857	−134.649		
1.62	6.7545	23.6700	60.4683	650.391	
1.68	7.2180	7.7250	−88.5833	−621.048	−3531.78
1.74	7.1458	−1.2033	−74.4025	59.0866	2267.12

Hint: The rationale for the estimation procedure of Example 4.8 is the near constancy of the relevant column in the divided difference table.

*4.30 Suppose f, f', f'', and f''' are continuous on $[a, b]$. Let P_2 be the quadratic that interpolates f at $\{x_i\}_{i=1}^3$, where $x_i = a + (i - 1)h$, and set $e_2(f, x) = f(x) - P_2(x)$. Since $e_2(f, x_i) = 0$, $1 \leq i \leq 3$, it follows from Rolle's theorem that e_2' has two distinct zeros, say y_1 and y_2.

(a) Let $w(x) = (x - y_1)(x - y_2)$ and define

$$\phi(t) = e_2'(f, t) - \frac{w(t)}{w(x_2)} e_2'(f, x_2)$$

Verify that $\phi(y_1) = \phi(y_2) = \phi(x_2) = 0$.

(b) Use Rolle's theorem and part (a) to show that $\phi''(\xi) = 0$ for some $\xi \in (x_1, x_3)$.

(c) From part (b) conclude that

$$e_2'(f, x_2) = f'''(\xi) \frac{h^2}{2}$$

(d) Use exercise 4.13 and part (c) to obtain

$$f'(x_2) - \frac{f(x_2 + h) - f(x_2 - h)}{2h} = f'''(\xi) \frac{h^2}{2}$$

which gives the error in the centered difference approximation to $f'(x_2)$.

(e) If M_3 is a bound for f''' on $[a, b]$, conclude from part (d) that

$$\left| f'(x_2) - \frac{f(x_2 + h) - f(x_2 - h)}{2h} \right| \leq \frac{M_3 h^2}{2}$$

4.31 Suppose $\tilde{f}_i = (1 - \rho_i) f(x_i)$, where $x_i = a + (i - 1)h$ and $h = (b - a)/N$. Assuming f has $N + 1$ continuous derivatives on $[a, b]$, show that

$$\max_{a \leq x \leq b} |f(x) - \tilde{P}_N(x)| \leq \frac{\rho}{2}(2^N + 1)M_0 + \frac{h^{N+1}}{4(N + 1)} M_{N+1}$$

where $\rho = \max_{1 \leq k \leq N+1} |\rho_k|$. Hint: See (4.20).

4.32 Suppose, as in Example 4.7 that quadratic interpolation is used in a table of values at equally spaced points. Let $\tilde{f}_i$ denote the floating point representation of $f(x_i)$ and assume that the relative error in $\tilde{f}_i$ is no more than u, the machine unit. Show that

$$\max_{x_i \leq x \leq x_{i+2}} |f(x) - \tilde{p}_2(x)| \leq 3uM_0 + \frac{\sqrt{3}h^3 M_3}{27}$$

4.33 Determine a cubic polynomial $Q_3(x)$ that interpolates f at a and b and satisifies $Q_3'(a) = f'(a)$, $Q_3'(b) = f'(b)$. The Q_3 is called a **Hermite cubic interpolate** of f. Hint: Try $Q(x) = A_0 + A_1(x - a) + A_2(x - a)^2 + A_3(x - a)^2(x - b)$ and determine the coefficients in terms of $f(a), f'(a), f(b),$ and $f'(b)$.

4.34 Suppose in Example 4.6 that the tabulated values of $f(x) = \cos x$ are correct only to four decimal digits. Use equation (4.20) to estimate the overall error. Use a four-place table for cosine and determine the polynomial interpolate suggested by Example 4.6. Determine the actual interpolation error at several intermediate points and compare with your estimated error. On the basis of your results would you say that (4.20) is a conservative error bound?

4.3 PIECEWISE POLYNOMIAL INTERPOLATION

In the previous section we argued that high degree polynomial interpolation in a table of values is frequently unsatisfactory. In Example 4.7 our approach was not to obtain a single polynomial interpolate for the entire table but rather to use different polynomial interpolates of low degree for various parts of the table. In this section we consider this approach more systematically. We examine two types of piecewise polynomial interpolation. The first type is "local" in the sense that the value of the interpolate at a point x^* is influenced only by the values of f near x^*. In the second type all of the tabular values of f influence the value of the interpolate at an intermediate point.

Piecewise Lagrange interpolation

The simplest example of piecewise Lagrange interpolation is linear interpolation in a table of values. Suppose f is given at $a = x_1 < x_2 < \cdots < x_{n+1} = b$. Let g_1 denote the function that interpolates f and is a polynomial of degree ≤ 1 on each subinterval $[x_i, x_{i+1}]$ and set $h_i = x_{i+1} - x_i$. Since g_1 is linear on $[x_i, x_{i+1}]$ and interpolates f, we must have

$$g_1(x) = f(x_i) + f[x_i, x_{i+1}](x - x_i) \qquad x \in [x_i, x_{i+1}]$$

Graphically g_1 looks like a "broken" line, as in Figure 4.1.

Since g_1 is a linear interpolate on $[x_i, x_{i+1}]$, we have from Theorem 4.3 the error formula

$$f(x) - g_1(x) = \frac{f''(z_i)}{2}(x - x_i)(x - x_{i+1}) \qquad x \in [x_i, x_{i+1}] \tag{4.21}$$

where $z_i \in (x_i, x_{i+1})$ depends on x. From this we obtain the following error bound (see also exercise 4.20).

THEOREM 4.5

Let g_1 be the piecewise linear Lagrange interpolate of f at $a = x_1 < x_2 < \cdots < x_{n+1} = b$ and set $h = \max_{1 \leq i \leq n} h_i$. If f'' is continuous on $[a, b]$ and $\max_{a \leq x \leq b} |f''(x)| \leq M_2$, then

$$\max_{a \leq x \leq b} |f(x) - g_1(x)| \leq \frac{M_2}{8} h^2 \tag{4.22}$$

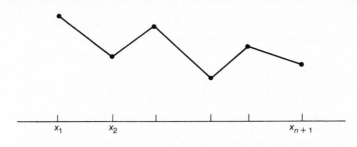

FIGURE 4.1

Observe that (in the absence of roundoff) the error $f(x) - g_1(x)$ can be made as small as we please by choosing h sufficiently small. Also note that the addition of more interpolation points poses no difficulty since g_1 is a simple linear function on each subinterval $[x_i, x_{i+1}]$.

This approach is easily extended to higher degree polynomials. For piecewise quadratic Lagrange interpolation we define $g_2(x)$ to be the function that interpolates f at $\{x_i\}_{i=1}^{n+1}$ and, for $i = 1, 3, 5, \ldots, n - 1$, is a polynomial of degree ≤ 2 on the subinterval $[x_i, x_{i+2}]$. Thus on each of the subintervals $[x_1, x_3]$, $[x_3, x_5]$, $\ldots$, $[x_{n-1}, x_{n+1}]$, g_2 is a different quadratic polynomial. Here we require n to be an even integer. The points $x_3, x_5, \ldots, x_{n-1}$ are called the **knots**—these are simply those interpolation points where g_2 changes from one polynomial to another. For piecewise linear Lagrange interpolation the knots are $x_2, x_3, \ldots, x_n$.

(handwritten margin note: WANT THE NUMBER OF SUBINTERVALS TO BE EVEN)

If x_i is a knot, then on each $[x_i, x_{i+2}]$ g_2 is the quadratic that interpolates f at x_i, x_{i+1}, and x_{i+2}. Therefore on $[x_i, x_{i+2}]$ g_2 is given by

$$g_2(x) = f(x_i) + f[x_i, x_{i+1}](x - x_i) + f[x_i, x_{i+1}, x_{i+2}](x - x_i)(x - x_{i+1})$$

and from Theorem 4.3 we have the error formula

$$f(x) - g_2(x) = \frac{f'''(w_i)}{3!}(x - x_i)(x - x_{i+1})(x - x_{i+2}) \qquad x \in [x_i, x_{i+2}]$$

$$(4.23)$$

where $w_i \in (x_i, x_{i+2})$ depends on x. From this we obtain (see Example 4.7 in the case of equal point spacing) the following error bound.

THEOREM 4.6

Let g_2 be the piecewise quadratic Lagrange interpolate of f at $a = x_1 < x_2 < \cdots < x_{n+1} = b$ (where n is even) and let $h = \max_{1 \leq i \leq n} h_i$. If f''' is continuous on $[a, b]$ and $\max_{a \leq x \leq b} |f'''(x)| \leq M_3$, then

$$\max_{a \leq x \leq b} |f(x) - g_2(x)| \leq \frac{M_3}{12} h^3 \qquad (4.24)$$

Proof

We have for $x \in [x_i, x_{i+2}]$

$$|f(x) - g_2(x)| \leq \frac{M_3}{3!} |(x - x_i)(x - x_{i+1})(x - x_{i+2})|$$

If $x \in [x_i, x_{i+1}]$, then $|(x - x_i)(x - x_{i+1})| \le h_i^2/4$, and for such x we also have $|x - x_{i+2}| \le h_i + h_{i+1}$. Therefore if $x \in [x_i, x_{i+1}]$

$$|(x - x_i)(x - x_{i+1})(x - x_{i+2})| \le \frac{h_i^2}{4}(h_i + h_{i+1}) \le \frac{h^3}{2}$$

Similarly for $x \in [x_{i+1}, x_{i+2}]$ we find

$$|(x - x_i)(x - x_{i+1})(x - x_{i+2})| \le \frac{h_{i+1}^2}{4}(h_i + h_{i+1}) \le \frac{h^3}{2}$$

and it follows that

$$\max_{a \le x \le b} |f(x) - g(x)| \le \max_{1 \le i \le n} \frac{M_3}{3!} \max_{x_i \le x \le x_{i+2}} |(x - x_i)(x - x_{i+1})(x - x_{i+2})|$$

$$\le \frac{h^3}{2} \frac{M_3}{3!}$$

which completes the proof.

For equally spaced points it is not difficult (with the aid of Example 4.7) to obtain a slightly better error bound. The error bound (4.24) shows why piecewise quadratic interpolation can be an excellent approximation device. As h is decreased (by adding interpolation points), the error decreases by a factor of h^3, which is quite respectable. However, the reader should be aware of the fact that the assumed smoothness of f is essential in the bound (4.24). If f''' does not exist but f'' is continuous, then the error decreases as h^2. If f is only continuously differentiable, then the error decreases by a factor of h.

EXAMPLE 4.10

It is required to find the number of equally spaced interpolation points such that the error in piecewise quadratic Lagrange interpolation for $f(x) = \sinh x$ on $[0, 1]$ is no more than 10^{-6}. By virtue of exercise 4.38 we have [since $f'''(x) = \cosh x$]

$$\max_{0 \le x \le 1} |\sinh x - g_2(x)| \le \frac{\sqrt{3}}{27} h^3 \max_{0 \le x \le 1} |\cosh x|$$

Since $(d/dx) \cosh x = \sinh x = (e^x - e^{-x})/2 > 0$, it follows that $\max_{0 \le x \le 1} |\cosh x| = \cosh 1 \doteq 1.54308$, and hence we choose $h(= 1/n)$ so that

$$\frac{\sqrt{3}}{27}(1.54308)h^3 \le 10^{-6}$$

Solving this inequality for h gives $h \doteq 2.16 \times 10^{-2}$, and hence about 47 $(= n + 1)$ points are required.

EXAMPLE 4.11

Consider the function $f(x) = \sqrt{x}$ on $[0.1, 2]$. We calculate $f'''(x) = 3x^{-5/2}/8$ and hence

$$\max_{0.1 \leq x \leq 2} |f'''(x)| = |f'''(0.1)| \doteq 118.6$$

It follows, for equal spacing, that the error in piecewise quadratic Lagrange interpolation is no more than 10^{-4} if h is chosen so that

$$\frac{\sqrt{3}}{27}(118.6)h^3 \leq 10^{-4}$$

This gives $h \simeq 2.36 \times 10^{-2}$, or about 81 equally spaced points.

Next we show that with unequal spacing we can obtain the same accuracy with far fewer points. Note that f''' is large near 0.1 but $|f'''(x)| \leq 1$ for $x \in [0.7, 2]$. Suppose we use equally spaced points with spacing h_1 on $[0.1, 0.7]$ and equally spaced points with spacing h_2 on $[0.7, 2]$. By the preceding analysis we need $h_1 \simeq 2.36 \times 10^{-2}$, which requires about 27 points on $[0.1, 0.7]$. For h_2 we want

$$\frac{\sqrt{3}h_2^3}{27} \leq 10^{-4}$$

and hence $h_2 \simeq 0.116$, which requires about 13 points on $[0.7, 2]$. Using this approach, we need roughly half the number of interpolation points as in the case of equal spacing throughout the interval.

By using a piecewise polynomial of degree $\leq k$, call it g_k, one can achieve approximations whose error bound contains a factor of h^{k+1} (provided f is sufficiently smooth). We shall not pursue this approach any further; rather we investigate a different type of piecewise polynomial interpolation that has become very popular of late, namely, cubic spline interpolation.

Before examining cubic splines, we point out a potential disadvantage of piecewise Lagrange interpolation. To illustrate this difficulty, consider the quadratic interpolate $g_2(x)$. Between each adjacent pair of knots g_2 is infinitely differentiable, but in general g_2' does not exist at the knots. This phenomenon is clearly illustrated in Figure 4.2. The graph of g_2 has sharp corners at the knots. Therefore if we want to use g_2' as an approximation to f', we are virtually assured of a poor approximation near the knots. On the other hand piecewise Lagrange interpolates are commonly used in numerical integration formulas whereby the integral

$$I(f) = \int_a^b f(x)\, dx$$

is approximated by

$$I_n(f) = \int_a^b g_k(x)\, dx$$

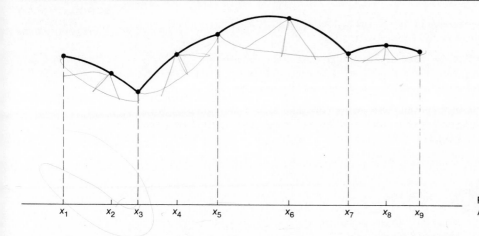

FIGURE 4.2
A typical g_2 interpolate

where g_k is a piecewise Lagrange interpolate of degree k for f at the points $a = x_1 < x_2 < \cdots < x_{n+1} = b$. We consider this approach in detail for $k = 1$ and 2 in Chapter 5.

EXERCISES

4.35 Show that g_1 can also be written as

$$g_1(x) = f(x_{i+1}) + f[x_i, x_{i+1}](x - x_{i+1}) \qquad x \in [x_i, x_{i+1}]$$

4.36 Integrate g_1 over $[x_i, x_{i+1}]$ to obtain

$$\int_{x_i}^{x_{i+1}} g_1(x)\, dx = \frac{h_i}{2} [f(x_i) + f(x_{i+1})]$$

This is the basis for the trapezoidal rule (Section 5.1).

4.37 Use (4.13) and exercise 4.36 to show that

$$\int_{x_i}^{x_{i+1}} f(x)\, dx = \frac{h_i}{2} [f(x_i) + f(x_{i+1})] + \int_{x_i}^{x_{i+1}} f[x_i, x_{i+1}, x](x - x_i)(x - x_{i+1})\, dx$$

4.38 For equally spaced points show that the error bound in Theorem 4.6 can be improved to $\sqrt{3}\, M_3 h^3/27$. Hint: See Example 4.7.

4.39 Estimate the number of equally spaced points required to approximate $f(x) = \sqrt{1 + x}$ on $[2, 3]$ with an absolute error of no more than 10^{-4} using piecewise quadratic Lagrange interpolation.

4.40 Show that g_2 on $[x_i, x_{i+2}]$ with i odd can also be expressed as

$$g_2(x) = f(x_{i+1}) + f[x_i, x_{i+1}](x - x_{i+1}) + f[x_i, x_{i+1}, x_{i+2}](x - x_i)(x - x_{i+1})$$

4.41 If x_i is a knot and $h_i = h_{i+1}$, show that

$$\int_{x_i}^{x_{i+2}} g_2(x)\, dx = \frac{h_i}{3} [f(x_i) + 4f(x_{i+1}) + f(x_{i+2})]$$

This is the basis for Simpson's rule (Section 5.1).

4.42 Suppose f is a polynomial of degree ≤ 2 and let g_2 be as in Theorem 4.6. Is it true that $g_2(x) = f(x)$ for all x? Justify your answer.

4.43 Determine the piecewise quadratic Lagrange interpolate for $f(x) = 1/(1 + 25x^2)$ on $[-1, 1]$ using interpolation points -1, -0.3, -0.6, -0.1, 0, 0.1, 0.6, 0.8, and 1. Estimate the error. Determine the actual error at several intermediate points.

4.44 Give an example of a continuous function that cannot be approximated with order h^3 accuracy using piecewise quadratic Lagrange interpolation. Hint: See the discussion following Theorem 4.6.

4.45 For equally spaced points ($h = h_i$, all i) express g_2 in terms of forward differences as

$$g_2(x) = f_i + \Delta f_i \tau + \Delta^2 f_i \frac{\tau(\tau - 1)}{2} \qquad x \in [x_i, x_{i+2}]$$

where $x = x_i + \tau h$. Hint: See exercise 4.15.

Cubic splines

Suppose f is tabulated at the points $a = x_1 < x_2 < \cdots < x_{n+1} = b$ and we attempt to determine a piecewise cubic function that interpolates these values of f. If n is a multiple of 4 and the knots are chosen to be $x_4, x_8, \ldots, x_{n-2}$, then there results the piecewise cubic Lagrange interpolate of f. As was previously observed, piecewise Lagrange interpolates are generally not differentiable at the knots.

The essential idea of cubic spline interpolation is to determine a piecewise cubic interpolate for f that is as smooth as possible on $[a, b]$. Suppose we choose the knots to be $x_2, x_3, \ldots, x_n$ and denote our desired piecewise cubic interpolate by $S(x)$. On each subinterval $[x_i, x_{i+1}]$, $1 \leq i \leq n$, we require S to be a cubic polynomial that satisfies $S(x_i) = f(x_i)$ and $S(x_{i+1}) = f(x_{i+1})$. We can write S on $[x_i, x_{i+1}]$ as

$$S(x) = A_{1,i} + A_{2,i}(x - x_i) + A_{3,i}(x - x_i)^2 + A_{4,i}(x - x_i)^3 \qquad (4.25)$$

where the four coefficients $\{A_{k,i}\}_{k=1}^4$ are to be determined. These four coefficients are uniquely specified by giving the value of S at four (distinct) points in $[x_i, x_{i+1}]$. Since we require S to interpolate f at only two points in $[x_i, x_{i+1}]$ (at the endpoints), we still have considerable freedom in constructing S. This additional freedom is used to make S smooth on $[a, b]$. A cubic spline interpolate S for f is required to satisfy

(i) $S(x_i) = f(x_i)$, $1 \leq i \leq n + 1$, where $a = x_1 < x_2 < \cdots < x_{n+1} = b$.

(ii) S is a cubic polynomial on $[x_i, x_{i+1}]$.

(iii) S is as smooth as possible on $[a, b]$; that is, for some $k \geq 0$ $S, S', S'',$ $\ldots, S^{(k)}$ are continuous on $[a, b]$.

Let us try to ascertain how large k in condition (iii) can be. Since S is completely specified on $[x_i, x_{i+1}]$ by the four coefficients $\{A_{k,i}\}_{k=1}^4$ and there are n such subintervals, we have a total of $4n$ coefficients to determine. Loosely speaking, this says we have $4n$ degrees of freedom in the determination of S. On the other hand conditions (i) and (iii) place constraints on S. Let us compare the number of constraints with the degrees of freedom. For each $i = 2, 3, \ldots, n$ condition (i) requires that the cubic polynomial on $[x_{i-1}, x_i]$ and the cubic polynomial on $[x_i, x_{i+1}]$ each interpolate f at x_i. These two constraints at each knot and the interpolation requirements at a and b give a total of $2(n-1) + 2 = 2n$ constraints for condition (i). If S satisfies (i), then S is automatically continuous on $[a, b]$. As for condition (iii), we require the derivatives to "match up" at the knots; that is, the jth derivative of S should have a common left-hand and right-hand limit at the knots. Thus condition (iii) can be expressed as, for $2 \le i \le n$

$$\lim_{x \to x_i^-} S^{(j)}(x) = \lim_{x \to x_i^+} S^{(j)}(x) \qquad j = 1, 2, \ldots, k$$

For each j this gives an additional $n - 1$ constraints on S. Thus there are a total of $2n + k(n-1)$ constraints, and we expect that the largest value of k, say k^*, satisfies

$$2n + k^*(n-1) \le 4n$$

We find $k^* = 2$, which results in $4n - 2$ constraints. In fact we shall demonstrate how to construct a piecewise cubic interpolate that satisfies (iii) with $k = 2$.

A function that satisfies conditions (i) to (iii) with $k = 2$ is called a **cubic spline interpolate** of f. More generally a **spline of order m** is a piecewise polynomial of degree $m - 1$ that is $m - 2$ times continuously differentiable. Here the order refers to the order of accuracy (see Theorem 4.7 at the end of this section). A cubic spline is therefore a spline of order 4 and a second-order spline is a continuous piecewise linear polynomial. Thus a piecewise linear Lagrange interpolate is also a **linear spline**.

In the determination of a cubic spline interpolate there are two fewer constraints than unknown coefficients, and we need two additional constraints to uniquely determine S. These two constraints consist in specifying S' or S'' at a and b. We postpone consideration of these **endpoint constraints** and proceed to the construction of S.

Let m_i denote the value of $S''(x_i)$. We show that S is completely determined by $\{m_i\}_{i=1}^{n+1}$ and $\{f(x_i)\}_{i=1}^{n+1}$. For clarity and convenience we denote the restriction of $S(x)$ to $[x_i, x_{i+1}]$ by $S_i(x)$, $f(x_i)$ by f_i, and $x_{i+1} - x_i$ by h_i. Since $S_i(x)$ is a cubic polynomial, $S_i''(x)$ is a linear polynomial whose values at x_i and x_{i+1} are m_i and m_{i+1}, respectively. Therefore

$$S_i''(x) = \frac{m_i}{h_i}(x_{i+1} - x) + \frac{m_{i+1}}{h_i}(x - x_i) \qquad x \in [x_i, x_{i+1}].$$

By integrating $S_i''(x)$ twice, we get two constants of integration, say c_i and d_i, so that

$$S_i(x) = \frac{m_i}{6h_i}(x_{i+1} - x)^3 + \frac{m_{i+1}}{6h_i}(x - x_i)^3 + \frac{c_i}{h_i}(x_{i+1} - x) + \frac{d_i}{h_i}(x - x_i)$$

$$(4.26)$$

In (4.26) we have chosen the particular form of the constants of integration for convenience in what follows. Note that we obtain the former equation by differentiating (4.26) two times. By condition (i) we want $S_i(x_i) = f_i$ and $S_i(x_{i+1}) = f_{i+1}$. Thus by substitution of x_i and x_{i+1} for x in (4.26) we find

$$f_i = \frac{m_i}{6} h_i^2 + c_i$$

$$f_{i+1} = \frac{m_{i+1}}{6} h_i^2 + d_i$$

We solve these two equations for c_i and d_i and substitute the resultant expressions into (4.26) to yield (after some manipulation)

$$S_i(x) = \frac{1}{h_i} \left\{ \frac{m_i}{6} (x_{i+1} - x)^3 + \frac{m_{i+1}}{6} (x - x_i)^3 \right\} + f_i$$
$$+ f[x_1, x_{i+1}](x - x_i) - \frac{h_i^2}{6} \left\{ (m_{i+1} - m_i) \frac{x - x_i}{h_i} + m_i \right\} \qquad (4.27)$$

This equation specifies S_i in terms of the unknown quantities m_i and m_{i+1}. The only requirement we have not considered is that S' be continuous at the knots. [In writing the equation for $S_i''(x)$, we implicitly used $\lim_{x \to x_i-} S''(x) = \lim_{x \to x_i+} S''(x) = m_i$.] From (4.27) we find the derivative of S_i and evaluate at x_i to yield

$$S_i'(x_i) = -\frac{m_i}{3} h_i - \frac{m_{i+1}}{6} h_i + f[x_i, x_{i+1}] \qquad (4.28)$$

Similarly, using (4.27) with i replaced by $i - 1$, we find

$$S_{i-1}'(x_i) = \frac{m_i}{3} h_{i-1} + \frac{m_{i-1}}{6} h_{i-1} + f[x_{i-1}, x_i] \qquad (4.29)$$

and for $2 \le i \le n$ we require $S_{i-1}'(x_i) = S_i'(x_i)$. Thus we equate (4.28) and (4.29) to yield

$$h_{i-1} m_{i-1} + 2(h_{i-1} + h_i)m_i + h_i m_{i+1} = 6\{f[x_i, x_{i+1}] - f[x_{i-1}, x_i]\} \qquad (4.30)$$

Since (4.30) must hold for $i = 2, 3, \ldots, n$, we have $n - 1$ linear equations in terms of the $n + 1$ unknowns $m_1, m_2, \ldots, m_{n+1}$. The determination of S reduces to the solution of a linear system of equations—a problem with which we are familiar.

At this point we need to place two additional constraints on S, which results in two additional equations. This gives rise to a square matrix problem, which can be solved by the methods of Chapter 3. We discuss in detail only two of the many possible ways of obtaining the additional equations and refer the reader to the exercises at the end of this section for alternate methods.

The first way of constraining S is to require $S'(a) = f'(a)$ and $S'(b) = f'(b)$. Of course this assumes that $f'(a)$ and $f'(b)$ are known. We use $S'(a) = f'(a)$ in (4.28) with $i = 1$ to give the equation

$$2h_1 m_1 + h_2 m_2 = 6\{f[x_1, x_2] - f'(a)\}$$

Similarly we use $S'(b) = f'(b)$ in (4.29) with $i = n + 1$ to give

$$h_n m_n + 2h_n m_{n+1} = 6\{f'(b) - f[x_n, x_{n+1}]\}$$

These two equations together with (4.30), for $2 \leq i \leq n$, give a square system of equations for the determination of S. In matrix form this system is

$$\begin{bmatrix} 2h_1 & h_1 & 0 \cdots \cdots \cdots \cdots \cdots \cdots \cdots \cdots 0 \\ h_1 & 2(h_1 + h_2) & h_2 \\ 0 \\ \vdots & & & & & 0 \\ 0 & & & h_{n-1} & 2(h_{n-1} + h_n) & h_n \\ 0 \cdots \cdots \cdots \cdots \cdots \cdots \cdots 0 & h_n & 2h_n \end{bmatrix} \begin{bmatrix} m_1 \\ m_2 \\ \vdots \\ m_n \\ m_{n+1} \end{bmatrix}$$

$$= 6 \begin{bmatrix} f[x_1, x_2] - f'(a) \\ f[x_2, x_3] - f[x_1, x_2] \\ \vdots \\ f[x_n, x_{n+1}] - f[x_{n-1}, x_n] \\ f'(b) - f[x_n, x_{n+1}] \end{bmatrix} \qquad (4.31)$$

This is a symmetric tridiagonal matrix problem whose coefficient matrix satisfies the hypotheses of Theorem 3.4. Thus the tridiagonal algorithm can be used for its solution.

For equally spaced interpolation points ($h_i = h$, all i) the system is particularly simple and can be written as

$$\begin{bmatrix} 2 & 1 & 0 & 0 \cdots \cdots \cdots 0 \\ 1 & 4 & 1 & 0 & & 0 \\ 0 & 1 & 4 & 1 & & \vdots \\ \vdots & & & & & 0 \\ 0 & & & 1 & 4 & 1 \\ 0 \cdots \cdots \cdots \cdots 0 & 1 & 2 \end{bmatrix} \begin{bmatrix} m_1 \\ m_2 \\ m_3 \\ \vdots \\ m_n \\ m_{n+1} \end{bmatrix} = 6h^{-2} \begin{bmatrix} \Delta f_1 - hf'(a) \\ \Delta^2 f_1 \\ \Delta^2 f_2 \\ \vdots \\ \Delta^2 f_{n-1} \\ hf'(b) - \Delta f_n \end{bmatrix} \qquad (4.32)$$

where $\Delta f_i = f_{i+1} - f_i$ and $\Delta^2 f_i = f_{i+2} - 2f_{i+1} + f_i$ (exercise 4.15).

The cubic spline that results from the endpoint constraints $f'(a) = S'(a)$ and $f'(b) = S'(b)$ is called the **complete cubic spline interpolate** of f and is denoted by $S_c(x)$.

If the derivative of f is unknown, we can require instead that S' approximately equal f' at the endpoints. One way to accomplish this is as follows. Let $P_a(x)$ denote the cubic polynomial that interpolates f at $\{x_i\}_{i=1}^4$. Then by (4.8) we have

$$P_a(x) = \sum_{j=1}^4 f[x_1, \ldots, x_j] \prod_{i=1}^{j-1} (x - x_i)$$

where $\prod_{i=1}^{0} (x - x_i) = 1$ by convention. Differentiation gives

$$P'_a(a) = f[x_1, x_2] - h_1 f[x_1, x_2, x_3]$$

$$+ h_1(h'_2 + h'_3)f[x_1, x_2, x_3, x_4] \quad (4.33)$$

Now we require that $S'(a) = P'_a(a)$, and hence we equate (4.33) to (4.28) with $i = 1$ to give

$$2h_1 m_1 + h_1 m_2 = 6h_1\{(h_1 + h_2)f[x_1, x_2, x_3, x_4] - f[x_1, x_2, x_3]\} \quad (4.34)$$

Similarly if $P_b(x)$ denotes the cubic polynomial that interpolates f at $\{x_{n+2-i}\}_{i=1}^{4}$ and we require that $P'_b(b) = S'(b)$, then there results the equation

$$h_n m_n + 2h_n m_{n+1} = 6h_n\{f[x_{n-1}, x_n, x_{n+1}]$$

$$+ (h_n + h_{n+1})f[x_{n-2}, x_{n-1}, x_n, x_{n+1}]\} \quad (4.35)$$

Equations (4.34), (4.35), and (4.30) for $2 \leq i \leq n$ give a tridiagonal system of equations for the determination of S. This system differs from (4.31) only in the first and last entries of the vector on the right-hand side. The cubic spline that results from the endpoint conditions $P'_a(a) = S'(a)$ and $P'_b(b) = S'(b)$ is called the **Lagrange cubic spline interpolate** of f and is denoted by $S_L(x)$.

Let us denote the tridiagonal coefficient matrix in (4.31) by **H**, the vector of unknowns by **m**, and the right-hand-side vector by **b**. Then S_c is determined by solving

$$\mathbf{Hm} = \mathbf{b}$$

To find S_L, we need only modify **b**. Specifically we define $\hat{\mathbf{b}}$ by $\hat{b}_i = b_i$ for $2 \leq i \leq n$, $\hat{b}_1 =$ right-hand side of (4.34), and $\hat{b}_{n+1} =$ right-hand side of (4.35). Then S_L is found by solving $\mathbf{Hm} = \hat{\mathbf{b}}$.

We summarize a procedure for computing the Lagrange cubic spline interpolate:

(a) Compute the divided differences $f[x_i, x_{i+1}]$ for $1 \leq i \leq n$ and form $\hat{\mathbf{b}}$.

(b) Set up the coefficient matrix **H** in a form suitable for application of the tridiagonal algorithm.

(c) Solve $\mathbf{Hm} = \hat{\mathbf{b}}$ by the tridiagonal algorithm, thereby obtaining $m_1, \ldots, m_{n+1}$.

(d) Determine the coefficients $\{A_{k,i}\}_{k=1}^{4}$ so that

$$S_i(x) = \sum_{k=1}^{4} A_{k,i}(x - x_i)^{k-1} \qquad x \in [x_i, x_{i+1}] \quad (4.36)$$

These coefficients are given by

$$A_{1,i} = f_i \qquad\qquad A_{3,i} = \frac{m_i}{2}$$

$$(4.37)$$

$$A_{2,i} = f[x_i, x_{i+1}] - \frac{h_i}{6}(m_{i+1} + 2m_i) \qquad A_{4,i} = \frac{m_{i+1} - m_i}{6h_i}$$

Then $S_L(z)$ can be evaluated by determining the index i for which $z \in$

$[x_i, x_{i+1}]$ and using (4.36) to compute $S_L(z)$. We leave it to the reader to establish the relationship between the coefficients $\{A_{k,i}\}$ and the m_i's given by (4.37).

The following is a subroutine for Lagrange cubic spline interpolation. In SPLINE the interpolation points and corresponding function values are stored in N1 dimensional arrays X and FX, respectively. The right-hand-side vector is stored in B. On return B contains the solution of the tridiagonal matrix problem $\mathbf{Hm} = \hat{\mathbf{b}}$ and array A contains the coefficients for S_L in the form (4.36).

```
      SUBROUTINE SPLINE(X,FX,N1,B,A)
C   X: ARRAY OF INTERPOLATION POINTS
C   FX: FUNCTION VALUES AT POINTS IN X
C   N1: NUMBER OF INTERPOLATION POINTS
C   B: ON ENTRY CONTAINS RIGHT HAND SIDE OF MATRIX
C      PROBLEM; ON RETURN CONTAINS SOLUTION OF MATRIX
C      PROBLEM
      REAL X(N1),FX(N1),B(N1),SD(41),SU(41),A(4,N1)
C   DETERMINE THE RIGHT HAND SIDE
      N=N1-1
      DO 2 J=1,N
      B(J)=(FX(J+1)-FX(J))/(X(J+1)-X(J))
2     CONTINUE
      DO 3 J=2,N
      JJ=N-J+2
      B(JJ)=6.*(B(JJ)-B(JJ-1))
3     CONTINUE
C   ENDPOINT CONDITIONS CORRESPONDING
C   TO S'(AA)=PA'(AA) AND S'(BB)=PB'(BB)
      W=B(2)/(X(3)-X(1))
      Z=B(3)/(X(4)-X(2))
      Z=(Z-W)/(X(4)-X(1))
      B(1)=(X(2)-X(1))*((X(3)-X(1))*Z-W)
      W=B(N-1)/(X(N)-X(N-2))
      Z=B(N)/(X(N1)-X(N-1))
      W=(Z-W)/(X(N1)-X(N-2))
      B(N1)=(X(N1)-X(N))*(Z+(X(N1)-X(N-1))*W)
C   SET UP COEFFICIENT MATRIX
      SD(1)=2.*(X(2)-X(1))
      SU(1)=SD(1)/2.
      DO 4 J=2,N
      SU(J)=X(J+1)-X(J)
      SD(J)=2.*(X(J+1)-X(J-1))
4     CONTINUE
      SD(N1)=2.*(X(N1)-X(N))
C   TRIDIAGONAL ALGORITHM FOR SOLUTION
C   OF LINEAR SYSTEM CORRESPONDING TO
C   EQNS. (4.30),(4.33),AND (4.34)
      DO 5 J=2,N1
      Q=SU(J-1)/SD(J-1)
      SD(J)=SD(J)-Q*SU(J-1)
      B(J)=B(J)-Q*B(J-1)
```

```
5     CONTINUE
      B(N1)=B(N1)/SD(N1)
      DO 6 K=1,N
        KK=N1-K
        B(KK)=(B(KK)-SU(KK)*B(KK+1))/SD(KK)
6     CONTINUE
C  DETERMINE THE COEFFICIENTS FOR S IN
C  THE FORM (4.36)
      DO 7 J=1,N
        DELTA=X(J+1)-X(J)
        A(4,J)=(B(J+1)-B(J))/(6.*DELTA)
        A(3,J)=B(J)/2.
        A(2,J)=(FX(J+1)-FX(J))/DELTA
        A(2,J)=A(2,J)-DELTA*(B(J+1)+2.*B(J))/6.
        A(1,J)=FX(J)
7     CONTINUE
      RETURN
      END
```

In order to modify this subroutine for other endpoint conditions, it is necessary to change the determination of B(1) and B(N1) in the eight lines of code
following the comment

C TO S'(AA) = PA'(AA) AND S'(BB) = PB'(BB)

For instance the determination of S_c results if these eight lines of code are
replaced by

B(1) = B(1) − 6.*DFX(1)

W = (FX(N1) − FX(N))/(X(N1) − X(N))

B(N1) = 6.*(DFX(2) − W)

where DFX is an array of dimension 2 that contains $f'(a)$ and $f'(b)$. The array
DFX must be added to the parameter list and declared in a dimension statement. The reader is asked to make these modifications in exercise 4.54.

The next subroutine, VALSPL, enables one to evaluate a cubic spline at
any given point. The subroutine accepts the N1 dimensional array X of interpolation points, the $4 \times N1$ array of coefficients A, and the point Z. On return,
the value of the spline at Z is SVAL.

```
      SUBROUTINE VALSPL(X,A,N1,Z,SVAL)
C  X: ARRAY OF INTERPOLATION POINTS
C  A: ARRAY OF SPLINE COEFFICIENTS
C  N1: NUMBER OF INTERPOLATION POINTS
C  Z: POINT AT WHICH VALUE OF SPLINE IS FOUND
C  SVAL: VALUE OF SPLINE AT Z
      DIMENSION X(N1),A(4,N1)
      INDX=1
      N=N1-2
C  USE LINEAR SEARCH TO FIND SUBINTERVAL INDEX
```

```
      DO 1 J=1,N
        IF(Z.LE.X(J+1)) GO TO 2
        INDX=INDX+1
  1    CONTINUE
  2    K=INDX
       W=Z-X(K)
       SVAL=A(1,K)+W*(A(2,K)+W*(A(3,K)+W*A(4,K)))
       RETURN
       END
```

In VALSPL, nested multiplication is used for the evaluation of the spline.

EXAMPLE 4.12

In the following program we compute S_L for $f(x) = 1/(1 + 25x^2)$, based on the points $x_i = -1 + 2(i - 1)/N$, $1 \le i \le N + 1$ for $N = 8, 16, \ldots, 80$. In each case we estimate the maximum interpolation error by computing

$$\max_{1 \le i \le 151} |f(z_i) - S_L(z_i)| \tag{4.38}$$

where $z_i = -1 + 2(i - 1)/150$. Examination of the output shows that the maximum error steadily decreases as the number of interpolation points increase. Compare this with Example 4.4!

```
       DIMENSION X(81),FX(81),SU(81),SD(81),B(81)
       DIMENSION A(4,81)
       F(X)=1./(1.+25.*X**2)
       AA=-1.0
       BB=1.0
       WRITE(6,99)
  99   FORMAT(' ',3X,'N',8X,'MA8 ERROR'//)
       DO 3 N=8,80,8
         N1=N+1
         H=(BB-AA)/FLOAT(N)
         DO 1 K=1,N1
           X(K)=AA+FLOAT(K-1)*H
           FX(K)=F(X(K))
  1      CONTINUE
         CALL SPLINE(X,FX,N1,B,A)
         ERR=0.0
         DO 2 J=1,151
           P=-1.+2.*FLOAT(J-1)/150.
           CALL VALSPL(X,A,N1,P,SVAL)
           DIFF=ABS(SVAL-F(P))
           IF(DIFF.GT.ERR)ERR=DIFF
  2      CONTINUE
         WRITE(6,101)N,ERR
  101    FORMAT(' ',2X,I3,3X,E14.7)
  3    CONTINUE
       STOP
       END
```

N	MAX ERROR
8	0.5584890E-01
16	0.3741622E-02
24	0.1917839E-02
32	0.6192923E-03
40	0.2783537E-03
48	0.1215935E-03
56	0.5942583E-04
64	0.3820658E-04
72	0.2711266E-04
80	0.2051890E-04

An analysis of the error in cubic spline interpolation is beyond the scope of the text. The error bounds in the following theorem were obtained by Hall and Meyer.

THEOREM 4.7

Suppose f is four times continuously differentiable on $[a, b]$ and $a = x_1 < x_2 < \cdots < x_{n+1} = b$. If S_c denotes the complete cubic spline interpolate of f at $\{x_i\}_{i=1}^{n+1}$, then

$$\max_{a \le x \le b} |f(x) - S_c(x)| \le \tfrac{5}{384} M_4 h^4 \qquad (4.39)$$

$$\max_{a \le x \le b} |f'(x) - S_c'(x)| \le \tfrac{1}{24} M_4 h^3 \qquad (4.40)$$

where $M_4 \ge \max_{a \le x \le b} |f^{(4)}(x)|$ and $h = \max_{1 \le i \le n} h_i$.

Thus for smooth f the complete cubic spline interpolate furnishes a fourth-order accurate approximation (the same order as the piecewise cubic Lagrange interpolate), and, in contrast to piecewise Lagrange interpolation, the derivative of f is approximated well by S_c' throughout the interval. Swartz and Varga in 1972 established error bounds analogous to those in Theorem 4.7 for the Lagrange cubic spline interpolate. Thus Theorem 4.7 is also valid for S_L (the Lagrange cubic spline interpolate) with different constants in the error bounds.

EXERCISES

4.46 Fill in the details involved in the derivation of (4.28) and (4.29).

4.47 Given the values $\{m_i\}_{i=1}^{n+1}$, one can use (4.27) for the evaluation of S_i. For computational purposes it is better to write S_i in the form (4.36) and use nested multiplication.

Show that (4.37) gives the coefficients $\{A_{k,i}\}_{k=1}^{4}$ for equation (4.36). Show that $S'(x_i) = A_{2,i}$ and $S''(x_i) = 2A_{3,i}$.

4.48 Write a subroutine based on differentiation of (4.36) for the evaluation of $S'(z)$. Call the subroutine DVALSP(X,A,N1,Z,DSVAL), where X, A, N1, and Z have the same meaning as in VALSPL. The value of $S'(z)$ should be returned in DSVAL. Use SPLINE and DVALSP to compute the derivative of S_L for $f(x) = \exp(-x)$ based on the interpolation points $x_i = (i-1)/20 - 1$, $1 \le i \le 21$. Evaluate $S'_L(z)$ at several intermediate points z and compare with $f'(z) = -\exp(-z)$. Discuss your results.

4.49 Integrate (4.27) over $[x_i, x_{i+1}]$ to obtain

$$\int_{x_i}^{x_{i+1}} S(x)\, dx = \frac{h_i}{2} (f_i + f_{i+1}) - \frac{h_i^3}{4!} (m_i + m_{i+1})$$

and hence conclude that

$$\int_a^b S(x)\, dx = \frac{1}{2} \sum_{i=1}^{n} h_i(f_i + f_{i+1}) - \frac{1}{4!} \sum_{i=1}^{n} h_i^3(m_i + m_{i+1})$$

This furnishes a reasonable procedure, called **cubic spline quadrature**, for the approximation of $\int_a^b f(x)\, dx$ when the choice of data points $\{x_i\}_{i=1}^{n+1}$ is not at your disposal.

4.50 Use hand computation to find the cubic spline interpolate of $f(x) = |x|$ at the points $x_1 = -2$, $x_2 = 0.5$, $x_3 = 0$, $x_4 = 1.5$, $x_5 = 2$ that satisfies $m_0 = m_5 = 0$. Determine the error at several intermediate points.

4.51 Determine the 4th-degree polynomial interpolate, in Newton form, for the data of exercise 4.50. Determine the error at the same intermediate points and compare the results.

4.52 The so-called **natural cubic spline interpolate** for f, which is denoted by S_N, satisfies the endpoint conditions $S''_N(a) = S''_N(b) = 0$. Give the linear system of equations for the determination of S_N. If $f''(a) \neq 0$ or $f''(b) \neq 0$, do you expect a good approximation of f by S_N on $[a, b]$? Try $f(x) = \cos x$ on $[0, \pi]$ with five equally spaced interpolation points.

4.53 Write a program to determine the natural cubic spline interpolate S_N of $f(x) = \cos x$ with interpolation points $\{(k-1)\pi/20\}_{k=1}^{21}$. Use VALSPL and DVALSP to compute

$$\max_{1 \le i \le 51} |\cos z_i - S_N(z_i)| = E_0$$

$$\max_{1 \le i \le 51} |\sin z_i + S'_N(z_i)| = E_1$$

where $z_i = (i-1)\pi/50$. Print E_0, E_1, and $\{m_i\}_{i=1}^{21}$.

4.54 Repeat exercise 4.53 for the complete cubic spline interpolate S_c of $f(x) = \cos x$. Compare the results with the bounds given in Theorem 4.7.

4.55 Repeat exercise 4.53 for the Lagrange cubic spline interpolate S_L of $f(x) = \cos x$. Compare the results with those of exercises 4.53 and 4.54.

4.56 For each of the cubic spline interpolates of exercises 4.53 to 4.55 perform spline quadrature (exercise 4.49) and discuss your results.

4.57 Linear search is not a very efficient procedure. Recode subroutine VALSPL to use binary search. Binary search is similar to the bisection method in that one chooses x_k, where $k = (i + j)/2$ or $k = (i + j + 1)/2$, to decide whether Z is in $[x_i, x_k]$ or in $[x_k, x_j]$. Test your revised code.

$$K = \frac{i + j}{2}$$

or

$$K = \frac{i + j + 1}{2}$$

4.58   Consider the experimental data

x_i	f_i	x_i	f_i	x_i	f_i
0	1.000	4.1	2.679	5.3	2.471
0.8	1.444	4.2	2.705	6.1	2.281
1.65	1.843	4.25	2.718	6.85	2.116
1.9	1.948	4.3	2.731	7.7	1.943
2.44	2.158	4.38	2.709	8.2	1.849
3.1	2.386	4.42	2.698	9.3	1.656
3.8	2.597	4.51	2.674	10.0	1.544

Plot the data and observe the sharp corner at 4.3. Use SPLINE and VALSPL to determine the cubic Lagrange spline interpolate for the data $\{(x_i, f_i)\}_{i=1}^{11}$, call it S_L^1. Determine another cubic Lagrange interpolate S_L^2 for the data $\{(x_i, f_i)\}_{i=11}^{21}$. Use an automatic plotter to plot S_L^1 and S_L^2 on the same graph. Discuss your results.

4.59   Use subroutines SPLINE and VALSPL to determine the cubic Lagrange spline interpolate for the data $\{(x_i, f_i)\}_{i=1}^{21}$ of exercise 4.58. Graph the resulting spline using an automatic plotter and compare with the graphs of S_L^1 and S_L^2. Discuss your results.

Basic splines

Consider the (common) situation where experimental data, say $\{y_i\}_{i=1}^{n+1}$, are obtained at equally spaced points $x_i = a + (i - 1)h$, $1 \le i \le n + 1$. Often it is the case that several experiments (for different experimental parameters) are performed with data gathered at the same points. For each set of data a polynomial or piecewise polynomial interpolate can be determined, thereby furnishing an approximation to the underlying function at intermediate points. If polynomial interpolation is used, then we recall (4.4), which gives the polynomial interpolate for data $\{y_i\}_{i=1}^{n+1}$ as

$$P_n(x) = \sum_{k=1}^{n+1} y_k L_k(x)$$

where L_k is a polynomial of degree n that depends only on $\{x_i\}_{i=1}^{n+1}$. Hence if $\{\hat{y}_i\}_{i=1}^{n+1}$ is another set of data at the same points, we immediately get the polynomial interpolate

$$\hat{P}_n(x) = \sum_{k=1}^{n+1} \hat{y}_k L_k(x)$$

The Lagrange representation for the polynomial interpolate has the desirable feature that the basic polynomials $\{L_k(x)\}_{k=1}^{n+1}$ can be determined once and for all and then used for any experiment at the same data points $\{x_i\}_{i=1}^{n+1}$.

In the preceding discussion we did not advocate the use of polynomial interpolation in Lagrange form. Our aim was to motivate the analogous question about cubic spline interpolation. Does there exist a set of **basic cubic splines**, say $\{\phi_k(x)\}_{k=0}^{m}$ for some integer m, such that every cubic spline interpolate based on $\{x_i\}_{i=1}^{n+1}$ has the representation

$$S(x) = \sum_{k=0}^{m} \alpha_k \phi_k(x) \qquad x \in [a, b] \tag{4.41}$$

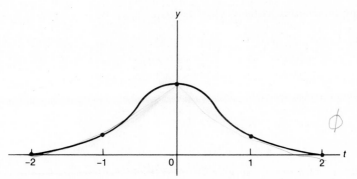

FIGURE 4.3

where $\{\alpha_k\}_{k=0}^{m}$ are constants? In (4.41) we want the basic cubic splines to depend only on the points $\{x_i\}_{i=1}^{n+1}$. Of course the answer to the question is yes. We refer to (4.41) as the **basic representation** of cubic splines.

Our approach is to simply define the ϕ_k's and show how the basic representation can be used. We remark that a basic representation also exists for unequal point spacing, but for simplicity we discuss only equal spacing. Consider the following piecewise cubic function:

$$\Phi(t) = \frac{1}{4} \begin{cases} (t+2)^3 & t \in [-2, -1] \\ 1 + 3(t+1) + 3(t+1)^2 - 3(t+1)^3 & t \in [-1, 0] \\ 1 + 3(1-t) + 3(1-t)^2 - 3(1-t)^3 & t \in [0, 1] \\ (2-t)^3 & t \in [1, 2] \\ 0 & |t| > 2 \end{cases}$$

whose graph is sketched in Figure 4.3. The following subroutine can be used to evaluate Φ at any point x. In writing the subroutine, we use the fact that Φ is an even function; that is, $\Phi(-x) = \Phi(x)$ for all x.

```
      SUBROUTINE BASIC(X,BVAL)
      IF(X.LT.2..AND.X.GT.-2.) GO TO 1
      BVAL=0.0
      RETURN
1     Z=-ABS(X)
      IF(Z.GT.-1.) GO TO 2
      BVAL=.25*(Z+2.)**3
      RETURN
2     CONTINUE
      Y=Z+1.
      BVAL=.25*(1.+3.*Y*(1.+Y*(1.-Y)))
      RETURN
      END
```

The value $\Phi(x)$ is returned in BVAL.

By differentiation of each polynomial piece we find

$$\Phi'(t) = \tfrac{3}{4} \begin{cases} (t+2)^2 & t \in (-2,-1) \\ 1 + 2(t+1) - 3(t+1)^2 & t \in (-1,0) \\ -1 - 2(1-t) + 3(1-t)^2 & t \in (0,1) \\ -(2-t)^2 & t \in (1,2) \\ 0 & |t| > 2 \end{cases}$$

$$\Phi''(t) = \tfrac{3}{4} \begin{cases} t+2 & t \in (-2,-1) \\ 1 - 3(t+1) & t \in (-1,0) \\ 1 - 3(1-t) & t \in (0,1) \\ 2 - t & t \in (1,2) \\ 0 & |t| > 2 \end{cases}$$

It is not difficult to see that both Φ' and Φ'' have common left-hand and right-hand limits at the knots -2, -1, 0, 1, and 2. For instance, Φ'' is continuous at -1, since

$$\lim_{t \to -1^+} \Phi''(t) = \lim_{t \to -1^+} \tfrac{3}{2}[1 - 3(t+1)] = \tfrac{3}{2} = \Phi''(-1)$$

and

$$\lim_{t \to -1^-} \Phi''(t) = \lim_{t \to -1^-} \tfrac{3}{2}(t+2) = \tfrac{3}{2} = \Phi''(-1)$$

It follows that Φ is a twice continuously differentiable piecewise cubic function; that is, Φ is a cubic spline. In Table 4.1 we give the values of Φ, Φ', and Φ'' at the knots.

For $i = 0, 1, \ldots, n+2$ we define the functions

$$\phi_i(x) = \Phi(h^{-1}(x - x_i)) \tag{4.42}$$

where, as before, $h = (b-a)/n$ and $x_i = a + (i-1)h$. Each ϕ_i is obtained from Φ by translating to the point x_i and then changing the scale of the independent variable (by a factor of h). The graph of $\phi_i(x)$ is given in Figure 4.4.

TABLE 4.1

	-2	-1	0	1	2
$\Phi(t)$	0	$\frac{1}{4}$	1	$\frac{1}{4}$	0
$\Phi'(t)$	0	$\frac{3}{4}$	0	$-\frac{3}{4}$	0
$\Phi''(t)$	0	$\frac{3}{2}$	-3	$\frac{3}{2}$	0

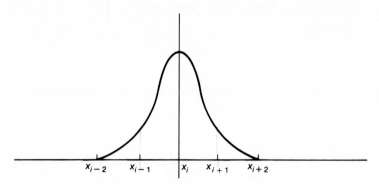

FIGURE 4.4

TABLE 4.2

$x =$	x_{i-2}	x_{i-1}	x_i	x_{i+1}	x_{i+2}
$\phi_i(x)$	0	$\dfrac{1}{4}$	1	$\dfrac{1}{4}$	0
$\phi_i'(x)$	0	$\dfrac{3}{4h}$	0	$-\dfrac{3}{4h}$	0
$\phi_i''(x)$	0	$\dfrac{3}{2h^2}$	$-\dfrac{3}{h^2}$	$\dfrac{3}{2h^2}$	0

An explicit formulation of ϕ_i is

$$\phi_i(x) = h^{-3} \begin{cases} (x - x_{i-2})^3 & x \in [x_{i-2}, x_{i-1}] \\ h^3 + 3h^2(x - x_{i-1}) \\ \quad + 3h(x - x_{i-1})^2 - 3(x - x_{i-1})^3 & x \in [x_{i-1}, x_i] \\ h^3 + 3h^2(x_{i+1} - x) \\ \quad + 3h(x_{i+1} - x)^2 - 3(x_{i+1} - x)^3 & x \in [x_i, x_{i+1}] \\ (x_{i+2} - x)^3 & x \in [x_{i+1}, x_{i+2}] \\ 0 & x \notin [x_{i+2}, x_{i+2}] \end{cases}$$

Derivatives of ϕ_i are easily computed by the chain rule. We have $\phi_i'(x) = h^{-1}\Phi'(h^{-1}(x - x_i))$ and $\phi_i''(x) = h^{-2}\Phi''(h^{-1}(x - x_i))$. Using these formulas for the derivatives together with Table 4.1, it is easy to evaluate ϕ_i, ϕ_i', and ϕ_i'' at the knots. Table 4.2 is useful when determining the coefficients $\{\alpha_k\}$ in the basic representation.

The following theorem provides the foundation for the basic representation of cubic splines.

THEOREM 4.8

Suppose S is any cubic spline function on $[a, b]$ with knots $\{x_i\}_{i=2}^n$, where

$x_i = a + (i-1)h$ and $h = (b-a)/n$. Then there exist unique coefficients $\{\alpha_k\}_{k=0}^{n+2}$ such that

$$S(x) = \sum_{k=0}^{n+2} \alpha_k \phi_k(x) \qquad x \in [a, b] \tag{4.43}$$

where ϕ_k is defined by (4.42).

The representation (4.43) is particularly appealing because each ϕ_k is given in terms of a single basic spline, namely Φ. In Theorem 4.8, S can be any cubic spline interpolate on $[a, b]$ with knots $\{x_i\}_{i=2}^n$. In particular suppose S_c is the complete cubic spline interpolate of f based on the points $\{x_i\}_{i=1}^{n+1}$. Then by (4.43) we have

$$S_c(x) = \sum_{k=0}^{n+2} \alpha_k \phi_k(x)$$

for some constants $\{\alpha_k\}_{k=0}^{n+2}$ that depend on $f'(a)$, $f'(b)$, and $\{f_i\}_{i=1}^{n+1}$. For each $i = 1, 2, \ldots, n+1$ we have

$$S_c(x_i) = \sum_{k=0}^{n+2} \alpha_k \phi_k(x_i) = f_i \tag{4.44}$$

and, since $\phi_k(x_i) = 0$ for $|k - i| > 2$, the interpolation equation (4.44) reduces to

$$\alpha_{i-1} \phi_{i-1}(x_i) + \alpha_i \phi_i(x_i) + \alpha_{i+1} \phi_{i+1}(x_i) = f_i$$

Use of Table 4.2 allows us to simplify these equations to

$$\alpha_{i-1} + 4\alpha_i + \alpha_{i+1} = 4f_i \qquad 1 \le i \le n+1 \tag{4.45}$$

For the endpoint condition $S_c'(a) = f'(a)$ we have

$$S_c'(a) = \sum_{k=0}^{n+2} \alpha_k \phi_k'(a) = f'(a)$$

and, since $\phi_k'(a) = 0$ for $k \ge 3$, we find (using Table 4.2 again)

$$\alpha_0 - \alpha_2 = -\tfrac{4}{3} h f'(a) \tag{4.46}$$

Similarly the condition $S_c'(b) = f'(b)$ gives the equation

$$-\alpha_n + \alpha_{n+2} = \tfrac{4}{3} h f'(b) \tag{4.47}$$

Equations (4.45), (4.46), and (4.47) give the following matrix problem for the determination of the coefficients in the basic representation of S_c:

$$
\begin{bmatrix}
1 & 0 & -1 & 0 \ldots\ldots\ldots\ldots\ldots 0 \\
1 & 4 & 1 & 0 \\
0 & 1 & 4 & 1 \\
\vdots & & \ddots & & & 0 \\
0 & & & 1 & 4 & 1 \\
0 \ldots\ldots\ldots\ldots\ldots 0 & -1 & 0 & 1
\end{bmatrix}
\begin{bmatrix}
\alpha_0 \\
\alpha_1 \\
\alpha_2 \\
\vdots \\
\alpha_{n+1} \\
\alpha_{n+2}
\end{bmatrix}
= 4
\begin{bmatrix}
-\dfrac{h}{3} f'(a) \\
f_1 \\
f_2 \\
\vdots \\
f_{n+1} \\
\dfrac{h}{3} f'(b)
\end{bmatrix}
$$

The coefficient matrix is almost tridiagonal—the (1, 3) and $(n + 3, n + 1)$ entries are not 0—and otherwise the system is at least as simple as (4.32), where S_c is determined by other means. In fact by a modification of the system we can produce a tridiagonal matrix problem. Specifically we add the next to last row to the last and add the second row to the first to get the equivalent system

$$
\begin{bmatrix}
2 & 4 & 0 \ldots\ldots\ldots\ldots .0 \\
1 & 4 & 1 \\
0 & 1 & 4 & 1 \\
\vdots & & \ddots & & 0 \\
0 & & & 1 & 4 & 1 \\
0 \ldots\ldots\ldots 0 & 4 & 2
\end{bmatrix}
\begin{bmatrix}
\alpha_0 \\
\alpha_1 \\
\alpha_2 \\
\vdots \\
\alpha_{n+1} \\
\alpha_{n+2}
\end{bmatrix}
= 4
\begin{bmatrix}
-\dfrac{h}{3} f'(a) + f_1 \\
f_1 \\
f_2 \\
\vdots \\
f_{n+1} \\
\dfrac{h}{3} f'(b) + f_{n+1}
\end{bmatrix}
\qquad (4.48)
$$

This tridiagonal system does not satisfy the hypotheses of Theorem 3.4, but the tridiagonal algorithm can be applied nonetheless (exercise 4.67). A similar analysis can be carried out for other cubic spline interpolates based on equally spaced points $\{x_i\}_{i=1}^{n+1}$. The only difference in the linear system of equations occurs in the endpoint conditions; that is, equations (4.45) hold for any such interpolate of f. The reader is asked to perform the analysis for S_L in exercise 4.65.

The subroutine CSPLIN uses the tridiagonal algorithm to solve (4.48) for the basic coefficients of the complete spline interpolate. The basic coefficients are returned in array B.

```
      SUBROUTINE CSPLIN(F,DF,AA,BB,N,B)
C  F,DF:EXTERNAL FUNCTION SUBPROGRAMS
C  AA,BB:ENDPOINTS OF INTERVAL WITH AA<BB
C  N:NUMBER OF INTERPOLATION POINTS=N+1
C  B:COEFFICIENTS FOR BASIC REPRESENTATION
      REAL B(43),U(43),L(42)
```

```
           H=(BB-AA)/FLOAT(N)
           N3=N+3
           N2=N+2
C   FORM THE RIGHT HAND SIDE
           B(1)=DF(AA)
           B(N3)=DF(BB)
           DO 1 J=2,N2
             X=AA+FLOAT(J-2)*H
             B(J)=4.*F(X)
     1     CONTINUE
           B(1)=-4.*H*B(1)/3.+B(2)
           B(N3)=4.*H*B(N3)/3.+B(N2)
C   TRIDIAGONAL ALGORITHM
           U(1)=2.
           L(1)=.5
           U(2)=2.
           DO 2 J=3,N2
             L(J-1)=1./U(J-1)
             U(J)=4.-L(J-1)
     2     CONTINUE
           L(N2)=4./U(N2)
           U(N3)=2.-L(N2)
C   FORWARD SUBSTITUTION
           DO 3 J=2,N3
             B(J)=B(J)-L(J-1)*B(J-1)
     3     CONTINUE
C   BACKSUBSTITUTION
           B(N3)=B(N3)/U(N3)
           N1=N+1
           DO 4 J=1,N1
             JN=N3-J
             B(JN)=(B(JN)-B(JN+1))/U(JN)
     4     CONTINUE
           B(1)=(B(1)-4.*B(2))/U(1)
C   ARRAY B NOW CONTAINS COEFFICIENTS
C   IN BASIC REPRESENTATION
           RETURN
           END
```

The subroutine SPLVAL is used to evaluate a spline given in the basic representation. In SPLVAL an index i (INDX in the program) is determined by linear search so that $[x_i, x_{i+1}]$ contains the point z where the spline is to be evaluated. Then $S(z)$ is given by

$$S(z) = \sum_{j=-1}^{2} \alpha_{i+j} \Phi(h^{-1}(z - x_{i+j}))$$

```
           SUBROUTINE SPLVAL(B,AA,BB,Z,N,SVAL)
C   B:COEFFICIENTS IN BASIC REPRESENTATION
C   AA,BB:INTERVAL ENDPOINTS WITH AA<BB
C   Z:POINT AT WHICH SPLINE IS EVALUATED
```

```
C  N: N+1=NUMBER OF INTERPOLATION POINTS
C  SVAL:VALUE OF SPLINE AT Z
       REAL B(43)
C  DETERMINE THE INDEX OF SUBINTERVAL
C  WHICH CONTAINS Z
       H=(BB-AA)/FLOAT(N)
       INDX=1
       NN=N-1
       DO 6 J=1,NN
         W=AA+H*FLOAT(J)
         IF(Z.LT.W) GO TO 7
         INDX=INDX+1
    6    CONTINUE
    7    T=(Z-AA)/H-FLOAT(INDX-1)
C  EVALUATE SPLINE AT Z
       SVAL=0.
       DO 8 K=1,4
         X=T+2.-FLOAT(K)
         CALL BASIC(X,BVAL)
         SVAL=SVAL+B(INDX+K-1)*BVAL
    8    CONTINUE
       RETURN
       END
```

EXAMPLE 4.13

We compute the basic representation of the complete spline interpolate for $f(x) = \log x$ on $[1, 3]$ based on 41 equally spaced points. Values of the spline and log function are printed at several intermediate points.

```
       EXTERNAL ALOG,DF                     Doesn't work
       REAL B(43),U(43),L(42),S(4)          Have to change
       N=40
       AA=1.0
       BB=3.0
       CALL CSPLIN(ALOG,DF,AA,BB,N,B)
       WRITE(6,99)
   99  FORMAT(' ',4X,'POINT',8X,'FUNCTION',8X,'SPLINE'//)
       DO 1 K=1,13
         Z=AA+FLOAT(K-1)*(BB-AA)/12.
         FZ=ALOG(Z)
         CALL SPLVAL(B,AA,BB,Z,N,SVAL)
         WRITE(6,88)Z,FZ,SVAL
    1    CONTINUE
   88  FORMAT(' ',2X,F9.6,2(3X,E14.7))
       STOP
       END
C
       REAL FUNCTION DF(X)
       DF=1./X
       RETURN
       END
```

POINT	FUNCTION	SPLINE
1.000000	0.0000000E 00	0.0000000E 00
1.166666	0.1541501E 00	0.1541498E 00
1.333333	0.2876818E 00	0.2876813E 00
1.500000	0.4054651E 00	0.4054642E 00
1.666666	0.5108252E 00	0.5108244E 00
1.833333	0.6061356E 00	0.6061346E 00
2.000000	0.6931471E 00	0.6931462E 00
2.166666	0.7731895E 00	0.7731887E 00
2.333333	0.8472977E 00	0.8472967E 00
2.500000	0.9162907E 00	0.9162902E 00
2.666666	0.9808290E 00	0.9808276E 00
2.833333	0.1041453E 01	0.1041452E 01
3.000000	0.1098612E 01	0.1098611E 01

Theorem 4.7 says that $S_c(x)$ is a fourth-order accurate approximation to $f(x)$. This can be illustrated computationally as follows. Let

$$E_N(f) = \max_{a \le x \le b} |S_c(x) - f(x)|$$

where S_c is the complete spline interpolate of f based on the $N + 1$ points $\{x_i = a + (i - 1)h\}_{i=1}^{N+1}$. If we treat the error bound of Theorem 4.7 as an equality, then

$$E_N(f) = Ch^4 \qquad (4.49)$$

where $C = 5M_4/384$ is independent of h. Based on (4.49) a simple calculation shows that the ratio

$$\frac{\log E_N - \log E_{2N}}{\log 2} \qquad (4.50)$$

should equal the exponent of h, that is, 4, in (4.49).

EXAMPLE 4.14

We use double-precision versions of subroutines CSPLIN, SPLVAL, and BASIC to estimate the ratio (4.50) for the function $f(x) = \sin 4x$ on $[0, \pi]$. The main program and output are

```
      DOUBLE PRECISION AA,BB,ERR,Z,DIFF,RATIO,P,FZ,SVAL,XLOG2
      DOUBLE PRECISION L(130),B(131),U(131),ERROR(8)
      EXTERNAL F,DF
      DOUBLE PRECISION F,DF
      DATA XLOG2/.6931472D0/
      WRITE(6,99)
99    FORMAT(' ',7X,'MAX ERROR',7X,' N '//)
      AA=0.D0
```

```
        BB=3.141592653589793D0
        DO 3 J=1,6
          N=4*2**(J-1)
          CALL CSPLIN(F,DF,AA,BB,N,B)
          ERR=0.D0
          DO 1 K=1,226
            Z=AA+DFLOAT(K-1)*(BB-AA)/225.D0
            FZ=F(Z)
            CALL SPLVAL(B,AA,BB,Z,N,SVAL)
            DIFF=DABS(FZ-SVAL)
            IF(DIFF.GT.ERR)ERR=DIFF
  1       CONTINUE
          ERROR(J)=ERR
          WRITE(6,88)ERR,N
  88      FORMAT(' ',4X,D14.7,5X,I3)
  3     CONTINUE
        WRITE(6,66)
  66    FORMAT(' ',2X,'THE ESTIMATES OF THE ORDER ARE :'/)
        DO 4 J=2,6
          RATIO=ERROR(J-1)/ERROR(J)
          P=DLOG(RATIO)/XLOG2
          N1=4*2**(J-2)
          N2=4*2**(J-1)
          WRITE(6,77) P,N1,N2
  77      FORMAT(' ',2X,'ORDER=',2X,F7.4,2X,'BASED ON N=',
     *I3,' AND N=',I3)
  4     CONTINUE
        STOP
        END
C
        DOUBLE PRECISION FUNCTION F(X)
        DOUBLE PRECISION X
        F=DSIN(4.D0*X)
        RETURN
        END
C
        DOUBLE PRECISION FUNCTION DF(X)
        DOUBLE PRECISION X
        DF=4.D0*DCOS(4.D0*X)
        RETURN
        END
```

MAX ERROR	N
0.8324204D 00	4
0.2281706D-01	8
0.1141807D-02	16
0.6305703D-04	32
0.3857110D-05	64
0.2421976D-06	128

THE ESTIMATES OF THE ORDER ARE :

ORDER=   5.1891   BASED ON N= 4 AND N= 8
ORDER=   4.3207   BASED ON N= 8 AND N= 16
ORDER=   4.1785   BASED ON N= 16 AND N= 32
ORDER=   4.0311   BASED ON N= 32 AND N= 64
ORDER=   3.9933   BASED ON N= 64 AND N=128

The values of E_N are decreasing, and the estimates of the order of accuracy (except for the first) are roughly equal to 4.

EXERCISES

4.60 Use (4.43) to show that

$$S'(x) = h^{-1} \sum_{i=0}^{n+2} \alpha_i \phi_i'(x)$$

and

$$S''(x) = h^{-2} \sum_{i=0}^{n+2} \alpha_i \phi_i''(x)$$

4.61 Use Table 4.2 and exercise 4.60 to show that

$$S''(x_i) = \tfrac{3}{2} h^{-2}(\alpha_{i+1} - 2\alpha_i + \alpha_{i-1})$$

4.62 Write a subroutine for the evaluation of Φ' called DBASIC(X,DVAL). Test your subroutine at the knots -2, -1, 0, 1, and 2 (Table 4.1) and at several intermediate points. Hint: Φ' is an odd function; that is, $\Phi'(-x) = -\Phi'(x)$ for all x.

4.63 Write a subroutine called DSPVAL (B,AA,BB,Z,N,DSVAL) for the evaluation of $S'(z)$ given by

$$S'(z) = h^{-1} \sum_{i=0}^{n+2} \alpha_i \phi_i'(z)$$

The parameters B, AA, BB, Z, and N are the same as in SPLVAL. Your subroutine should be similar to SPLVAL with the role of BASIC replaced by DBASIC.

4.64 Suppose S is a cubic spline interpolate for f given in basic form, that is, (4.43). Let $m_i = S''(x_i)$ for $1 \leq i \leq n + 1$.

(a) Show that $m_i = \tfrac{3}{2} h^{-2}(\alpha_{i-1} - 2\alpha_i + \alpha_{i+1})$, $1 \leq i \leq n + 1$.
(b) Use part (a) to show that

$$\frac{h^3}{4!} \sum_{i=1}^{n} (m_i + m_{i-1}) = \frac{h}{16} (\alpha_0 - \alpha_2 - \alpha_n + \alpha_{n+2})$$

(c) Use part (b) and exercise 4.49 to show that

$$\int_a^b S(x)\,dx = \frac{h}{2} \sum_{i=1}^{n} [f_i + f_{i+1}] - \frac{h}{16}(\alpha_0 - \alpha_2 - \alpha_n + \alpha_{n+2})$$

4.65 Modify subroutine CSPLIN to perform Lagrange spline interpolation. Call the subroutine LSPLIN(F,AA,BB,N,B), where AA, BB, N, and B are the same as in

CSPLIN. Instead of supplying an **EXTERNAL** function subprogram, let F be an array of dimension N + 3 that contains the function values, that is, $F(1) = f(x_1)$, $F(2) = f(x_2)$, and so on. Hint: In forming the right-hand-side vector, you need only replace DF(AA) by $P'_a(a)$ and DF(BB) by $P'_b(b)$. Test your subroutine by repeating exercise 4.55 for the basic representation.

4.66 For the natural cubic spline interpolate show that the endpoint conditions $S''(a) = S''(b) = 0$ give the equations

$$\alpha_0 - 2\alpha_1 + \alpha_2 = 0$$

$$\alpha_n - 2\alpha_{n+1} + \alpha_{n+2} = 0$$

These equations together with (4.45) determine S_N. Modify this system, using elementary row operations, to obtain a tridiagonal system for the determination of S_N. Does the tridiagonal algorithm apply to your system? Support your answer.

4.67 Show that the tridiagonal algorithm can be applied to (4.48); in other words, show that, when Algorithm 3.6 is applied to (4.48), $u_i \neq 0$ for each i.

4.68 Use subroutines CSPLIN, SPLVAL, DSPVAL, BASIC, and DBASIC to determine the complete cubic spline interpolate for $f(x) = \sinh x$ on $[0, 5]$ with $h = 0.1$. Print the values of S_c and S'_c at $z_j = \frac{1}{2}(x_j + x_{j+1})$, $1 \leq j \leq n$. Compare with the values of f and f'.

4.69 Use subroutines LSPLIN, SPLVAL, DSPVAL, BASIC, and DBASIC to determine the Lagrange cubic spline interpolate for $f(x) = \arctan x$ on $[0, 2]$ with $h = 0.05$. Compute and print

$$E_0 = \max_{1 \leq i \leq 101} |S_L(z_i) - f(z_i)|$$

$$E_1 = \max_{1 \leq i \leq 101} |S'_L(z_i) - f'(z_i)|$$

where $z_i = 0.02(i - 1)$.

In the next three exercises we develop the basic representation for **linear spline interpolation**.

4.70 Define $L(t)$ by

$$L(t) = \begin{cases} 1 + t & \text{if } -1 < t \leq 0 \\ 1 - t & \text{if } 0 < t < 1 \\ 0 & \text{if } |t| \geq 1 \end{cases}$$

and let $\psi_i(x) = L(h^{-1}(x - x_i))$, where as before $x_i = a + (i - 1)h$ and $h = (b - a)/n$. Sketch the graphs of L and ψ_i.

4.71 Show that

$$\psi_i(x_j) = \begin{cases} 1 & \text{if } i = j \\ 0 & \text{if } i \neq j \end{cases}$$

and hence conclude that

$$\psi(x) = \sum_{i=1}^{n+1} y_i \psi_i(x)$$

satisfies $\psi(x_j) = y_j$, $1 \leq j \leq n + 1$.

4.72 Show that the piecewise linear Lagrange interpolate of f at $\{x_i\}_{i=1}^{n+1}$ can be written as

$$g_1(x) = \sum_{i=1}^{n+1} f(x_i)\psi_i(x)$$

4.73 Determine the number of equally spaced points necessary to approximate $\sinh x$ on $[-2, 2]$ by a complete cubic spline interpolate with an error of no more than 10^{-6}. Hint: See Theorem 4.7. For this choice of h how well does S_c' approximate $\cosh x$ on $[-2, 2]$? Write a program to determine S_c and S_c'. Compute and print

$$E_0 = \max_{1 \le i \le 201} |S_c(z_i) - \sinh z_i|$$

$$E_1 = \max_{1 \le i \le 201} |S_c'(z_i) - \cosh z_i|$$

Where $z_i = -2 + .02(i - 1)$. Compare E_0 and E_1 with the bounds in Theorem 4.7.

Project 5

The purpose of this project is to test the conditioning and obtainable accuracy of Lagrange cubic spline interpolation when used for integration and differentiation. As a test function, use

$$g(x) = \begin{cases} 1 & \text{if } x = 0 \\ \dfrac{\sin x}{x} & \text{if } x \ne 0 \end{cases}$$

We have tabulated values of g (rounded to seven digits) in the following table.

x	$f(x)$	x	$f(x)$	x	$f(x)$
0	1.000000	1.05	0.8261174	2.1	0.4110521
0.15	0.9962542	1.2	0.7766992	2.25	0.3458103
0.3	0.9850674	1.35	0.7227580	2.4	0.2814430
0.45	0.9665901	1.5	0.6649967	2.55	0.2186995
0.6	0.9410708	1.65	0.6041606	2.7	0.1582888
0.75	0.9088517	1.8	0.5410265	2.85	0.1008695
0.9	0.8703632	1.95	0.4763896	3.00	0.0470400

Use subroutines LSPLIN (exercise 4.65), BASIC, SPLVAL, DSPVAL (exercise 4.63), and DBASIC (exercise 4.62) to compute

(a) S_L for g based on the tabular values; print

$$\max_{0 \le i \le 101} |S_L(z_i) - g(z_i)| = E_0$$

where $z_i = 0.03(i - 1)$.

(b) $\int_0^3 S_L(x)\, dx$, using the result of exercise 4.64, part c.

(c) $S_L'(x_i)$, $2 \le i \le 20$, where $x_i = 0.15(i - 1)$. Print these values and the estimated maximum error

$$E_1 = \max_{0 \le i \le 101} |S_L'(z_i) - g'(z_i)|$$

Now repeat steps (a), (b), and (c) with the tabular values rounded to six, four, and then two digits. Compare the results as the precision of the data is decreased. Discuss your results. Is differentiation or integration more sensitive to errors in the data?

NOTES AND COMMENTS

Sections 4.1 and 4.2

The development of polynomial interpolation and finite differences dates back to the early seventeenth century, when Harriot and Briggs (at Oxford) studied the problems of oceanic navigation. The book by Goldstine (1977) gives a detailed historical account of this and other subjects. The books by Davis (1963) and Cheney (1966) are excellent sources on the modern theory of interpolation and approximation.

Example 4.4 is essentially due to Runge and was published in the early 1900s. The Runge "phenomena" disappears when the interpolation points are suitably chosen (exercise 4.18). The book by Conte and de Boor (1980, pp. 235–245) gives a nice discussion of the Chebyshev points and their relevance to interpolation and approximation.

Section 4.3

Piecewise Lagrange interpolation provides the foundation for several of the most popular methods of numerical integration (Section 5.1). Spline functions were introduced by Schoenberg in the 1940s, and subsequently there has been an "explosion" of research on splines and their applications. In the book by de Boor (1978), an excellent source for further study, the practical applications (through FORTRAN programs) of piecewise polynomials are developed.

Splines have had a significant impact on the relatively new method of finite elements for the numerical solution of partial differential equations. A large portion of the recent research in numerical analysis is devoted to piecewise polynomial functions and finite element methods. The books by Strang and Fix (1973) and Prenter (1975) are among the most accessible to (advanced) undergraduate students.

NUMERICAL INTEGRATION AND DIFFERENTIATION

5

In this chapter we consider the numerical solution of the two fundamental problems of elementary calculus, namely, the integration and differentiation of a given function.

Recall the standard approach for evaluating definite integrals. Given f on $[a, b]$, the problem is to compute the definite integral

$$I(f) = \int_a^b f(x)\, dx \tag{5.1}$$

If F is an antiderivative for f, that is, $F'(x) = f(x)$, then the fundamental theorem of calculus says that $I(f) = F(b) - F(a)$. For instance, we have

$$\int_0^1 \frac{1}{1 + x^2}\, dx = \arctan 1 - \arctan 0 = \frac{\pi}{4}$$

since $\arctan x$ is an antiderivative for $1/(1 + x^2)$.

Unfortunately it is not always a simple matter to find an antiderivative. Consider the problem of determining the arc length of the graph of some function g on $[a, b]$. The arc length L is given by the integral

$$L = \int_a^b \{1 + [g'(x)]^2\}^{1/2}\, dx$$

This integral is typically difficult or impossible to evaluate by means of a simple antiderivative.

The numerical methods we consider for (5.1) consists in replacing f by a simple function p that approximates f on $[a, b]$. There results a formula of the form

$$I(p) \equiv I_N(f) = \sum_{i=1}^{N} w_i f(x_i)$$

which is called a numerical integration or **quadrature** formula. Here $x_1 < x_2 < \cdots < x_N$ are called the quadrature **nodes** and $\{w_i\}_{i=1}^{N}$ are the **weights**. The most common choice for p is a polynomial or piecewise polynomial interpolate of f. We begin by developing the trapezoidal and Simpson rules using piecewise Lagrange interpolates.

5.1 TWO BASIC INTEGRATION RULES

Suppose f is known at the points $a = x_1 < x_2 < \cdots < x_{N+1} = b$. We approximate $I(f)$ by $I(g_k)$, where g_k is the piecewise Lagrange interpolate of degree k for f. For $k = 1$ and $k = 2$, there results the trapezoidal and Simpson rules, respectively. For $k = 1$ we have

$$I(g_1) = \int_a^b g_1(x)\, dx = \sum_{i=1}^{N} \int_{x_i}^{x_{i+1}} g_1(x)\, dx$$

Since $g_1(x) = f_i + f[x_i, x_{i+1}](x - x_i)$ on $[x_i, x_{i+1}]$, it follows that

$$\int_{x_i}^{x_{i+1}} g_1(x)\, dx = \frac{h_i}{2}(f_i + f_{i+1})$$

where $h_i = x_{i+1} - x_i$ and $f_i = f(x_i)$. Therefore,

$$I(g_1) = \frac{1}{2} \sum_{i=1}^{N} h_i(f_i + f_{i+1})$$

gives the familiar **trapezoidal rule**, which we henceforth denote by $T_N(f)$. For equally spaced nodes, that is, $h_i = h = (b - a)/N$, the formula may be written more conveniently as

$$T_N(f) = \frac{h}{2}\left[f(a) + 2 \sum_{i=2}^{N} f_i + f(b) \right] \tag{5.2}$$

EXAMPLE 5.1

The following is a simple program for the trapezoidal rule (5.2) applied to $f(x) = e^{-x^2}$ on $[0, 1]$. Values for $T_N(f)$ are printed for $N = 10, 15, \ldots, 100$. The correct value is 0.74682413.

```
      EXTERNAL F
      A=0.
      B=1.
      WRITE(6,88)
   88 FORMAT(' ',3X,'N',4X,'TRAP RULE'/)
      DO 2 N=10,100,5
      CALL TRAPEZ(F,A,B,N,TNF)
```

```
        WRITE(6,101)N,TNF
  2     CONTINUE
101     FORMAT(' ',2X,I3,2X,F11.7)
        STOP
        END


        SUBROUTINE TRAPEZ(F,A,B,N,VALU)
C  F:EXTERNAL FUNCTION, THE INTEGRAND
C  A,B:ENDPOINTS OF INTERVAL WITH A<B
C  N:N+1=NUMBER OF FUNCTION VALUES
C  VALU:VALUE OF TRAP RULE
        H=(B-A)/FLOAT(N)
        VALU=.5*H*(F(A)+F(B))
        IF(N.EQ.1) RETURN
        SUM=0.0
        N1=N-1
        DO 3 J=1,N1
          X=A+H*FLOAT(J)
          SUM=SUM+F(X)
  3     CONTINUE
        VALU=VALU+H*SUM
        RETURN
        END
C
        REAL FUNCTION F(X)
        F=EXP(-X**2)
        RETURN
        END
```

N	TRAP RULE
10	0.7462101
15	0.7465512
20	0.7466703
25	0.7467239
30	0.7467526
35	0.7467701
40	0.7467813
45	0.7467883
50	0.7467936
55	0.7467976
60	0.7468013
65	0.7468035
70	0.7468059
75	0.7468066
80	0.7468083
85	0.7468093
90	0.7468107
95	0.7468112
100	0.7468117

When using the trapezoidal rule, it is advantageous to know how many nodes are sufficient for a desired accuracy. In what follows, we obtain a bound for the **discretization error** $I(f) - T_N(f)$. This bound can be used to determine N such that $I(f) - T_N(f)$ is within a prescribed error tolerance.

For each subinterval $[x_i, x_{i+1}]$, g_1 is the linear interpolate of f at x_i, x_{i+1}. Thus by (4.13)

$$f(x) = g_1(x) + f[x_i, x_{i+1}, x](x - x_i)(x - x_{i+1})$$

and it follows that

$$\int_{x_i}^{x_{i+1}} f(x)\, dx = \int_{x_i}^{x_{i+1}} g_1(x)\, dx + \int_{x_i}^{x_{i+1}} f[x_i, x_{i+1}, x](x - x_i)(x - x_{i+1})\, dx$$

Since $(x - x_i)(x - x_{i+1}) \le 0$ on $[x_i, x_{n+1}]$, an application of the integral mean value theorem gives

$$\int_{x_i}^{x_{i+1}} f[x_i, x_{i+1}, x](x - x_i)(x - x_{i+1})\, dx$$

$$= f[x_i, x_{i+1}, \xi_i] \int_{x_i}^{x_{i+1}} (x - x_i)(x - x_{i+1})\, dx$$

for some point $\xi_i \in (x_i, x_{i+1})$. A simple calculation shows that

$$\int_{x_i}^{x_{i+1}} (x - x_i)(x - x_{i+1})\, dx = -\frac{h_i^3}{6}$$

and hence

$$\int_{x_i}^{x_{i+1}} f(x)\, dx = \frac{h_i}{2}(f_i + f_{i+1}) - \frac{h_i^3}{6} f[x_i, x_{i+1}, \xi_i]$$

If f'' is continuous on $[a, b]$, then by (4.14)

$$f[x_i, x_{i+1}, \xi_i] = \frac{f''(\gamma_i)}{2} \qquad \text{some } \gamma_i \in [x_i, x_{i+1}]$$

so that

$$\int_{x_i}^{x_{i+1}} f(x)\, dx = \frac{h_i}{2}(f_i + f_{i+1}) - \frac{h_i^3}{12} f''(\gamma_i) \tag{5.3}$$

By summation of (5.3) we obtain

$$I(f) = T_N(f) - \frac{1}{12} \sum_{i=1}^{N} h_i^3 f''(\gamma_i)$$

Moreover, if M_2 is a bound for f'' on $[a, b]$ and $h = \max_{1 \le i \le N} h_i$, then it follows that

$$|I(f) - T_N(f)| \le \frac{1}{12} \sum_{i=1}^{N} h_i^3 |f''(\gamma_i)|$$

$$\le \frac{h^3}{12} M_2 N \tag{5.4}$$

This is the required bound for the discretization error.

We summarize our error analysis for $T_N(f)$ with equal spacing (the case of primary interest) in the following theorem.

THEOREM 5.1

Assume that f is twice continuously differentiable on $[a, b]$ and M_2 is a bound on $[a, b]$ for f''. Let $x_i = a + (i - 1)h$, where $h = (b - a)/N$. Then

$$|T_N(f) - I(f)| \le \frac{M_2 h^2 (b - a)}{12} \tag{5.5}$$

This error bound shows that, as $h \to 0$, the error tends to zero like a multiple of h^2. Consequently we say that the trapezoidal rule is **second-order accurate** or, equivalently, is a **second-order method**. The bound (5.5) indicates that a large value of M_2, that is, large second derivative, dictates a relatively small choice of h whereas a relatively large choice of h is feasible when M_2 is small.

EXAMPLE 5.2

We demonstrate how the trapezoidal rule could be used to generate a table of values of $g(x) = \mathrm{erf}(x)$ on $[0, 1]$. It is required that each tabular entry be accurate to within 10^{-4}. Recall from Example 4.5 that

$$\mathrm{erf}(x) = \frac{2}{\sqrt{\pi}} \int_0^x e^{-t^2}\, dt$$

Let $f(t) = 2e^{-t^2}/\sqrt{\pi}$; then for any $b \in (0, 1]$ we approximate $I(f) = \int_0^b f(t)\, dt = \mathrm{erf}(b)$ by

$$T_N(f) = \frac{h}{2}\left[f(0) + 2 \sum_{i=2}^N f_i + f(b) \right]$$

where $h = b/N$. In order to determine a bound for f'' on $[0, b]$, we calculate

$$f''(t) = \frac{4}{\sqrt{\pi}}\, e^{-t^2}(2t^2 - 1)$$

$$f'''(t) = -\frac{8te^{-t^2}}{\sqrt{\pi}}\, (4t^2 - 3)$$

The only critical point of f'' (that is, where $f''' = 0$) in $(0, 1)$ is $\sqrt{3}/2$. Therefore,

$$\max_{0 \le t \le 1} |f''(t)| = \max\left\{ |f''(0)|, \left|f''\left(\frac{\sqrt{3}}{2}\right)\right|, |f''(1)| \right\}$$

$$= |f''(0)| = \frac{4}{\sqrt{\pi}}$$

For the required accuracy it suffices to choose h such that $bh^2/3\sqrt{\pi} \le 10^{-4}$. Solving for $h\, (= b/N)$ gives $h \le 0.02306/\sqrt{b}$ or, equivalently, $N \ge 43.366 b^{3/2}$.

Next we consider approximating $I(f)$ by $I(g_2)$, thereby obtaining son's rule. We assume that N is an even integer and, for odd i, $x_{i+1} = (x_i + x_{i+2})/2$. Recalling that $x_3, x_5, \ldots, x_{N-1}$ are the knots for g_2, we write

$$I(g_2) = \int_a^b g_2(x)\, dx = \sum_{\substack{i=1 \\ i \text{ odd}}}^{N-1} \int_{x_i}^{x_{i+2}} g_2(x)\, dx$$

By exercise 4.41

$$\int_{x_i}^{x_{i+2}} g_2(x)\, dx = \frac{h_i}{3}(f_i + 4f_{i+1} + f_{i+2})$$

where we have used the fact that x_{i+1} is the midpoint of $[x_i, x_{i+2}]$, that is, $h_i = h_{i+1}$. Summation of the last equation gives **Simpson's rule**:

$$S_N(f) = I(g_2) = \frac{h_1}{3}(f_1 + 4f_2 + f_3) + \frac{h_3}{3}(f_3 + 4f_4 + f_5) + \cdots$$
$$+ \frac{h_{N-1}}{3}(f_{N-1} + 4f_N + f_{N+1}) \qquad (5.6)$$

Since (in this section) the case of equal node spacing is of primary interest, we write Simpson's rule in a more convenient form. For $h_i = h$ we notice that each odd subscripted term $3 \le i \le N-1$ in (5.6) appears twice in the sum. Therefore we regroup the terms as follows:

$$S_N(f) = \frac{h}{3}(f_1 + f_{N+1}) + \frac{2h}{3}\sum_{i=1}^{N/2-1} f_{2i+1} + \frac{4h}{3}\sum_{i=1}^{N/2} f_{2i}$$

or, equivalently,

$$S_N(f) = \frac{h}{3}\left\{ f(a) + 2\sum_{i=1}^{N/2-1} f_{2i+1} + 4\sum_{i=1}^{N/2} f_{2i} + f(b) \right\} \qquad (5.7)$$

This form of Simpson's rule is used in the FORTRAN subroutine SIMPSN, given in the next example.

EXAMPLE 5.3

Recall that

$$\int_0^1 \frac{1}{1+x^2}\, dx = \frac{\pi}{4} = 0.78539816\ldots$$

We compute S_N for even N between 2 and 34. The value of S_{12} is accurate to six decimal digits, and, because of roundoff, the succeeding values of S_N offer no improvement in accuracy.

```
EXTERNAL F
A=0.
B=1.
WRITE(6,77)
```

```
   77  FORMAT(' ',3X,'N',4X,'SIMP RULE'/)
       DO 5 N=2,34,2
         CALL SIMPSN(F,A,B,N,SIMP)
         WRITE(6,100)N,SIMP
    5  CONTINUE
  100  FORMAT(' ',2X,I2,2X,F11.7)
       STOP
       END
C
       SUBROUTINE SIMPSN(F,A,B,N,SIMP)
C  F:EXTERNAL FUNCTION,THE INTEGRAND
C  A,B:ENDPOINTS OF INTERVAL WITH A<B
C  N:N+1=NUMBER OF EQUALLY SPACED POINTS
C  N: MUST BE AN EVEN INTEGER
       H=(B-A)/FLOAT(N)
       SIMP=H*(F(A)+F(B)+4.*F(B-H))/3.
       IF(N.EQ.2) RETURN
       K=N/2
       K1=K-1
       DO 2 J=1,K1
         X=A+H*FLOAT(2*J-1)
         SIMP=SIMP+H*(4.*F(X)+2.*F(X+H))/3.
    2  CONTINUE
       RETURN
       END
C
       REAL FUNCTION F(X)
       F=1./(1.+X**2)
       RETURN
       END
```

N	SIMP RULE
2	0.7833331
4	0.7853918
6	0.7853979
8	0.7853980
10	0.7853979
12	0.7853981
14	0.7853979
16	0.7853979
18	0.7853981
20	0.7853979
22	0.7853980
24	0.7853981
26	0.7853979
28	0.7853979
30	0.7853977
32	0.7853976
34	0.7853978

The discretization error for Simpson's rule cannot be obtained in the same straightforward manner as that given for $T_N(f)$. In Section 5.2 we shall demonstrate, by means of the Euler-Maclaurin formula, that

$$\int_{x_i}^{x_{i+2}} f(x) \, dx = \int_{x_i}^{x_{i+2}} g_2(x) \, dx - \frac{h_i^5}{90} f^{(4)}(\mu_i) \tag{5.8}$$

where $f^{(4)}$ is assumed to be continuous on $[a, b]$ and μ_i is some point in $[x_i, x_{i+2}]$. Equation (5.8) shows that one should have h_i relatively small on that portion of $[a, b]$ where $f^{(4)}$ is large, whereas h_i could be relatively large where $f^{(4)}$ is small. We explore the possibility of nonuniform node spacing more fully in Section 5.3. For the time being we examine only uniform node spacing. Thus for $h_i = h$ we sum (5.8) to give

$$I(f) = S_N(f) - \frac{h^5}{90} \sum_{\substack{i=1 \\ i \text{ odd}}}^{N-1} f^{(4)}(\mu_i)$$

from which the following theorem is a direct consequence.

THEOREM 5.2

Suppose f is four times continuously differentiable on $[a, b]$. Let N be even and $h = (b - a)/N$. Then $S_N(f)$, given by (5.7), satisfies

$$|I(f) - S_N(f)| \leq \frac{h^4(b - a)}{180} M_4 \tag{5.9}$$

where $M_4 \geq \max_{a \leq x \leq b} |f^{(4)}(x)|$.

In comparison with the trapezoidal rule, it is seen that Simpson's rule is **fourth-order accurate**. For this reason Simpson's rule has been and continues to be a popular method for numerical integration. However, the reader should not conclude that $S_N(f)$ is always a better choice than $T_N(f)$.

EXAMPLE 5.4

The **sine integral function** arises frequently in scientific applications. This function is defined by

$$\text{Si}(x) = \int_0^x \frac{\sin t}{t} \, dt$$

Values of $\text{Si}(x)$ are tabulated in many mathematical handbooks. We use (5.9) to estimate the number of points required in Simpson's rule to find $\text{Si}(1)$ within 10^{-6}. The fourth derivative of $f(t) = (\sin t)/t$ could be calculated directly; however, we can estimate $f^{(4)}(t)$ using the Taylor series expansion for $\sin t$ about 0. We have

$$\sin t = \sum_{k=0}^{\infty} (-1)^k \frac{t^{2k+1}}{(2k+1)!} \qquad \text{all } t$$

and hence

$$f(t) = \sum_{k=0}^{\infty} (-1)^k \frac{t^{2k}}{(2k+1)!}$$

By termwise differentiation we find

$$f^{(4)}(t) = \sum_{k=2}^{\infty} (-1)^k \frac{2k(2k-1)(2k-2)(2k-3)}{(2k+1)!} t^{2k-4}$$

which is an alternating series for $t > 0$. It follows that (exercise 1.48)

$$|f^{(4)}(t) - \tfrac{1}{5}| \leq \frac{t^2}{14}$$

and hence

$$\max_{0 \leq t \leq 1} |f^{(4)}(t)| \leq \tfrac{1}{5} + \tfrac{1}{14} \leq 0.272 = M_4$$

By (5.9) we choose h so that $h^4 M_4/180 \leq 10^{-6}$ or, equivalently,

$$h^4 \leq \tfrac{1.8}{0.272} 10^{-4}$$

Solving for h $(= 1/N)$ gives $h \leq 0.1603$, and thus $N \geq 6.234$. Since N must be even, the choice $N = 8$ is sufficient.

EXERCISES

5.1   Use subroutine TRAPEZ to compute $T_N(f)$ for each of the following integrals. Let $E_N(f) = I(f) - T_N(f)$. Print $T_N(f)$ and $E_N(f)$ for $N = 8, 16, 32, 64, 128, 256$ in each case.

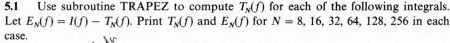

(a) $\displaystyle\int_0^1 \frac{e^x}{1 + e^x} dx$ (b) $\displaystyle\int_{0.01}^2 x \log x \, dx$ (c) $\displaystyle\int_{-1}^{1.99} \frac{1}{x^2 - 4} dx$

(d) $\displaystyle\int_0^\pi \cos 2x \, dx$ (e) $\displaystyle\int_1^4 \frac{1}{x} dx$ (f) $\displaystyle\int_0^{\pi/4} e^{-x} \sin x \, dx$

The values of these integrals, rounded to six decimal digits, are (a) 0.620115, (b) 0.386550, (c) -1.77189, (d) 0.0, (e) 1.38629, and (f) 0.177603.

5.2   Compute $T_N(f)$ for $f(x) = \sin x$ on $[0, \pi]$ with $N = 10, 20, 40, 80$. Print $E_N(f) = |T_N(f) - I(f)|$ and the ratios E_{20}/E_{10}, E_{40}/E_{20}, E_{80}/E_{40}. According to Theorem 5.1, these ratios should approximately equal $\tfrac{1}{4}$. Verify this! Compare the computed ratios with that predicted by Theorem 5.1.

5.3   Repeat exercise 5.2 for $f(x) = 2 + \sin x$ on $[-\pi/4, \pi/4]$.

5.4   Compute the arc length of the graph of $f(x) = \tan x$ on $[0, \pi/4]$ using the trapezoidal rule. Use a program that computes $T_{8N}(f)$ for $N = 1, 2, \ldots$ and terminate the computation when

$$|T_{8N}(f) - T_{8(N-1)}(f)| \leq 10^{-6}$$

5.5   Determine the number of nodes such that $T_N(f)$ approximates $\int_0^2 e^x \sin x \, dx$ with an absolute error of no more than 10^{-6}. Hint: Use Theorem 5.1.

5.6   Compute $S_N(f)$ for $N = 4, 8, 16, 32, 64$, where $f(x) = 1/(x \log x)$ on $[2, 6]$. Print $E_N(f) = |I(f) - S_N(f)|$ and the ratios E_8/E_4, E_{16}/E_8, E_{32}/E_{16}, E_{64}/E_{32}. According to Theorem 5.2, these ratios should be approximately equal to $\frac{1}{16}$. Discuss your results. Hint: $\log (\log x)$ is an antiderivative for f.

5.7   Repeat exercise 5.4 using Simpson's rule.

5.8   Repeat exercise 5.1 using subroutine SIMPSN.

5.9   In Section 5.2 it is demonstrated that for even N

$$I(f) - T_N(f) \simeq \tfrac{1}{3}[T_N(f) - T_{N/2}(f)]$$

and

$$I(f) - S_N(f) \simeq \tfrac{1}{15}[S_N(f) - S_{N/2}(f)]$$

These give a posteriori error estimates for the trapezoidal and Simpson rules, respectively. Verify these estimates empirically by performing the following experiment. Let $f(x) = 4/(1 + x^2)$ on $[0, 1]$ and for $N = 4, 8, 16, 32, 64$ print the actual and estimated errors for each method. Discuss your results. Note: $I(f) = \pi$.

5.10   For even N show that $S_N(f) = [4T_N(f) - T_{N/2}(f)]/3$.

5.11   Determine the largest value of k such that
(a) $I(x^n) = T_N(x^n)$ $\quad n = 0, 1, \ldots, k$
(b) $I(x^n) = S_N(x^n)$ $\quad n = 0, 1, \ldots, k$
Hint: Use Theorems 5.1 and 5.2.

*5.12   Suppose $I_N(f)$ is a numerical method for $I(f)$ of the form

$$I_N(f) = w_1 f(x_1) + w_2 f(x_2) + \cdots + w_N f(x_N)$$

and $I_N(x^n) = I(x^n)$ for $n = 0, 1, 2, \ldots, k$. Does it follow that $I_N(p) = I(p)$ for all polynomials p of degree $\leq k$? Justify your answer.

5.13   Suppose $I_N(f) = \sum_{i=1}^N w_i f(x_i)$ is a numerical method for $\int_0^1 f(x) \, dx$. Show that this method, when applied to $\int_a^b g(t) \, dt$, is

$$(b - a) \sum_{i=1}^N w_i g(t_i)$$

where $t_i = a + x_i(b - a)$. Hint: Use the substitution $t = a + x(b - a)$ in the integral $\int_a^b g(t) \, dt$ and apply I_N to the result.

5.14   Let $I_3(f) = w_1 f(a) + w_2 f((a + b)/2) + w_3 f(b)$. Determine w_1, w_2, and w_3 such that

$$I_3(x^n) = I(x^n) \quad n = 0, 1, 2$$

Hint: Consider $[a, b] = [0, 1]$ and use the result of exercise 5.13. Is it also true that $I_3(x^3) = I(x^3)$? Does I_3 give a familiar integration rule?

5.15   Let $I_4(f) = w_1 f'(a) + w_2 f(a) + w_3 f(b) + w_4 f'(b)$ and determine $\{w_i\}_{i=1}^4$ such that

$$I_4(x^n) = I(x^n) \quad n = 0, 1, 2, 3$$

Hint: Consider $[a, b] = [0, 1]$ and use the result of exercise 5.13.

5.16 Let $I(f) = \int_0^1 f(x)\,dx$. For $h \neq \frac{1}{2}$ find $\{w_i\}_{i=1}^4$ such that $I_4(f) = w_1 f(0) + w_2 f(h) + w_3 f(1-h) + w_4 f(1)$ satisfies

$$I_4(x^n) = I(x^n) \qquad n = 0, 1, 2, 3$$

Note: The resulting rule is called the Simpson 3/8 rule.

5.2 THE EULER-MACLAURIN FORMULA AND APPLICATIONS

In this section we give the famous Euler-Maclaurin formula and discuss some of its many applications in numerical computation. This formula enables us to obtain a formula for the discretization error in Simpson's rule [see (5.8)] as well as a new integration formula called the corrected trapezoidal rule and provides the foundation for a powerful integration technique called Romberg integration.

Derivation of the formula

In order to derive the Euler-Maclaurin formula, we introduce a sequence of polynomials with some special properties. Specifically we show how to construct polynomials $q_0(t), \ldots, q_k(t)$ such that

(i)   degree of $q_j = j$

(ii) $q'_{j+1}(t) = q_j(t) \qquad j \geq 0$ (5.10)

(iii) $\int_0^1 q_j(t)\,dt = 0 \qquad j \geq 1$

First note that conditions (ii) and (iii) give

$$\int_0^1 q_j(t)\,dt = \int_0^1 q'_{j+1}(t)\,dt = q_{j+1}(1) - q_{j+1}(0) = 0 \qquad j \geq 1$$

Hence we require $q_{j+1}(1) = q_{j+1}(0)$ for $j \geq 1$. Let $q_0(t) = 1$; then by (ii) we have $q_1(t) = t + c_1$ for some constant c_1. Since $\int_0^1 q_1(t)\,dt = \frac{1}{2} + c_1$, condition (iii) requires that $c_1 = -\frac{1}{2}$ and hence $q_1(t) = t - \frac{1}{2}$. To determine $q_2(t)$, we use (ii) again to get $q_2(t) = t^2/2 - t/2 + c_2$ for some constant c_2. Condition (iii) requires that $0 = \int_0^1 q_2(t)\,dt = -\frac{1}{12} + c_2$, and thus $c_2 = \frac{1}{12}$ or $q_2(t) = t^2/2 - t/2 + \frac{1}{12}$. Continuing in this manner, we find:

$$q_0 = 1$$

$$q_1(t) = t - \frac{1}{2}$$

$$q_2(t) = \frac{t^2}{2} - \frac{t}{2} + \frac{1}{12}$$

$$q_3(t) = \frac{t^3}{3!} - \frac{t^2}{4} + \frac{t}{12}$$

$$q_4(t) = \frac{t^4}{4!} - \frac{t^3}{12} + \frac{t^2}{24} - \frac{1}{720}$$

$$q_5(t) = \frac{t^5}{5!} - \frac{t^4}{48} + \frac{t^3}{72} - \frac{t}{720}$$

and so on

It is possible to show that $q_j(0) = 0$ for all odd integers $j \geq 3$, that is,

$$q_{2i+1}(1) = q_{2i+1}(0) = 0 \qquad i \geq 1$$

Now suppose that F is a given function with k continuous derivatives on $[0, 1]$. Using integration by parts gives

$$\int_0^1 F(t)\, dt = \int_0^1 F(t)q_0(t)\, dt = F(t)q_1(t)\,|_0^1 - \int_0^1 F'(t)q_1(t)\, dt$$

$$= \tfrac{1}{2}[F(0) + F(1)] - \int_0^1 F'(t)q_1(t)\, dt$$

By applying integration by parts again, we find

$$\int_0^1 F'(t)q_1(t)\, dt = F'(t)q_2(t)\,|_0^1 - \int_0^1 F''(t)q_2(t)\, dt$$

$$= q_2(0)[F'(1) - F'(0)] - \int_0^1 F''(t)q_2(t)\, dt$$

so that

$$\int_0^1 F(t)\, dt = \tfrac{1}{2}[F(0) + F(1)] - q_2(0)[F'(1) - F'(0)] + \int_0^1 F''(t)q_2(t)\, dt$$

By continuing this process of repeated integration by parts, we obtain as our result the following theorem.

THEOREM 5.3

Suppose F has k continuous derivatives on $[0, 1]$ and $\{q_j\}_{j=0}^k$ is the sequence of polynomials satisfying (5.10) with $q_0(t) = 1$. Then

$$\int_0^1 F(t)\, dt = \tfrac{1}{2}[F(0) + F(1)] + \sum_{j=1}^{k-1} (-1)^j q_{j+1}(0)[F^{(j)}(1) - F^{(j)}(0)]$$

$$+ (-1)^k \int_0^1 F^{(k)}(t)q_k(t)\, dt$$

The reader should keep in mind that in this theorem $q_3(0) = q_5(0) = \cdots = 0$, that is, $q_{2i+1}(0) = 0$ for $i \geq 1$.

The Euler-Maclaurin formula follows from Theorem 5.3 by a particular choice of F, defined as follows. Suppose f is a given function on $[a, b]$ having k continuous derivatives. Let $x_i = a + (i - 1)h$ and $h = (b - a)/N$. Define F on $[0, 1]$ by

$$F(t) = f(x_i + th)$$

Then $F(0) = f(x_i) = f_i$ and $F(1) = f_{i+1}$. Moreover, by the chain rule

$$F'(t) = hf'(x_i + th)$$

$$F''(t) = h^2 f''(x_i + th)$$

$$\cdots\cdots\cdots\cdots\cdots\cdots\cdots$$

$$F^{(k)}(t) = h^k f^{(k)}(x_i + th)$$

To evaluate $\int_0^1 F(t)\, dt$ in terms of f, we make the substitution $x = x_i + th$:

$$\int_0^1 F(t)\, dt = \int_0^1 f(x_i + th)\, dt = \frac{1}{h} \int_{x_i}^{x_{i+1}} f(x)\, dx$$

Therefore Theorem 5.3, with this choice of F, gives

$$\int_{x_i}^{x_{i+1}} f(x)\, dx = \frac{h}{2}(f_i + f_{i+1}) + \sum_{j=1}^{k-1} (-1)^j h^{j+1} q_{j+1}(0)[f^{(j)}(x_{i+1}) - f^{(j)}(x_i)]$$

$$+ (-1)^k h^{k+1} \int_0^1 f^{(k)}(x_i + th) q_k(t)\, dt \qquad (5.11)$$

By summation of the last equation from $i = 1$ to $i = N$, we obtain the following Euler-Maclaurin formula.

THEOREM 5.4

Suppose f has k, $k \geq 2$, continuous derivatives on $[a, b]$; then

$$I(f) = T_N(f) - h^2 q_2(0)[f'(b) - f'(a)] - h^4 q_4(0)[f'''(b) - f'''(a)] + \cdots$$

$$+ (-1)^{k-1} h^k q_k(0)[f^{(k-1)}(b) - f^{(k-1)}(a)] + R_k^N(f) \qquad (5.12)$$

where

$$R_k^N(f) = (-1)^k h^{k+1} \sum_{i=1}^{N} \int_0^1 f^{(k)}(x_i + th) q_k(t)\, dt$$

In (5.12) the first few values of $q_{2j}(0)$ are $q_2(0) = \frac{1}{12}$, $q_4(0) = -\frac{1}{720}$, and $q_6(0) = \frac{1}{30,240}$. Recall that $q_3(0) = q_5(0) = \cdots = 0$.

EXERCISES

5.17 Use (5.10) to derive $q_3(t)$, $q_4(t)$, and $q_5(t)$.

5.18 Verify the formula

$$\int_0^1 F(t)\, dt = \frac{1}{h} \int_{x_i}^{x_{i+1}} f(x)\, dx$$

for $f(x) = x^2$. Hint: Recall that $F(t) = f(x_i + th)$.

5.19 Use integration by parts to show that for $j \geq 1$

$$\int_0^1 F^{(j-1)}(t) q_{j-1}(t)\, dt = q_j(0)[F^{(j-1)}(1) - F^{(j-1)}(0)] - \int_0^1 F^{(j)}(t) q_j(t)\, dt$$

Hint: Use (5.10).

5.20   (a) Let $[a, b] = [0, N]$ and $h = 1$ in (5.12) to obtain

$$\sum_{n=0}^{N} f(n) = \int_0^N f(x)\, dx + \tfrac{1}{2}[f(0) + f(N)] + \tfrac{1}{12}[f'(N) - f'(0)]$$

$$- \tfrac{1}{720}[f'''(N) - f'''(0)] - R_4^N(f)$$

(b) Take $f(x) = x$ in part (a) and conclude that

$$\sum_{n=1}^{N} n = \frac{N(N + 1)}{2}$$

(c) Take $f(x) = x^2$ in part (a) and conclude that

$$\sum_{n=1}^{N} n^2 = \frac{(2N + 1)(N + 1)N}{6}$$

(d) Find a formula for $\sum_{n=1}^{N} n^3$.

Some consequences of the formula

The Euler-Maclaurin formula has numerous applications, of which we examine only three. It immediately comes to mind that, for $f'(a)$ and $f'(b)$ known, the **corrected trapezoidal rule**

$$CT_N(f) = T_N(f) - \frac{h^2}{12}\, [f'(b) - f'(a)]$$

gives an improved approximation (fourth-order) to $I(f)$. Indeed the Euler-Maclaurin formula says that

$$I(f) - CT_N(f) = -\frac{h^4}{720}\, [f'''(b) - f'''(a)] + R_4^N(f) \tag{5.13}$$

Note also that even if $f'(a)$ and $f'(b)$ are not known but satisfy $f'(a) = f'(b)$, then $CT_N(f) = T_N(f)$. For such functions $T_N(f)$ is fourth-order accurate and compares favorably with Simpson's rule.

In order to determine an error bound for $CT_N(f)$, we could use (5.13); however, it is more convenient to use (5.11) with $k = 4$. In fact, by summing (5.11) with $k = 4$, we obtain (5.13). Thus we write

$$\int_{x_i}^{x_{i+1}} f(x)\, dx = \frac{h}{2}\,(f_i + f_{i+1}) - \frac{h^2}{12}\,(f'_{i+1} - f'_i) + \frac{h^4}{720}\,(f'''_{i+1} - f'''_i)$$

$$+ h^5 \int_0^1 f^{(4)}(x_i + th)q_4(t)\, dt$$

It is not difficult to show (using the substitution $x = x_i + th$) that

$$\int_0^1 f^{(4)}(x_i + th)\, dt = h^{-1} \int_{x_i}^{x_{i+1}} f^{(4)}(x)\, dx = h^{-1}(f'''_{i+1} - f'''_i)$$

and hence

$$\int_{x_i}^{x_{i+1}} f(x)\, dx = \frac{h}{2}(f_{i+1} + f_i) - \frac{h^2}{12}(f'_{i+1} - f'_i) + h^5 \int_0^1 f^{(4)}(x_i + th)Q_4(t)\, dt$$

where $Q_4(t) = q_4(t) + \frac{1}{720} = t^2(t-1)^2/24$. Since Q_4 is of one sign on $[0, 1]$, an application of the integral mean value theorem gives

$$\int_{x_i}^{x_{i+1}} f(x)\, dx = \frac{h}{2}\,(f_i + f_{i+1}) - \frac{h^2}{12}\,(f'_{i+1} - f'_i) + \frac{h^5}{720}\,f^{(4)}(x_i + \xi h) \qquad (5.14)$$

where we have used the fact that $\int_0^1 Q_4(t) = \frac{1}{720}$. Summation of (5.14) gives

$$I(f) = CT_N(f) + \frac{h^5}{720}\sum_{i=1}^N f^{(4)}(x_i + \xi h)$$

Finally, if M_4 is a bound for $f^{(4)}$ on $[a, b]$, we obtain the error bound

$$|I(f) - CT_N(f)| \le \frac{h^5}{720}\sum_{i=1}^N |f^{(4)}(x_i + \xi h)|$$

$$\le \frac{h^5}{720}\,M_4 N = \frac{h^4(b-a)}{720}\,M_4 \qquad (5.15)$$

EXAMPLE 5.5

To illustrate the improved accuracy of $CT_N(f)$ over $T_N(f)$, we consider the results of Example 5.1. For $N = 20$ (that is, $h = 0.05$) we computed $T_{20}(f) = 0.7466703$, whereas $I(f) = 0.74682413\ldots$. We have $f'(x) = -2xe^{-x^2}$ and hence $f'(0) = 0, f'(1) = -2/e$. Therefore we obtain

$$CT_{20}(f) = T_{20}(f) + \frac{(0.05)^2}{6e} = 0.7468235\ldots$$

which is a significant improvement over $T_{20}(f)$.

A second application of Theorem 5.4 is a formula for the error in Simpson's rule. The Euler-Maclaurin formula does not involve $S_N(f)$ directly. However, by exercise 5.10, Simpson's rule results from a linear combination of trapezoidal rules—one for h and the other for $2h$. Actually we want to derive a local error formula [see (5.8)] for Simpson's rule. Toward this end we sum (5.14) from $i = j$ to $i = j + 1$ and obtain

$$\int_{x_j}^{x_{j+2}} f(x)\, dx = \frac{h}{2}\,(f_j + 2f_{j+1} + f_{j+2}) - \frac{h^2}{12}\,(f'_{j+2} - f'_j)$$

$$+ \frac{h^5}{720}\,[f^{(4)}(x_j + \xi h) + f^{(4)}(x_{j+1} + \gamma h)] \qquad (5.16)$$

for some $\xi, \gamma \in [0, 1]$. Next we write (5.14) with $i = j$ and h replaced by $2h$, that is, for some $\alpha \in [0, 1]$:

$$\int_{x_j}^{x_{j+2}} f(x)\, dx = h(f_{j+2} + f_j) - \frac{h^2}{3}\,(f'_{j+2} - f'_j) + \frac{2h^5}{45}\,f^{(4)}(x_j + 2\alpha h) \qquad (5.17)$$

By subtracting $\frac{1}{3}$ times equation (5.17) from $\frac{4}{3}$ times (5.16), we obtain

$$\int_{x_j}^{x_{j+2}} f(x)\, dx = \frac{h}{3}\,(f_j + 4f_{j+1} + f_{j+2}) + \frac{h^5}{90}\left[\frac{f^{(4)}(\xi_j) - 8f^{(4)}(\alpha_j) + f^{(4)}(\gamma_j)}{6}\right]$$

where $\xi_j = x_j + \xi h$, $\alpha_j = x_j + 2\alpha h$, and $\gamma_j = x_j + \gamma h$. For h sufficiently small the continuity of $f^{(4)}$ implies that

$$f^{(4)}(\xi_j) \simeq f^{(4)}(\alpha_j) \simeq f^{(4)}(\gamma_j)$$

and hence we obtain the estimate

$$\int_{x_j}^{x_{j+2}} f(x)\, dx \simeq \frac{h}{3}(f_j + 4f_{j+1} + f_{j+2}) - \frac{h^5}{90} f^{(4)}(\alpha_j)$$

which is essentially (5.8). This local error estimate for Simpson's rule is particularly important when nonuniform mode spacing is used. It plays an essential role in the method of adaptive Simpson quadrature.

EXERCISES

5.21 Given an example of a function f for which $CT_N(f) = T_N(f)$.

5.22 Modify subroutine TRAPEZ to compute the corrected trapezoidal rule. Call the new subroutine CTRAP(F,DF,A,B,N,VALU), where DF is an external subprogram for $f'(x)$. All the remaining parameters in the list should have the same function as in TRAPEZ. Use CTRAP to compute $CT_{100}(f)$ for $f(x) = e^{-x^2}$ on $[0, 1]$.

5.23 Determine the largest value of k for which

$$CT_N(x^n) = I(x^n) \qquad n = 0, 1, \ldots, k$$

5.24 Suppose $f'(a)$, $f'(b)$, $f'''(a)$, and $f'''(b)$ are known. Give an integration rule that is sixth-order accurate.

5.25 Assume that $f'(a)$ and $f'(b)$ are readily available. On the basis of Theorem 5.2 and equation (5.15), which of the two rules $S_N(f)$ and $CT_N(f)$ is preferable? Why?

5.26 Compare the number of function evaluations in $T_N(f)$, $CT_N(f)$, and $S_N(f)$.

5.27 Determine the number of points required to approximate $I(f) = \int_{0.1}^{10} (1/x)\, dx$ within 10^{-6} using (a) the trapezoidal rule, (b) the corrected trapezoidal rule, and (c) Simpson's rule.

5.28 Compute as well as you can the value of

$$\int_{0.1}^{2} \frac{(\log x)^2}{x}\, dx$$

by $I_N(f)$, $S_N(f)$, and $CT_N(f)$. The correct value is

$$I(f) = \tfrac{1}{3}(\log x)^3 |_{0.1}^{2} = 4.18036540$$

Compare the number of function evaluations in each case. Suggestion: Proceed as in exercise 5.4.

5.29 Discuss the relative advantages and disadvantages of $T_N(f)$, $S_N(f)$, and $CT_N(f)$. Pay particular attention to accuracy, efficiency, and the class of integrands (that is, how smooth) for which these rules can be expected to perform well.

Romberg integration

Suppose that, for some N, $T_N(f)$ has been computed. By the Euler-Maclaurin formula we know (for smooth f) that the error $I(f) - T_N(f)$ has an expansion in terms of even powers of h. Can this knowledge be used to obtain an improved approximation to $I(f)$?

The procedure of Romberg integration provides a systematic method for obtaining successively higher order approximations to $I(f)$. The basis for this procedure is (5.12). Suppose k is even in (5.12), say $k = 2p$. Then for some coefficients $K_2, K_4, \ldots, K_{2p}$,

$$I(f) - T_N(f) = K_2 h^2 + \cdots + K_{2p} h^{2p} + R_{2p}^N(f) \qquad (5.18)$$

In fact the coefficients are given by

$$K_{2j} = -q_{2j}(0)[f^{(2j-1)}(b) - f^{(2j-1)}(a)] \qquad 1 \le j \le p$$

but this particular formula for K_{2j} is of little importance. As far as Romberg integration is concerned, the important feature of (5.18) is that the coefficients are independent of h. For simplicity of notation we write (5.18) as

$$I(f) - T_N(f) = K_2 h^2 + K_4 h^4 + O(h^6) \qquad (5.19)$$

where the remaining terms in (5.18) have been lumped together in $O(h^6)$ (big oh of h^6). A quantity $g(h)$ is said to be $O(h^p)$, that is, $g(h) = O(h^p)$, if $g(h)/h^p$ is bounded for $|h|$ sufficiently small. Loosely speaking, this means that $g(h)$ tends to zero at least as fast as does h^p for $h \to 0$.

If N is even, we may replace h by $2h$ in (5.19) to yield

$$I(f) - T_{N/2}(f) = 2^2 K_2 h^2 + 2^4 K_4 h^4 + O(h^6) \qquad (5.20)$$

On comparing (5.19) with (5.20), observe that 4 times (5.19) has the same leading term, namely $4K_2 h^2$, as does (5.20). Therefore if equation (5.20) is subtracted from 4 times (5.19), the leading error terms cancel and there results

$$3I(f) - [4T_N(f) - T_{N/2}(f)] = -12K_4 h^4 + O(h^6)$$

Dividing through by 3 gives

$$I(f) - \tfrac{1}{3}[4T_N(f) - T_{N/2}(f)] = K_4^1 h^4 + O(h^6) \qquad (5.21)$$

where K_{2j}^1, $2 \le j \le p$, denotes the new coefficients in (5.21). Equation (5.21) says that the linear combination of $T_N(f)$ and $T_{N/2}(f)$ given by

$$T_{N/2}^1(f) = T_N(f) + \frac{T_N(f) - T_{N/2}(f)}{3}$$

is a fourth-order accurate approximation to $I(f)$. If, in computing $T_N(f)$, we have the foresight to choose N even and to store the values $\{f_j\}_{j=1}^{N+1}$ in memory, then $T_{N/2}^1(f)$ can be computed with little additional computational effort. Note also that $T_{N/2}^1(f) = S_N(f)$ (exercise 5.10).

p. 200

EXAMPLE 5.6

In Example 5.1 we computed

$$T_{10}(f) = 0.7462101 \qquad \text{and} \qquad T_{20}(f) = 0.7466703$$

whereas $I(f) = 0.74682413\ldots$. The approximations $T_{10}(f)$ and $T_{20}(f)$ are accurate only to three decimal digits; however,

$$T^1_{10}(f) = \frac{4(0.7466703) - 0.7462101}{3} = 0.7468237$$

is in error by only 0.43×10^{-6}.

The process described for eliminating the leading error term in the expansion for $T_N(f)$ can also be applied to $T^1_{N/2}(f)$. The starting point is equation (5.21), which we rewrite as

$$I(f) - T^1_{N/2}(f) = K^1_4 h^4 + K^1_6 h^6 + O(h^8) \tag{5.22}$$

By analogy with our previous discussion we write down (5.22) with h replaced by $2h$ (assuming N is divisible by 4) and subtract the resultant equation from 16 times (5.22). This gives

$$15I(f) - [16T^1_{N/2}(f) - T^1_{N/4}(f)] = -48K^1_6 h^6 + O(h^8)$$

or, equivalently,

$$I(f) - \tfrac{1}{15}[16T^1_{N/2}(f) - T^1_{N/4}(f)] = K^2_6 h^6 + O(h^8)$$

where, as before, K^2_{2j}, $3 \leq j \leq p$, denotes the new coefficients. Therefore

$$T^2_{N/4}(f) = T^1_{N/2}(f) + \frac{T^1_{N/2}(f) - T^1_{N/4}(f)}{15}$$

provides a sixth-order approximation to $I(f)$. Notice that the computation of $T^2_{N/4}(f)$ requires $T_N(f)$, $T_{N/2}(f)$, and $T_{N/4}(f)$.

This process can be continued to produce higher order approximations provided the error expansion at each stage consists of known, even powers of h. In what follows, we construct a so-called **T table** for the computation of $T^1_{N/2}$, $T^2_{N/4}$, $T^3_{N/8}$, and so on. This table is similar to a divided difference table, as will soon be apparent. The format of the table is:

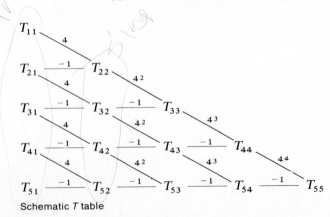

Schematic *T* table

The first column entries are values of the trapezoidal rule and are given by

$$T_{j1} = T_{N/2^{j-1}}(f) \qquad j \geq 1$$

The second column entries are values of Simpson's rule and are given by

$$T_{j2} = T^1_{N/2^{j-1}}(f) = S_{N/2^{j-2}}(f) \qquad j \geq 2$$

and in general

$$T_{ij} = T^{j-1}_{N/2^{i-1}}(f) \qquad i \geq j \geq 1$$

where, by convention, $T^0_{N/2^{i-1}}(f) \equiv T_{N/2^{i-1}}(f)$.

Given the first column entries, the remaining tabular values can be computed column by column using the general formula

$$T_{ij} = T_{i-1,j-1} + \frac{T_{i-1,j-1} - T_{i,j-1}}{4^{j-1} - 1} \qquad i \geq j \geq 2 \tag{5.23}$$

The denominator constants in (5.23) are given in the T table. For instance (5.23) gives

$$T_{22} = T_{11} + \frac{T_{11} - T_{21}}{3} = T^1_{N/2}(f)$$

$$T_{53} = T_{42} + \frac{T_{42} - T_{52}}{15} = T^2_{N/16}(f)$$

The diagonal entries in the T table are of primary interest and represent $T_N(f)$, $T^1_{N/2}(f)$, $T^2_{N/4}(f)$, $T^3_{N/8}(f)$, and so forth.

In order to compute the entries in a T table, it is convenient to modify subroutine TRAPEZ (Example 5.1) to accept an array of function values FX of dimension N1 = N + 1 and an index K. The index K specifies the number of values in FX to be used in a trapezoidal rule. Specifically the subroutine returns $T_{N/2^K}(f)$; that is, $K = 0$ gives $T_N(f)$, $K = 1$ gives $T_{N/2}(f)$, and so on. It is required that $N/2^K$ be an even integer. The following is a double-precision version of the modified subroutine:

```
      SUBROUTINE TRAP(FX,N1,H,K,VALU)
C  FX:ARRAY OF INTEGRAND VALUES
C  N1:NUMBER OF VALUES IN FX
C  H:EQUAL POINT SPACING FOR VALUES IN FX
C  K:SPECIFIES TRAP RULE FOR N/2**K,WHERE N=N1-1
C  THE RATIO (N1-1)/2**K MUST BE AN EVEN INTEGER
C  VALU: VALUE OF SPECIFIED TRAP RULE
      DOUBLE PRECISION FX(N1),DELTA,VALU,SUM,H
      M=2**K
      DELTA=H*DFLOAT(M)
      VALU=.5D0*DELTA*(FX(1)+FX(N1))
      N=N1-1
      IF(M.EQ.N) RETURN
      NK=M+1
      SUM=0.D0
      DO 1 J=NK,N,M
        SUM=SUM+FX(J)
```

```
1     CONTINUE
      VALU=VALU+DELTA*SUM
      RETURN
      END
```

EXAMPLE 5.7

In the following program we compute the first four rows of the T table for

$$I(f) = \int_0^\pi \sin x \, dx$$

The first column entries are $\{T_{64/2^j}\}_{j=0}^3$, and we observe that $T_{44} = T_8^3(f)$ gives the correct value of $I(f)$ to 12 decimal digits.

```
      DOUBLE PRECISION FX(65),T(4,4),A,B,H,X,VALU,P
      A=0.D0
      B=3.141592653589793D0
      N=64
      H=(B-A)/DFLOAT(N)
      N1=N+1
      DO 1 K=1,N1
        X=A+H*DFLOAT(K-1)
        FX(K)=DSIN(X)          ← TEST IF ITS ZERO
1     CONTINUE                   DEFINE IT IF IT IS
      DO 2 J=1,4                 EQUAL TO ZERO
        JJ=J-1
        CALL TRAP(FX,N1,H,JJ,VALU)
        T(J,1)=VALU
2     CONTINUE
      WRITE(6,88)
88    FORMAT(' ',/3X,'THE T TABLE IS:'//)
      DO 3 J=2,4
        J1=J-1
        P=DFLOAT(4**J1)
        DO 3 I=J,4
          T(I,J)=(P*T(I-1,J-1)-T(I,J-1))/(P-1.D0)
3     CONTINUE
      DO 4 I=1,4
        WRITE(6,100) (T(I,J),J=1,I)
4     CONTINUE
100   FORMAT(' ',4(1X,F15.11))
      STOP
      END
```

THE T TABLE IS:

```
1.99959838864
1.99839336097    2.00000006453
1.99357034377    2.00000103337    1.99999999994
1.97423160195    2.00001659105    1.99999999619    2.00000000000
```

For a suitable N our analysis says that the diagonal entries in the T table T_N, $T_{N/2}^1$, $T_{N/4}^2$, ... should be second-, fourth-, sixth-, ... order approximations to $I(f)$ if f is smooth. This does not necessarily mean that $T_{N/2}^1(f)$ is more accurate than $T_N(f)$ or that $T_{N/4}^2(f)$ is more accurate than $T_{N/2}^1(f)$. In order to understand the possible difficulty, let us examine $T_{N/2}^1(f)$ more closely. This term was defined in such a way as to eliminate the leading term in the error expansion (5.18). However, it may happen that the leading term $K_2 h^2$ is not dominant; that is, it is possible that, for the range of h being used, the coefficients $\{K_{2j}\}$ are such that the error in $T_N(f)$ is being controlled by $K_4 h^4 + O(h^6)$ rather than by $K_2 h^2$. In this case $T_{N/2}^1(f)$ is unlikely to be more accurate than $T_N(f)$. How can we be assured that $T_{N/2}^1(f)$ is more accurate than $T_N(f)$? Fortunately this question can be answered by performing a simple computational test. We derive this test only for $T_{N/2}^1(f) = T_{22}$ and simply state the test for $T_{N/2^{j-1}}^{j-1}(f) = T_{jj}, j \geq 2$.

If the leading term in (5.18) is also the dominant one, then

$$I(f) - T_N(f) \simeq K_2 h^2$$

and

$$I(f) - T_{N/2}(f) \simeq 4K_2 h^2$$

Therefore, upon subtraction, we obtain the estimate

$$T_{N/2}(f) - T_N(f) \simeq -3K_2 h^2 \tag{5.24}$$

Replacing h by $2h$ in (5.24) gives

$$T_{N/4}(f) - T_{N/2}(f) \simeq -3K_2(2h)^2$$

and hence the ratio

$$R_1 = \frac{T_{N/4}(f) - T_{N/2}(f)}{T_{N/2}(f) - T_N(f)} = \frac{T_{31} - T_{21}}{T_{21} - T_{11}}$$

should be $\simeq 4$. Consequently if the computed ratio R_1 is $\simeq 4$, we have assurance that $K_2 h^2$ is dominant and that $T_{N/2}^1(f) = T_{22}$ is an improvement over $T_N(f) = T_{11}$.

In general the test is as follows. For $j \geq 1$, compute the ratio

$$R_j = \frac{T_{j+2, j} - T_{j+1, j}}{T_{j+1, j} - T_{jj}}$$

If $R_j \simeq 4^j$, then accept $T_{j+1, j+1}$ as an improvement over T_{jj}.

EXAMPLE 5.8

For $f(x) = (\sin x)/x$ on $[0, \pi]$, we compute the diagonal T table entries $\{T_{jj}\}_{j=1}^6$ for $N = 64$. Computations are carried out in double precision so as to minimize roundoff error. Examination of the ratios $\{R_j\}_{j=1}^3$ indicates that T_{22} is an improvement over T_{11} and that T_{33} is more accurate than T_{22}.

```
TN(F)= 0.18518731351100 01    N= 64
TN(F)= 0.1851681372414D 01    N= 32
TN(F)= 0.1850914140365D 01    N= 16
TN(F)= 0.1847842306445D 01    N=  8
TN(F)= 0.1835508123281D 01    N=  4
TN(F)= 0.1785398163397D 01    N=  2
```

THE DIAG ENTRIES IN T TABLE ARE:

```
      1.85187313511
      1.85193705601
      1.85193705198
      1.85193705198
      1.85193705198
      1.85193705198
```

THE RATIOS ARE:

```
      4.0009
     16.0304
     64.6077
```

In addition to testing that $T_{j+1,\,j+1}$ is more accurate than T_{jj}, it is also possible to use the T table to estimate the error $I(f) - T_{j+1,\,j+1}$. We give the details of the derivation only for $j = 1$. Write down (5.22) with h replaced by $2h$ and subtract (5.22) from the resultant equation to yield

$$T^1_{N/2}(f) - T^1_{N/4}(f) = 15K^1_4 h^4 + O(h^6)$$

Therefore by neglecting the higher order terms, we obtain

$$K^1_4 h^4 \simeq \tfrac{1}{15}[T^1_{N/2}(f) - T^1_{N/4}(f)]$$

Substitution of this estimate for $K^1_4 h^4$ into (5.22) gives the error estimate

$$I(f) - T^1_{N/2}(f) \simeq \tfrac{1}{15}[T^1_{N/2}(f) - T^1_{N/4}(f)]$$

or, equivalently,

$$I(f) - T_{22} \simeq \tfrac{1}{15}[T_{22} - T_{32}]$$

In general the a posteriori error estimate is

$$I(f) - T_{jj} \simeq \frac{T_{jj} - T_{j+1,\,j}}{4^j - 1} \qquad j \geq 1 \tag{5.25}$$

EXAMPLE 5.9

For the diagonal entries in the T table of Example 5.8, we compute the error estimate (5.25) for $1 \leq j \leq 5$:

```
FOR T(1,1)   ERREST= 0.6392089872D-04
FOR T(2,2)   ERREST=-0.4028082623D-08
.FOR T(3,3)   ERREST= 0.1942777509D-11
FOR T(4,4)   ERREST=-0.4629847703D-14
FOR T(5,5)   ERREST= 0.4948794714D-16
```

In applying the method of Romberg integration, it is essential that the integrand be smooth. Theorem 5.4 says that the error expansion (5.18) is valid provided f is $2p$ times continuously differentiable. If $p \leq 1$, no improvement should be expected in T_{22}; if $p \leq 2$, no improvement is expected in T_{33}; and so on. When applicable, the method of Romberg integration can be an efficient and highly accurate technique.

It is also possible to show that the method of Romberg integration is **well-conditioned**. A small perturbation of the function (integrand) values does not cause a substantial change in the diagonal T table entries. In fact we have the following theorem.

THEOREM 5.5

Suppose $\tilde{f}_i = f(x_i) + \varepsilon_i$ and let $\varepsilon = \max_{1 \leq i \leq N+1} |\varepsilon_i|$. Let $\{T_{jj}\}$ and $\{\tilde{T}_{jj}\}$ denote the diagonal T table entries based on function values $\{f(x_i)\}_{i=1}^{N+1}$ and $\{\tilde{f}_i\}_{i=1}^{N+1}$, respectively. Then for each $j \geq 1$,

$$|T_{jj} - \tilde{T}_{jj}| \leq (b - a)\varepsilon$$

For $j = 1, 2$, this theorem says that the trapezoidal and Simpson rules satisfy

$$|T_N(f) - T_N(\tilde{f})| \leq (b - a)\varepsilon$$

and

$$|S_N(f) - S_N(\tilde{f})| \leq (b - a)\varepsilon$$

Notice that the bound $(b - a)\varepsilon$ is independent of h, the node spacing. The quantity ε_i could represent roundoff or measurement error in $f(x_i)$; however, Theorem 5.5 does not take into account any additional roundoff error due to floating point arithmetic.

In computing Simpson's rule, the overall error (ignoring that due to floating point arithmetic) consists of the discretization error plus the error due to inexact data. Thus the error is

$$I(f) - S_N(\tilde{f}) = I(f) - S_N(f) + S_N(f) - S_N(\tilde{f})$$

and consequently, by Theorems 5.2 and 5.5,

$$|I(f) - S_N(\tilde{f})| \leq |I(f) - S_N(f)| + |S_N(f) - S_N(\tilde{f})|$$

$$\leq \frac{M_4(b - a)h^4}{180} + \varepsilon(b - a)$$

If $\tilde{f}$ represents the value of f obtained from a function subprogram, then it may be reasonable to suppose that $\tilde{f}_i = f_i(1 + \rho_i)$, where $|\rho_i| \le u$, where u is the machine unit. Then the bound is

$$|I(f) - S_N(\tilde{f})| \le \frac{M_4(b-a)h^4}{180} + M_0 u(b-a)$$

where M_0 is a bound for f on $[a, b]$. A similar error bound holds for $I(f) - T_N(\tilde{f})$ using Theorems 5.1 and 5.5.

EXERCISES

5.30 Use a program like that in Example 5.7 to compute the T table for $f(x) = x^2 e^x$ on $[0, 2]$. Use single precision and print the entire table for $N = 32$. Print the absolute error in each diagonal entry.

5.31 Repeat the previous exercise in double precision and discuss the effect of round-off error.

5.32 Use hand calculation to compute the ratios $\{R_j\}_{j=1}^3$ from the output in exercises 5.30 and 5.31. Discuss your results.

5.33 Suppose $\{T_{128/2^j}(f)\}_{j=0}^6$ have been computed. If these values are used for Romberg integration, how many rows and columns are in the T table? How many function evaluations are required in the Romberg method?

5.34 Consider a T table with five rows and columns. Explain why the calculation of R_4 and R_5 is impossible. Hint: See exercise 5.32.

5.35 If $T_{33} = T_{43}$, what is the value of T_{44}?

5.36 Show that $T_{33} = (64T_{11} - 20T_{21} + T_{31})/45$.

*5.37 For $j \ge 2$, show that $T_{jj} = \alpha_1 T_{11} + \alpha_2 T_{21} + \cdots + \alpha_j T_{j1}$ where the coefficients satisfy $\alpha_1 + \alpha_2 + \cdots + \alpha_j = 1$.

5.38 Use the program of Example 5.7 as the basis for writing a subroutine to implement Romberg integration. Call the subroutine ROM(FX,M,H,T,R) and design it to accept an array FX of M function values with point spacing H. The subroutine returns the T table in T and ratios in R. Make certain that $M - 1 = 2^k$ (for some k), in which case T is $(k - 1) \times (k - 1)$ and R is of dimension $k - 3$. Test your subroutine on the problem of exercise 5.30.

5.39 From the following data calculate $\int_{2.1}^{3.6} f(x)\, dx$ as best you can using subroutine ROM. Which diagonal T table entry do you consider most accurate? Why? The tabular values are correct to the digits shown.

x_i	$f(x_i)$	x_i	$f(x_i)$
2.1	1.688249	2.9	2.241773
2.2	1.764731	3.00	2.302585
2.3	1.838961	3.1	2.361797
2.4	1.911023	3.2	2.419479
2.5	1.981001	3.3	2.475698
2.6	2.048982	3.4	2.530517
2.7	2.115050	3.5	2.583998
2.8	2.179287	3.6	2.636196

5.40 In the text we observed that the T table could be generated column by column using (5.23). Show that (5.23) can also be used to generate the table row by row. Give an algorithm for this approach. Is there any advantage to this method? Hint: If R_j is not approximately equal to 4^j, then the computation of T_{ii}, $i \geq j + 1$, is unwarranted.

5.41 Consider the integral $I = \int_0^1 (\sin x / \sqrt{x})\, dx$.

(a) Would it be appropriate to apply Romberg integration to I? Hint: Is the integrand smooth?

(b) Make the change of variables $x = t^2$ to show that $I = 2 \int_0^1 \sin t^2\, dt$.

Why is this integral better suited for Romberg integration? Use subroutine ROM to compute I and estimate the error.

5.42 For which of the following integrals would you expect Romberg integration to perform poorly? Why?

(a) $\displaystyle\int_0^2 \frac{1}{x - \sqrt{3}}\, dx$

(b) $\displaystyle\int_0^1 f(x)\, dx$ where $f(x) = \begin{cases} \dfrac{\cos x - 1}{x} & \text{if } x \neq 0 \\ 0 & \text{if } x = 0 \end{cases}$

(c) $\displaystyle\int_{-1}^2 e^{-x}|\sin \pi x|\, dx$

(d) $\displaystyle\int_{-1}^1 \sqrt{1 - x^2}\, dx$

(e) $\displaystyle\int_0^1 f(x)\, dx$ where $f(x) = \begin{cases} \dfrac{e^x - 1}{x} & \text{if } x \neq 0 \\ 0 & \text{if } x = 0 \end{cases}$

(f) $\displaystyle\int_0^1 x^{1/3}(1 + e^{-x})\, dx$

5.43 Show that Simpson's rule $S_N(f)$ satisfies (for smooth f) an error formula of the form

$$I(f) - S_N(f) = C_4 h^4 + C_6 h^6 + O(h^8)$$

Hint: $T_{22} = S_N(f)$.

5.44 Suppose L is approximated by $L(h)$ and the error formula

$$L - L(h) = C_1 h + C_3 h^3 \qquad \text{as } h \to 0$$

is valid. Determine a linear combination of $L(h)$ and $L(2h)$ that is third-order accurate. Can you find a linear combination of $L(h)$, $L(2h)$, and $L(4h)$ that gives L exactly?

5.45 Make the substitution $x = t^3$ in the integral of exercise 5.42, part (f), to obtain a more suitable integral for Romberg integration. Evaluate as best you can by using subroutine ROM.

5.46 Show that

$$|T_N(f) - T_N(\tilde{f})| \leq \varepsilon(b - a)$$

where ε, f, and $\tilde{f}$ are as in Theorem 5.5. Hint: Use (5.2) for $T_N(f)$ and $T_N(\tilde{f})$. Estimate the difference. Conclude that

$$|I(f) - T_N(\tilde{f})| \le \frac{M_2 h^2 (b - a)}{12} + \varepsilon(b - a)$$

by using Theorem 5.1.

5.3 ADAPTIVE SIMPSON QUADRATURE

All of the methods in Sections 5.1 and 5.2 for approximating $I(f)$ are implemented using uniformly spaced nodes. These methods are easy to program and perform quite satisfactorily if the integrand f is reasonably well-behaved. However, if f exhibits bad behavior on some portion (possibly small) of the interval, then such methods are not very efficient. This is because, for a given error tolerance, the choice of h is dictated by the worst behavior of f on the interval. All of these methods satisfy an error bound of the form

$$|I(f) - I_N(f)| \le Ch^p M_p$$

where I_N denotes the particular method and M_p is a bound for $f^{(p)}$. The bad behavior of f typically manifests itself in a large value for M_p. Consequently h must be relatively small in order to satisfy the error tolerance. It is important to realize that M_p is large even if $|f^{(p)}(x)|$ is large only on a small portion of the interval of integration.

EXAMPLE 5.10

We use Simpson's rule for the integral

$$I(f) = \int_{0.01}^{4} \frac{1}{x} \, dx$$

and require the absolute error to be no more than 10^{-2}. We consider two ways of applying Simpson's rule in order to illustrate the shortcomings of uniform node spacing. In the first way we determine h for $S_N(f)$ applied to $I(f)$, and in the second we split $I(f)$ into

$$I(f) = \int_{0.01}^{0.5} \frac{1}{x} \, dx + \int_{0.5}^{4} \frac{1}{x} \, dx = I^1(f) + I^2(f)$$

and apply Simpson's rule separately to each integral.
    Since $f^{(4)}(x) = 4!/x^5$, we find

$$M_4 = \max_{0.01 \le x \le 4} |f^{(4)}(x)| = |f^{(4)}(0.01)| = 24 \times 10^{10}$$

$$M_4^1 = \max_{0.01 \le x \le 0.5} |f^{(4)}(x)| = M_4$$

$$M_4^2 = \max_{0.5 \le x \le 4} |f^{(4)}(x)| = |f^{(4)}(0.5)| = 768$$

By Theorem 5.2 we choose h such that

$$\frac{3.99h^4(24 \times 10^{10})}{180} \le 10^{-2}$$

which gives $h \simeq 0.00117$. This requires approximately 3410 uniformly spaced nodes for Simpson's rule applied to $I(f)$.

In the second case let h_1 and h_2 denote the spacing for Simpson's rule applied to $I^1(f)$ and $I^2(f)$, respectively. We use an error tolerance of 0.5×10^{-2} in each to insure an overall error of no more than 10^{-2}. Then h_1 and h_2 are chosen such that

$$\frac{0.49h_1^4(24 \times 10^{10})}{180} \le 0.5 \times 10^{-2}$$

$$\frac{3.5h_2^4(768)}{180} \le 0.5 \times 10^{-2}$$

It follows that $h_1 \simeq 0.00166$ and $h_2 \simeq 0.1352$. Therefore approximately 295 nodes are required for $I^1(f)$ and 26 nodes for $I^2(f)$. In the first case the bad behavior of f near $x = 0.01$ results in an extremely fine node spacing throughout the interval, whereas a fine spacing is required in the second case only on $[0.01, 0.5]$. Clearly the second approach is significantly more efficient (though not entirely satisfactory since 321 nodes are required).

Of course the integral in this example is easily evaluated using calculus but nonetheless clearly illustrates the need for methods that are not restricted to uniform node spacing. A judicious choice of unevenly spaced nodes can result in a considerably more efficient application of Simpson's rule. The approach given in Example 5.10 requires a priori knowledge of the behavior of $f^{(4)}$—knowledge that is generally not available. Another criticism of this approach is that it requires a separate program for each different integral.

Adaptive quadrature is a procedure that is designed to use a given quadrature formula whose nodes are automatically selected so as to produce an approximation to $I(f)$ within a specified error tolerance and with "optimal" efficiency. Because of its simplicity and order of accuracy, we describe such a procedure based on Simpson's rule. However, the reader should be made aware than more accurate rules are available.

Suppose that we are given $f(x)$ on $[a, b]$ and an absolute error tolerance ε. It is required to approximate $I(f)$ by some Simpson rule within the specified tolerance. The adaptive procedure that we next develop requires that (i) the local errors in Simpson's rule be able to be estimated computationally and (ii) there be a strategy for choosing local error tolerances ε^i such that the overall error specification is satisfied.

Suppose $a = t_1 < t_2 < \cdots < t_k = b$ and on each subinterval $[t_i, t_{i+1}]$ we compute a Simpson rule approximation, call it $S^i(f)$, to

$$I^i(f) = \int_{t_i}^{t_{i+1}} f(x)\, dx$$

Assume that $E^i(f) = I^i(f) - S^i(f)$ can be estimated, say $|E^i(f)| \simeq \text{ERR}^i$, where ERR^i is computable. How small should ERR^i be in order that

$$|I(f) - \sum_{i=1}^{k-1} S^i(f)| \leq \varepsilon$$

By the triangle inequality we have

$$|I(f) - \sum_{i=1}^{k-1} S^i(f)| = \left| \sum_{i=1}^{k-1} [I^i(f) - S^i(f)] \right|$$

$$\leq \sum_{i=1}^{k-1} |I^i(f) - S^i(f)| \simeq \sum_{i=1}^{k-1} \text{ERR}^i$$

Since ERR^i estimates only that portion of the error due to $[t_i, t_{i+1}]$, a natural requirement is

$$\frac{\text{ERR}^i}{\varepsilon} = \frac{t_{i+1} - t_i}{b - a}$$

Moreover if $\text{ERR}^i \leq \varepsilon(t_{i+1} - t_i)/(b - a)$, then $\sum_{i=1}^{k-1} \text{ERR}^i \leq \varepsilon$ and we are reasonably assured that the overall error tolerance is satisfied. Therefore we use the strategy of choosing $S^i(f)$ such that

$$\text{ERR}^i \leq \frac{t_{i+1} - t_i}{b - a} \varepsilon$$

Next we demonstrate how the local error estimate ERR^i can be obtained. Consider the approximation on $[t_i, t_{i+1}]$. Let $4h_i = t_{i+1} - t_i$ and $x_j = t_i + (j - 1)h_i$, $1 \leq j \leq 5$. Define the following Simpson rule approximations:

$$\tilde{S}^i(f) = 2\frac{h_i}{3}[f(x_1) + 4f(x_3) + f(x_5)]$$

$$S_R^i(f) = \frac{h_i}{3}[f(x_3) + 4f(x_4) + f(x_5)]$$

$$S_L^i(f) = \frac{h_i}{3}[f(x_1) + 4f(x_2) + f(x_3)]$$

$$S^i(f) = S_R^i(f) + S_L^i(f)$$

By (5.8) we can estimate the **local errors** in $S^i(f)$ and $\tilde{S}^i(f)$ as

$$I^i(f) - \tilde{S}^i(f) = -\frac{(2h_i)^5}{90} f^{(4)}(\mu_i) \tag{5.26}$$

$$I^i(f) - S^i(f) = -\frac{h_i^5}{90}[f^{(4)}(\xi_i) + f^{(4)}(\gamma_i)] \tag{5.27}$$

where $\mu_i \in [x_1, x_5]$, $\xi_i \in [x_1, x_3]$, and $\gamma_i \in [x_3, x_5]$. Now we make the crucial assumption that $f^{(4)}(x)$ does not vary much over $[t_i, t_{i+1}] = [x_1, x_5]$. Then by subtracting (5.27) from (5.26) we obtain

$$S^i(f) - \tilde{S}^i(f) \simeq -\frac{2}{90} f^{(4)}(2^4 - 1)h_i^5 \tag{5.28}$$

where $f^{(4)}$ denotes the assumed constant value of the fourth derivative. From (5.28) we find that

$$-\tfrac{2}{90}f^{(4)}h_i^5 \simeq \frac{S^i(f) - \tilde{S}^i(f)}{2^4 - 1}$$

which when substituted on the right-hand side of (5.27) gives the error estimate

$$I^i(f) - S^i(f) \simeq \frac{S^i(f) - \tilde{S}^i(f)}{15}$$

Therefore the requisite absolute error estimate for $E^i(f)$ is $\mathrm{ERR}^i = |S^i(f) - \tilde{S}^i(f)|/15$.

The adaptive Simpson procedure essentially consists in applying S^i and $\tilde{S}^i$ to each subinterval $[t_i, t_{i+1}]$. If $S^i(f)$ satisfies

$$\frac{|S^i(f) - \tilde{S}^i(f)|}{15} \leq \varepsilon\,\frac{t_{i+1} - t_i}{b - a} \tag{5.29}$$

then $S^i(f)$ is accepted as a sufficiently accurate approximation to $I^i(f)$ and we move on to the next interval, $[t_{i+1}, t_{i+2}]$, for processing. If (5.29) is not satisfied, then we halve the interval and repeat the process on the left-hand half $[t_i, (t_i + t_{i+1})/2]$. In this way we sweep out the interval from left to right until the entire interval has been covered. We illustrate the ideas in the following simple example.

EXAMPLE 5.11

Let $I(f) = \int_0^1 x^{3/2}\,dx$ and $\varepsilon = 10^{-4}$. We begin with the points $t_1 = 0$ and $t_2 = 1$. Then $h_1 = 0.25$ and we calculate

$$S_L^1(f) = \tfrac{0.25}{3}[f(0) + 4f(0.25) + f(0.5)] = 0.0711294$$

$$S_R^1(f) = \tfrac{0.25}{3}[f(0.5) + 4f(0.75) + f(1)] = 0.329302$$

$$S^1(f) = S_R^1(f) + S_L^1(f) = 0.400431$$

$$\tilde{S}^1(f) = \tfrac{0.5}{3}[f(0) + 4f(0.5) + f(1)] = 0.402369$$

Then we compute $|S^1(f) - \tilde{S}^1(f)|/15 = 1.292 \times 10^{-4} > 10^{-4}$. Since the local error criterion is not met, we halve the interval and repeat the process on the left-hand half. Now we have $t_1 = 0$, $t_2 = 0.5$, and $t_3 = 1$. For the subinterval $[0, 0.5]$ we compute $S^1(f)$ and $\tilde{S}^1(f)$. We now have $h_1 = 0.125$, and we note that $\tilde{S}^1(f)$ is the same as our previously computed $S_L^1(f)$. Also we find for $[0, 0.5]$ the new values

$$S_L^1(f) = \tfrac{0.125}{3}[f(0) + 4f(0.125) + f(0.25)] = 0.012574$$

$$S_R^1(f) = \tfrac{0.125}{3}[f(0.25) + 4f(0.375) + f(0.5)] = 0.058213$$

and hence

$$S^1(f) = 0.070787$$

Again we compute

$$\tfrac{1}{15}|S^1(f) - \tilde{S}^1(f)| = 0.228 \times 10^{-4} < 0.5 \times 10^{-4}$$

and since criterion (5.29) is satisfied, we accept 0.070787 as a satisfactory approximation to $\int_0^{0.5} x^{3/2}\, dx$. It remains to consider $[t_2, t_3] = [0.5, 1]$. We observe that $\tilde{S}^2(f)$ is the same as our first computed $S_R^1(f)$. Also for $[0.5, 1]$ we compute the new values

$$S_R^2(f) = \tfrac{0.125}{3}[f(0.75) + 4f(0.875) + f(1)] = 0.205144$$

$$S_L^2(f) = \tfrac{0.125}{3}[f(0.5) + 4f(0.625) + f(0.75)] = 0.124146$$

$$S^2(f) = 0.32929$$

Then

$$\tfrac{1}{15}|S^2(f) - \tilde{S}^2(f)| = 0.8 \times 10^{-6} < 0.5 \times 10^{-4}$$

and hence we accept $S^2(f) = 0.32929$ as a satisfactory approximation to $\int_{0.5}^1 x^{3/2}\, dx$. The overall Simpson approximation to $I(f)$ is

$$0.070787 + 0.32929 = 0.400077$$

and if we sum the error estimates, we find that the absolute error is expected to be no worse than $0.228 \times 10^{-4} + 0.8 \times 10^{-6} \simeq 0.229 \times 10^{-4}$. The exact value of the integral is 0.4, and the actual absolute error is 0.77×10^{-4}.

When implementing the adaptive Simpson procedure in a FORTRAN program, it is necessary to put an upper limit on the number of times an interval can be bisected. In our version of this procedure, called ADPSIM, the upper limit is called LEVMAX and has been set equal to 10. Therefore in ADPSIM the smallest subinterval over which $S^i(f)$ can be applied is $(b - a)/2^{10}$. If this limit is exceeded, then the most recent approximation $S^i(f)$ is accepted by default and added to the accumulated value (VALU in the program) of Simpson approximations. A flag IFLG is then updated by 1 in order to keep track of the number of default approximations in VALU. Whenever criterion (5.29) is not satisfied, it is necessary to bisect the interval and repeat the computation of S^i (SIMP2) and $\tilde{S}^i$ (SIMP1) for the left-hand half of the interval. In this event all of the information for the right-hand half of the subinterval is saved in arrays FKEEP, XKEEP, and SIMPR, which contain function values, nodes, and $S_R^i(f)$, respectively. These quantities are needed later as $[a, b]$ is swept out from left to right. We note that the first dimension of FKEEP and XKEEP and the dimension of SIMPR must be $\geq$ LEVMAX. The termination criteria for the procedure is, cease computation when the interval $[a, b]$ has been covered by subintervals $[t_i, t_{i+1}]$ over which a Simpson approximation $S^i(f)$ has been accepted (either normally or by default). The subroutine returns four quantities: VALU, ERREST, IFLG, and NOFUN. ERREST is an estimate for the absolute error in VALU, and NOFUN is the number of function evaluations required in computing VALU.

In this program we attempt to provide sufficient documentation so that the reader can follow the flow of the computation as a subinterval is tested for acceptance. All of the Simpson computations are done using two "working" arrays F and X, which contain function values and nodes for the current subinterval. Note that the specification TOL = 0.0 is permissible. When this specification is used, the value produced by ADPSIM is the best (assuming IFLG = 0) that can be obtained on your machine; that is, the machine unit (exercise 1.32) is used instead of TOL.

```
      SUBROUTINE ADPSIM(FUN,A,B,TOL,ERREST,VALU,IFLG,NOFUN)
C     FUN:USER SUPPLIED FUNCTION SUBPROGRAM,THE INTEGRAND
C     A,B:LIMITS OF INTEGRATION WITH A<B
C     TOL:USER SUPPLIED ERROR TOLERANCE:TOL=0.0 IS OK
C     ERREST:ESTIMATED ABSOLUTE ERROR BOUND FOR VALU
C     VALU:FINAL APPROXIMATION TO THE DEFINITE INTEGRAL
C     IFLG:NUMBER OF SUBINTERVALS WHERE ERROR CRITERION
C          HAS NOT BEEN SATISFIED
C     NOFUN:NUMBER OF FUNCTION EVALUATIONS
      REAL FKEEP(10,3),XKEEP(10,3),X(5),F(5),SIMPR(10)
C  INITIALIZATION
      VALU=0.0
      IFLG=0
      JFLG=0
      ERREST=0.0
C  PROTECT AGAINST UNREASONABLE TOLERANCE
      UNIT=1.
   17 UNIT=.5*UNIT
      U=1.+UNIT
      IF(U.GT.1.) GO TO 17
      TOL2=TOL+UNIT
      TOL1=TOL2*15./(B-A)
      LEVMAX=10
C  LEVMAX:MAXIMUM NUMBER OF TIMES A SUBINTERVAL
C         CAN BE BISECTED
      H=(B-A)/4.
      DO 1 J=1,5
        X(J)=A+H*FLOAT(J-1)
        F(J)=FUN(X(J))
   1  CONTINUE
      NOFUN=5
      LEVEL=1
C  SAVE INFORMATION FOR RIGHT HALF OF SUBINTERVAL
   6  DO 2 K=1,3
        FKEEP(LEVEL,K)=F(K+2)
        XKEEP(LEVEL,K)=X(K+2)
   2  CONTINUE
      H=(X(5)-X(1))/4.
      SIMPR(LEVEL)=H/3.*(F(3)+4.*F(4)+F(5))
      IF(JFLG.LE.0) SIMP1=2.*H/3.*(F(1)+4.*F(3)+F(5))
      SIMPL=H/3.*(F(1)+4.*F(2)+F(3))
      SIMP2=SIMPL+SIMPR(LEVEL)
      DIFF=ABS(SIMP1-SIMP2)
C  TEST FOR ACCEPTABLE VALUE ON CURRENT SUBINTERVAL
      IF(DIFF.LE.TOL1*4.*H) GO TO 9
      LEVEL=LEVEL+1
```

```
          SIMP1=SIMPL
C  HAS MAXIMUM NUMBER OF BISECTIONS BEEN EXCEEDED?
          IF(LEVEL.LE.LEVMAX) GO TO 12
          IFLG=IFLG+1
C  ACCEPT APPROXIMATION FOR CURRENT SUBINTERVAL
   9      LEVEL=LEVEL-1
          JFLG=0
          VALU=VALU+SIMP2
          ERREST=ERREST+DIFF/15.
C  TERMINATION CRITERION:HAS ENTIRE INTERVAL BEEN COVERED?
          IF(LEVEL.LE.0) RETURN
C  UPDATE X,F FOR NEXT SUBINTERVAL
          DO 3 J=1,3
            JJ=2*J-1
            F(JJ)=FKEEP(LEVEL,J)
            X(JJ)=XKEEP(LEVEL,J)
   3      CONTINUE
          GO TO 5
C  UPDATE X,F FOR BISECTED SUBINTERVAL
   12     JFLG=1
          F(5)=F(3)
          F(3)=F(2)
          X(5)=X(3)
          X(3)=X(2)
   5      DO 4 J=1,2
            JJ=2*J
            X(JJ)=.5*(X(JJ+1)+X(JJ-1))
            F(JJ)=FUN(X(JJ))
   4      CONTINUE
          NOFUN=NOFUN+2
          GO TO 6
          END
```

EXAMPLE 5.12

The following is a sample calling (main) program and output for ADPSIM
applied to the integral

$$\int_{0.1}^{2} \frac{1}{x}\,dx = \log 20 = 2.995732274\ldots$$

```
          EXTERNAL G
          A=0.1
          B=2.0
          TOL=1.E-06
          CALL ADPSIM(G,A,B,TOL,ERROR,RESULT,NFLG,NUM)
          WRITE(6,101) RESULT
   101    FORMAT(' ','THE VALUE OF THE INTEGRAL=',E14.7/)
          WRITE(6,102) ERROR
   102    FORMAT(' ','WITH A PREDICTED ERROR BOUND OF  ',E14.7/)
          WRITE(6,99) NUM
    99    FORMAT(' ','THE FUNCTION WAS EVALUATED ',I3,' TIMES'/)
          IF(NFLG.GT.0) WRITE(6,88) NFLG
```

```
88  FORMAT(' ','WARNING:RESULT MAY BE UNRELIABLE.FLAG=',I4)
    STOP
    END
C

    REAL FUNCTION G(X)
    G=1./X
    RETURN
    END
```

THE VALUE OF THE INTEGRAL= 0.2995719E 01

WITH A PREDICTED ERROR BOUND OF   0.3558891E-06

THE FUNCTION WAS EVALUATED 133 TIMES

In this section we have attempted to illustrate the basic ideas in adaptive quadrature methods. Our subroutine ADPSIM does not contain some of the more sophisticated features present in state-of-the-art programs, such as roundoff error controls, relative error estimation, and detection of singularities in the integrand. As a consequence we recommend that, in practice, the reader use only well-tested and clearly documented library routines for adaptive quadrature.

EXERCISES

5.47 Consider the integral $\int_{0.1}^2 x^{5/2} \log x \, dx$, whose value to seven digits is 1.317256. Compute $T_{500}(f)$ and $S_{500}(f)$ and use ADPSIM with TOL = 0.0. Compare the results for accuracy and efficiency.

5.48 Consider the integral $\int_0^2 e^{-5x} \sin kx \, dx$. For $k = 1, 5, 10$, and 20, use ADPSIM with TOL = 0.0 to compute the approximate values of the integrals. As k increases, the integrand becomes more oscillatory and therefore more "difficult." How does NOFUN change as k increases?

5.49 Apply ADPSIM to $\int_0^{2\pi} |\sin x| \, dx$ with TOL $= 10^{-6}$. Note: $f(x) = |\sin x|$ is not differentiable on $[0, 2\pi]$. Does adaptive Simpson quadrature perform reasonably well nonetheless?

5.50 Without doing any computation, explain why the integral $\int_0^\pi (2 + \cos 8x) \, dx$ would "fool" subroutine ADPSIM.

5.51 Apply ADPSIM with TOL = 0.0 to the integral of exercise 5.50.

5.52 Would the integral $\int_0^{3.2} (2 + \cos 8x) \, dx$ fool ADPSIM?

5.53 Let
$$f(x) = \begin{cases} x^2 + \cos 4\pi x & 0 \le x \le 1 \\ 2e^{1-x} & 1 < x \le 3 \end{cases}$$

(a) Evaluate $I(f)$ using calculus.
(b) Apply ADPSIM to $I(f)$ with TOL $= 10^{-6}$.
(c) Split $I(f)$ into $\int_0^1 f(x) \, dx + \int_1^3 f(x) \, dx$ and use ADPSIM on each integral with TOL $= 0.5 \times 10^{-6}$. Compare the results with those of part (b).

5.54   Of the following problems which would be difficult for ADPSIM? Why?

(a) $\displaystyle\int_0^1 \frac{e^x}{\sqrt{x}}\,dx$ $\qquad$ TOL $= 10^{-5}$ $\quad$ (b) $\displaystyle\int_2^{10} \log(1+x^4)\,dx$ $\qquad$ TOL $= 10^{-5}$

(c) $\displaystyle\int_0^2 \frac{1}{x-\sqrt{3}}\,dx$ $\qquad$ TOL $= 10^{-1}$ $\quad$ (d) $\displaystyle\int_0^3 \frac{\sqrt{x}}{1+x^4}\,dx$ $\qquad$ TOL $= 10^{-6}$

(e) $\displaystyle\int_{-80}^{1000} x^{5/2} \cos x\,dx$ $\qquad$ TOL $= 10^{-6}$ $\quad$ (f) $\displaystyle\int_0^6 |x+2 \cos x|\,dx$ $\qquad$ TOL $= 10^{-5}$

(g) $\displaystyle\int_\pi^{2\pi} (6 - \sin 8x)\,dx$ $\qquad$ TOL $= 10^{-3}$

Try to answer without actually using ADPSIM. As a last resort apply the subroutine and report your findings.

5.55   Suppose ADPSIM were applied to $f(x)$ on $[0, 2]$ where

$$f(x) = \begin{cases} \dfrac{\cos x - 1}{x} & \text{if } x \ne 0 \\[2mm] 0 & \text{if } x = 0 \end{cases}$$

Explain why it would not be appropriate to use the function subprogram

```
REAL FUNCTION F(X)
F = (COS(X) - 1.)/X
RETURN
END
```

Hint: Consider subtractive cancellation for $x \simeq 0$. Show how Taylor's theorem can be used to give an alternating series for $f(x)$. How many terms in the series are required for an absolute error tolerance of 10^{-7} for $0 \le x \le 0.25$?

5.4   A BRIEF INTRODUCTION TO GAUSSIAN QUADRATURE

One measure of the effectiveness of an integration rule $I_N(f)$ is its **precision**. By this we mean the index k such that

$$I_N(x^n) = I(x^n) \qquad n = 0, 1, \ldots, k$$

but $I_N(x^{k+1}) \ne I(x^{k+1})$. Thus for the trapezoidal rule the precision is 1 whereas Simpson's rule has precision 3.

Consider an integration rule of the form

$$I_N(f) = A_1 f(\xi_1) + \cdots + A_N f(\xi_N) \tag{5.30}$$

where the nodes $\{\xi_i\}_{i=1}^N$ and weights $\{A_i\}_{i=1}^N$ are to be chosen so as to give the highest possible precision. In the early 1800s Gauss considered this situation for $[-1, 1]$, that is, for $I_N(f)$ applied to

$$I(f) = \int_{-1}^1 f(x)\,dx \tag{5.31}$$

Gauss demonstrated that there is a unique choice of nodes and weights such that $I_N(f)$ has precision $2N - 1$. Moreover this is the best (highest) possible precision. For $N = 2$ it is not difficult to derive the rule as follows. For $I_2(f)$ to have precision 3, we require that

$$I_2(x^n) = I(x^n) \qquad n = 0, 1, 2, 3$$

This gives the system of equations

$$A_1 + A_2 = 2$$
$$A_1 \xi_1 + A_2 \xi_2 = 0$$
$$A_1 \xi_1^2 + A_2 \xi_2^2 = \tfrac{2}{3} \qquad\qquad (5.32)$$
$$A_1 \xi_1^3 + A_2 \xi_2^3 = 0$$

Observe that (5.32) is a nonlinear system of equations in the four unknowns A_1, A_2, ξ_1, and ξ_2. At first glance it is not clear whether this system of equations has a solution or, if it does, whether that solution is unique. In fact Gauss proved that (5.32) is uniquely solvable.

We note that the weights cannot be zero. Indeed if $A_1 = 0$, then $A_2 = 2$ and (5.32) requires that $2\xi_2 = 0$ and $2\xi_2^2 = \tfrac{2}{3}$, which is clearly impossible. Similar considerations show that the nodes cannot be zero. To solve the system, we proceed as follows. Multiply the first equation by ξ_1 and subtract the result from the second equation to obtain

$$A_2(\xi_2 - \xi_1) = -2\xi_1 \qquad\qquad (5.33)$$

Next multiply the second equation by ξ_1 and subtract the result from the third equation to yield

$$A_2 \xi_2(\xi_1 - \xi_2) = \tfrac{2}{3} \qquad\qquad (5.34)$$

Substitution of (5.33) into the left-hand side of (5.34) gives $2\xi_1 \xi_2 = \tfrac{2}{3}$ or $\xi_1 \xi_2 = \tfrac{1}{3}$. By subtracting ξ_1^2 times the second equation from the last, we find

$$A_2 \xi_2(\xi_1^2 - \xi_2^2) = 0$$

and since $A_2 \xi_2 \neq 0$, we must have $\xi_1^2 = \xi_2^2$. From $\xi_1 \xi_2 = \tfrac{1}{3}$ and $\xi_1^2 = \xi_2^2$, it follows that $\xi_1 = 1/\sqrt{3} = -\xi_2$. Using this in the first two equations of (5.32) gives $A_1 = A_2 = 1$. Consequently the precision 3, two-point Gaussian quadrature rule for (5.31) is

$$I_2(f) = f\left(-\frac{1}{\sqrt{3}}\right) + f\left(\frac{1}{\sqrt{3}}\right) \qquad\qquad (5.35)$$

This two-point rule is not restricted to $[-1, 1]$ since any finite interval $[c, d]$ can be transformed into $[-1, 1]$ by a simple linear change of variables. Indeed by

$$x(t) = \frac{d - c}{2} t + \frac{d + c}{2} \qquad\qquad (5.36)$$

we obtain

$$\int_c^d g(x)\, dx = \frac{d - c}{2} \int_{-1}^1 g(x(t))\, dt$$

Then if we let $f(t) = (d - c)g(x(t))/2$ and apply (5.35), there results the following two-point rule for $[c, d]$:

$$\int_c^d g(x) \, dx \simeq \frac{d - c}{2} \left[g\left(x\left(-\frac{1}{\sqrt{3}} \right) \right) + g\left(x\left(\frac{1}{\sqrt{3}} \right) \right) \right]$$

where $x(\pm 1/\sqrt{3})$ is given by (5.36).

EXAMPLE 5.13

We use the two-point Gauss formula to estimate

$$\int_0^1 e^{-x^2} \, dx = 0.74682413\ldots$$

We have $x(t) = (t + 1)/2$, and hence

$$\int_0^1 e^{-x^2} \, dx \simeq \tfrac{1}{2}[e^{-(\sqrt{3}-1)^2/12} + e^{-(\sqrt{3}+1)^2/12}] = 0.7465875\ldots$$

This is a much better approximation than the value $T_{15}(f)$ computed in Example 5.1.

In order to obtain a more accurate approximation than $I_2(f)$, we can proceed in one of two ways. We either (a) determine $I_N(f)$ for some $N > 2$ or (b) split the given interval $[a, b]$ into subintervals of uniform length and sum the results of I_2 applied to each subinterval. The latter approach is called the two-point **composite Gaussian quadrature** and is derived as follows.

Given f on $[a, b]$, let $h = (b - a)/M$ and $x_i = a + (i - 1)/h$. Then we write

$$I(f) = \int_a^b f(x) \, dx = \sum_{i=1}^M \int_{x_i}^{x_{i+1}} f(x) \, dx \tag{5.37}$$

For the ith integral in the sum we make the change of variables $x(t) = (ht + x_i + x_{i+1})/2$ to obtain

$$\int_{x_i}^{x_{i+1}} f(x) \, dx = \frac{h}{2} \int_{-1}^1 f\left(\frac{ht}{2} + x_{i+1/2} \right) dt$$

where $x_{i+1/2} = (x_i + x_{i+1})/2 = \text{midpoint of } [x_i, x_{i+1}]$. An application of (5.35) gives

$$\int_{x_i}^{x_{i+1}} f(x) \, dx \simeq \frac{h}{2} [f(z_i^1) + f(z_i^2)]$$

where $z_i^1 = x_{i+1/2} - h/(2\sqrt{3})$ and $z_i^2 = x_{i+1/2} + h/(2\sqrt{3})$. Finally we obtain the composite rule

$$\int_a^b f(x) \, dx \simeq I_2^M(f) = \frac{h}{2} \sum_{i=1}^M [f(z_i^1) + f(z_i^2)] \tag{5.38}$$

EXAMPLE 5.14

In the following simple program we compute the composite two-point Gaussian quadrature rule for $f(x) = e^{-x^2}$ on $[0, 1]$ with $M = 1, 2, \ldots, 6$. The value obtained for $M = 6$ requires 12 function evaluations and gives a more accurate approximation than $T_{100}(f)$ does in Example 5.1.

```
        ROOT3=SQRT(3.0)
        A=0.0
        B=1.0
        WRITE(6,77)
  77    FORMAT(' ','THE COMPOSITE GAUSS 2-POINT RULE GIVES:')
        DO 1 N=1,6
          H=(B-A)/FLOAT(N)
          P=H/(2.*ROOT3)
          SUM=0.0
          DO 2 J=1,N
            Z1=A+(FLOAT(J)-.5)*H-P
            Z2=Z1+P*2.
            SUM=SUM+F(Z1)+F(Z2)
  2       CONTINUE
          VALU=SUM*H/2.
          WRITE(6,88) VALU,N
  1     CONTINUE
  88    FORMAT(' ',/3X,'INTEGRAL = ',E14.7,'  FOR N=',I2)
        STOP
        END
C
        REAL FUNCTION F(X)
        F=EXP(-X**2)
        RETURN
        END
```

```
THE COMPOSITE GAUSS 2-POINT RULE GIVES:

    INTEGRAL =  0.7465944E 00  FOR N= 1

    INTEGRAL =  0.7468028E 00  FOR N= 2

    INTEGRAL =  0.7468195E 00  FOR N= 3

    INTEGRAL =  0.7468224E 00  FOR N= 4

    INTEGRAL =  0.7468228E 00  FOR N= 5

    INTEGRAL =  0.7468233E 00  FOR N= 6
```

As was previously mentioned, a more accurate approximation can also be obtained (for smooth f) by determining the N-point Gaussian rule, $N > 2$.

Using the same method as for $I_2(f)$ would require the solution of the nonlinear system of equations

$$I_N(x^n) = \int_{-1}^{1} x^n \, dx \qquad n = 0, 1, \ldots, 2N - 1$$

for the $2N$ unknowns $\{A_i\}_{i=1}^{N}$ and $\{\xi_i\}_{i=1}^{N}$. This is by no means a simple task. Fortunately there is another method for obtaining Gaussian rules, one that applies in a more general setting than (5.31).

Consider an integral of the form

$$I(f) = \int_{a}^{b} w(x)f(x) \, dx \qquad \qquad (5.39)$$

where w is a positive continuous function on (a, b) called a **weight function**. Given such a weight function, it can be shown that there is a sequence of polynomials $\{p_n(x)\}$ such that

(i)   degree of $p_n = n$

(ii) $\int_{a}^{b} w(x)p_n(x)p_m(x) \, dx$ is $\begin{cases} = 0 \text{ if } m \neq n \\ \neq 0 \text{ if } m = n \end{cases}$

(iii) p_n has n real distinct zeros $\xi_1^n, \xi_2^n, \ldots, \xi_n^n$ in (a, b).

These polynomials are said to be **orthogonal** [because of condition (ii)] with respect to the weight function w. For each different weight function and interval (a, b) there is a different sequence of orthogonal polynomials. These polynomials and their zeros have been tabulated for many weight functions that arise in applications. The significance of these polynomials with regard to Gaussian quadrature lies in the following fact.

THEOREM 5.6

Let $I_N(f)$ be of the form (5.30). Then

$$I_N(x^n) = \int_{a}^{b} x^n w(x) \, dx \qquad n = 0, 1, \ldots, 2N - 1$$

if and only if $\{\xi_i\}_{i=1}^{N}$ are the zeros of $p_N(x)$ and the weights are given by

$$A_i = \int_{a}^{b} \Phi_i^2(x)w(x) \, dx$$

where

$$\Phi_i(x) = \prod_{\substack{j=1 \\ j \neq i}}^{N} \left(\frac{x - \xi_j}{\xi_i - \xi_j} \right)$$

Moreover the error formula

$$\int_{a}^{b} w(x)f(x) \, dx - I_N(f) = C_N f^{(2N)}(\gamma)$$

holds for some constant C_N and some $\gamma \in (a, b)$.

TABLE 5.1 Gauss-Legendre
weights and nodes

N	ξ_i	A_i
2	$\pm 1/\sqrt{3}$	1.0
3	$\pm\sqrt{3/5}$	0.555556
	0	0.888889
4	± 0.861136	0.347855
	± 0.339981	0.652145
5	± 0.906180	0.236927
	± 0.538469	0.478629
	0.0	0.568889

The simplest application of Theorem 5.6 is $w(x) \equiv 1$ and $[a, b] = [-1, 1]$, that is, our original problem (5.31). In this case the orthogonal poly-nomials are the **Legendre polynomials** and are usually denoted by $\{L_n(x)\}$. These polynomials are defined by the general formula

$$L_n(x) = \frac{1}{2^n n!} \frac{d^n}{dx^n} (x^2 - 1)^n \qquad n \geq 1$$

and $L_0(x) = 1$. The first four Legendre polynomials are

$$L_0(x) = 1$$

$$L_1(x) = x$$

$$L_2(x) = \tfrac{3}{2}(x^2 - \tfrac{1}{3})$$

$$L_3(x) = \tfrac{5}{2}(x^3 - \tfrac{3}{5}x)$$

In Table 5.1 we give the nodes and weights for $I_N(f)$ with $w \equiv 1$ and $[a, b] = [-1, 1]$. The values are rounded to six digits.

In Theorem 5.6 the interval (a, b) is not required to be finite. For example, if $w(x) = e^{-x}$ on $(0, +\infty) = (a, b)$, the resulting orthogonal poly-nomials are the **Laguerre polynomials**, defined by the general formula

$$Q_n(x) = e^x \frac{d^n}{dx^n} (x^n e^{-x}) \qquad n \geq 0$$

The first few Laguerre polynomials are

$$Q_0(x) = 1$$

$$Q_1(x) = 1 - x$$

$$Q_2(x) = x^2 - 4x + 2$$

$$Q_3(x) = 6 - 18x + 9x^2 - x^3$$

Table 5.2 gives the nodes and weights (to six correct digits) for $I_N(f)$ applied to $\int_0^\infty e^{-x} f(x)\, dx$.

TABLE 5.2 Gauss-Laguerre weights and nodes

N	ξ_i	A_i
2	0.585786	0.146447
	3.414214	0.853553
3	0.415775	0.711093
	2.294280	0.278518
	6.289945	0.0103893
4	0.322548	0.603154
	1.745761	0.357419
	4.536620	0.0388879
	9.395071	0.000539295

Our final example of orthogonal polynomials and corresponding quadrature rule is particularly appealing because all of the weights are equal (for a given N). The **Chebyshev polynomials** are defined by

$$T_n(x) = \cos(n \arccos x) \qquad n \geq 0$$

These polynomials are orthogonal with respect to $w(x) = 1/(1 - x^2)^{1/2}$ on $(-1, 1)$. The first few Chebyshev polynomials are

$$T_0(x) = 1$$

$$T_1(x) = x$$

$$T_2(x) = 2x^2 - 1$$

$$T_3(x) = 4x^3 - 3x$$

and have the attractive feature that the zeros can be given explicitly by formula. The zeros of $T_n(x)$ are

$$\xi_i^n = \cos\left(\frac{2i - 1}{2n}\pi\right) \qquad 1 \leq i \leq n$$

The N-point Gauss-Chebyshev quadrature formula is

$$\int_{-1}^{1} \frac{f(x)}{(1 - x^2)^{1/2}} \, dx \simeq \frac{\pi}{N} \sum_{i=1}^{N} f\left(\cos\left(\frac{2i - 1}{2N}\pi\right)\right)$$

and this formula has precision $2N - 1$, that is, it gives the exact result for $f(x) = x^n$, $n = 0, 1, \ldots, 2N - 1$.

The Gaussian rules have several features that distinguish them from other quadrature formulas. In general the nodes are irrational numbers and are unevenly spaced. This makes their use somewhat inconvenient since the nodes and weights must be looked up in a table. The most comprehensive tables for the Gaussian rules are given in Stroud and Secrest (1966). On the other hand, for a given smooth f and error tolerance, these rules require significantly fewer function evaluations than most other methods. Moreover, many of the Gaussian rules are very effective in treating certain types of improper integrals. Specifically, if the interval of integration is infinite or if the integrand has a singularity at an endpoint, then Gaussian quadrature may be the method of choice.

EXERCISES

5.56 (a) Mimic the approach given in the text to derive the following composite Gauss-Legendre four-point rule:

$$\int_a^b f(x)\,dx \simeq \frac{h}{2}\sum_{j=1}^M \left\{\sum_{i=1}^4 A_i f(z_i^j)\right\} = I_4^M(f)$$

where $z_i^j = x_{j+1/2} + h\xi_i/2$ and $\{A_i\}_{i=1}^4$ and $\{\xi_i\}_{i=1}^4$ are the appropriate weights and nodes in Table 5.1.

(b) Write a simple program to implement the composite rule of part (a). Test your program on $f(x) = x^7$, $x \in [0, 1]$, for $M = 1, 2$. The result should be exact except for the roundoff error in the weights, nodes, and floating point arithmetic.

5.57 Apply the program of exercise 5.56 to the integral

$$I = \int_0^1 \frac{1}{1 + x^2}\,dx = \frac{\pi}{4}$$

Try several values of M (≤ 10) and compare the results with Example 5.3.

5.58 Show that the rule

$$\int_{-1}^1 f(x)\,dx \simeq \frac{1}{9}[5f(-\sqrt{0.6}) + 8f(0) + 5f(\sqrt{0.6})]$$

has precision 5.

5.59 According to Theorem 5.6, the two-point Gauss-Legendre rule (5.35) satisfies

$$I(f) - I_2(f) = C_2 f^{(4)}(\gamma)$$

for some constant C_2 and $\gamma \in [-1, 1]$. Show that $C_2 = \frac{1}{135}$ by considering $f(x) = x^4$.

5.60 Use the result of exercise 5.59 to establish the following estimates for (5.38):

*(a) Show that

$$\int_{x_i}^{x_{i+1}} f(x)\,dx - \frac{h}{2}[f(z_i^1) + f(z_i^2)] = \left(\frac{h}{2}\right)^5 \frac{f^{(4)}(\gamma_i)}{135}$$

for some $\gamma_i \in [x_i, x_{i+1}]$. Hint: Apply the chain rule to $f(ht/2 + x_{i+1/2})$.

(b) Sum the equation of (a) to conclude that

$$\left| I(f) - \frac{h}{2}\sum_{i=1}^M [f(z_i^1) + f(z_i^2)]\right| \le \frac{h^4(b-a)M_4}{4320}$$

where M_4 is a bound for $f^{(4)}(x)$ on $[a, b]$.

(c) Use part (b) to determine h such that the composite two-point Gauss-Legendre quadrature $I_2^M(f)$ applied to $\int_2^8 \log x\,dx$ is accurate to within 10^{-6}.

*5.61 Suppose $\tilde{f}(\xi_i) = f(\xi_i) + \varepsilon_i$ and (5.30) has precision $2N - 1$ as an approximation to (5.39). If $\varepsilon = \max_{1 \le i \le N} |\varepsilon_i|$, show that

$$|I_N(f) - I_N(\tilde{f})| \le \varepsilon \sum_{i=1}^{N} A_i = \varepsilon \int_a^b w(x)\,dx$$

Hint: $I_N(1) = I(1)$ and by Theorem 5.6 the weights $\{A_i\}$ are all positive.

5.62 Apply the four-point Gauss-Laguerre formula to

$$I = \int_0^\infty \frac{1}{x^3 + 1}\,dx = \frac{2\pi}{3\sqrt{3}}$$

Hint: $I = \int_0^\infty e^{-x}\left(\frac{e^x}{x^3 + 1}\right) dx$

5.63 Determine a suitable change of variables to transform the integral

$$I = \int_0^\infty \frac{e^{-3x^2}}{(1 + x^2)^{1/2}}\,dx$$

into a form suitable for Gauss-Laguerre quadrature.

5.64 Find the exact value of the integral $I = \int_0^\infty t^3 e^{-t}\,dt$ by a suitable Gaussian quadrature.

5.65 Apply the two-point Gauss-Chebyshev quadrature to $I = \int_{-1}^1 (1 - x^2)^{1/2}\,dx$. Hint:

$$I = \int_{-1}^1 \frac{1 - x^2}{(1 - x^2)^{1/2}}\,dx$$

5.66 Show that the change of variables $x(t) = (b - a)t/2 + (b + a)/2$ transforms

$$\int_a^b \frac{f(x)}{[(x - a)(b - x)]^{1/2}}\,dx \quad \text{into} \quad \int_{-1}^1 \frac{f(x(t))}{(1 - t^2)^{1/2}}\,dt$$

5.67 Use exercise 5.66 to apply the N-point Gauss-Chebyshev quadrature to the integral

$$I = \int_0^{1/3} \frac{6x}{[x(1 - 3x)]^{1/2}}\,dx = \frac{\pi}{\sqrt{3}}$$

Can you obtain the exact result for some N?

5.5 TRICKS OF THE TRADE

In applications it frequently happens that problems arise in a form that is not suited for solution by a standard numerical method. It then becomes necessary to transform or modify the problem to a more suitable form. This is more of an art than a science, and it is difficult to give a systematic discussion of these special tricks. We are content to give several examples that illustrate some of the basic ideas.

The standard methods of Sections 5.1 and 5.2 either are not applicable or perform poorly on a given problem for (normally) two reasons: (1) the interval of integration is infinite or (2) the integrand or one of its lower order derivatives does not exist at one or more points in the interval of integration.

Perhaps the simplest difficulty to remedy is the case where the integrand or one of its lower order derivatives has a jump discontinuity at a point. For example, the integral

$$I_1 = \int_0^1 e^{-x^2} |\sin 2\pi x| \ dx$$

can be rewritten as

$$I_1 = \int_0^{0.5} e^{-x^2} \sin 2\pi x \ dx - \int_{0.5}^1 e^{-x^2} \sin 2\pi x \ dx$$

and each of these integrals can be treated by any of the standard quadrature methods, for example, by Romberg integration. For I_1 the difficulty is the jump discontinuity in the first derivative of the integrand at $x = 0.5$.

Similarly it is clear that

$$I_2 = \int_1^4 f(x) \ dx \qquad \text{where } f(x) = \begin{cases} e^x/\sqrt{x} & 1 \le x \le 2.7 \\ e^x \sin 2x & 2.7 < x \le 4 \end{cases}$$

should be split into

$$I_2 = \int_1^{2.7} \frac{e^x}{\sqrt{x}} \ dx + \int_{2.7}^4 e^x \sin 2x \ dx$$

and each integral treated separately.

In some situations a change of variables can be used to alleviate the difficulty. Consider an integral of the form

$$\int_a^b \frac{f(x)}{(x-a)^p} \ dx$$

where p is a rational number in $(0, 1)$ and f is smooth. Then the substitution $x(t) = a + t^{1/p}$ may result in a more manageable integral. As an example we make the substitution $x(t) = 2 + t^3$ in the integral

$$I_3 = \int_2^{10} \frac{\log x}{(x-2)^{1/3}} \ dx$$

and obtain

$$I_3 = 3 \int_0^2 t \log(2 + t^3) \ dt$$

The singularity at $x = 2$ has been removed, and the new form of I_3 has a smooth integrand.

A change of variables can also be used to transform an infinite interval of integration into a finite one. For $a > 0$, an integral of the form $\int_a^\infty f(x) \ dx$ results in

$$\int_0^{1/a} \frac{f(1/t)}{t^2} \ dt$$

when the substitution $x = 1/t$ is used. Here we must exercise caution because a singularity may be introduced at $t = 0$. Consider the integral

$$I_4 = \int_1^\infty \frac{e^{-x}}{x} \, dx$$

which, for $x = 1/t$, becomes

$$I_4 = \int_0^1 \frac{e^{-1/t}}{t} \, dt$$

Note that $e^{-1/t}/t$ is indeterminate at $t = 0$. However, by virtue of the power series for $e^{1/t}$, we have

$$\frac{e^{-1/t}}{t} = \frac{1}{te^{1/t}} = \frac{1}{t\left(1 + \dfrac{1}{t} + \dfrac{1}{2!\,t^2} + \cdots\right)}$$

$$= \frac{1}{t + 1 + \dfrac{1}{2!\,t} + \dfrac{1}{3!\,t^2} + \cdots} \tag{5.40}$$

and hence $\lim_{t \to 0+} e^{-1/t}/t = 0$. Moreover we can split the interval and write

$$I_4 = \int_0^\varepsilon \frac{1}{te^{1/t}} \, dt + \int_\varepsilon^1 \frac{e^{-1/t}}{t} \, dt$$

For t in $(0, \varepsilon)$ we have, by (5.40), the estimate

$$\frac{1}{te^{1/t}} < n!\, t^{n-1}$$

so that for $n = 6$ we find

$$\int_0^\varepsilon \frac{1}{te^{1/t}} \, dt < \int_0^\varepsilon 6!\, t^5 \, dt = 5!\, \varepsilon^6$$

For $\varepsilon = 0.025$, this gives the estimate

$$\int_0^\varepsilon \frac{1}{te^{1/t}} \, dt < 0.293 \times 10^{-7}$$

and hence I_4 is estimated by

$$I_4 \simeq \int_{0.025}^1 \frac{e^{1/t}}{t} \, dt$$

which can be evaluated by standard methods.

The substitution $t = e^{-x}$, which transforms an integral of the form

$$\int_0^\infty f(x) \, dx$$

into

$$\int_0^1 \frac{f(-\log t)}{t} \, dt$$

can sometimes be useful. In this regard we should point out that $-\log t = w(t)$ on $(0, 1)$ is a weight function for which Gaussian quadrature nodes and weights are known and tabulated.

Another technique that can be used to handle a singularity involves the use of power series. To illustrate the idea, we consider the integral

$$I_5 = \int_0^{\pi/2} \frac{\cos x - 1}{\sqrt{x}} \, dx$$

whose integrand is indeterminate at $x = 0$. As in the case of I_4 we split the interval and write

$$I_5 = \int_0^\varepsilon \frac{\cos x - 1}{\sqrt{x}} \, dx + \int_\varepsilon^{\pi/2} \frac{\cos x - 1}{\sqrt{x}} \, dx$$

From the power series for $\cos x$ we find

$$\cos x - 1 = -\frac{x^2}{2!} + \frac{x^4}{4!} - \frac{x^5}{5!} + \cdots$$

and hence

$$\frac{\cos x - 1}{\sqrt{x}} = -\frac{x^{3/2}}{2!} + \frac{x^{7/2}}{4!} - \frac{x^{11/2}}{6!} + \cdots$$

This is an alternating series for $x > 0$, and hence if the series is truncated after the third term, the error is bounded by $x^{15/2}/(8!)$. Therefore the estimate

$$\int_0^\varepsilon \frac{\cos x - 1}{\sqrt{x}} \, dx \simeq \int_0^\varepsilon \left(-\frac{x^{3/2}}{2!} + \frac{x^{7/2}}{4!} - \frac{x^{11/2}}{6!} \right) dx$$

is in error by at most

$$\int_0^\varepsilon \frac{x^{15/2}}{8!} \, dx = \frac{2\varepsilon^{17/2}}{8! \, 17}$$

For $\varepsilon = 0.1$, we find $2\varepsilon^{17/2}/(8! \, 17) \simeq 10^{-14}$ and hence

$$I_5 \simeq \int_0^{0.1} \left(-\frac{x^{3/2}}{2!} + \frac{x^{7/2}}{4!} - \frac{x^{11/2}}{6!} \right) dx + \int_{0.1}^{\pi/2} \frac{\cos x - 1}{\sqrt{x}} \, dx$$

The first integral can be evaluated exactly, and the second can be treated by any standard quadrature formula.

EXERCISES

5.68 The integrand in $\int_0^2 \sin \sqrt{x} \, dx$ is not differentiable at $x = 0$. Remove the singularity by an appropriate change of variables.

5.69 Use integration by parts to show that

$$\int_0^\pi \frac{\sin x}{\sqrt{x}} \, dx = -2 \int_0^\pi \sqrt{x} \cos x \, dx$$

Integrate by parts again to find

$$\int_0^\pi \frac{\sin x}{\sqrt{x}}\, dx = \frac{4\pi^{3/2}}{3} - \frac{4}{3}\int_0^\pi x^{3/2} \sin x\, dx$$

If Simpson's rule is to be used, would it be worthwhile to integrate by parts one or two more times? Justify your answer.

5.70 Make the substitution $x = t^2$ after the first integration by parts in exercise 5.69 to get

$$\int_0^\pi \frac{\sin x}{\sqrt{x}}\, dx = -4 \int_0^{\sqrt{\pi}} t^2 \cos t^2\, dt$$

Is this form suitable for a standard quadrature formula? If so, why?

5.71 Write the integral

$$I = \int_0^\infty \frac{1}{1 + x^2}\, dx$$

as

$$I = \int_0^1 \frac{1}{1 + x^2}\, dx + \int_1^\infty \frac{1}{1 + x^2}\, dx$$

Transform the second integral into one with a finite interval of integration.

5.72 Transform

$$\int_{-\infty}^\infty \frac{1}{1 + x^2}\, dx$$

into the sum of integrals over finite intervals. Hint: Use exercise 5.71 and the fact that the integrand is even.

5.73 Use integration by parts to show that

$$\int x^n \log x\, dx = \frac{x^{n+1}}{n + 1}\left(\log x - \frac{1}{n + 1}\right) \qquad n = 0, 1, 2, \ldots$$

*5.74 Write the integral $I = \int_0^\pi \sin x \log x\, dx$ as

$$I = \int_0^\varepsilon \sin x \log x\, dx + \int_\varepsilon^\pi \sin x \log x\, dx$$

Use the power series for $\sin x$ and the result of exercise 5.73 to show that

$$I_\varepsilon \equiv \int_0^\varepsilon \sin x \log x\, dx = \sum_{n=0}^\infty \frac{(-1)^n \varepsilon^{2n+2}}{(2n + 2)!}\left(\log \varepsilon - \frac{1}{2n + 2}\right)$$

$$= \frac{\varepsilon^2}{2!}(\log \varepsilon - \tfrac{1}{2}) - \frac{\varepsilon^4}{4!}(\log \varepsilon - \tfrac{1}{4}) + \frac{\varepsilon^6}{6!}(\log \varepsilon - \tfrac{1}{6}) + \cdots$$

For $\varepsilon = 0.1$, this is an alternating series whose first two terms give an accurate value for I_ε, that is,

$$\int_0^{0.1} \sin x \log x\, dx \simeq \frac{10^{-2}}{2!}(\log 0.1 - \tfrac{1}{2}) - \frac{10^{-4}}{4!}(\log 0.1 - \tfrac{1}{4})$$

The integral $\int_{0.1}^\pi \sin x \log x\, dx$ can be handled by standard methods.

5.75   The integral

$$\int_0^1 \frac{e^{-x}}{(1-x)^{1/2}} \, dx$$

has a singularity at $x = 1$. Show that the singularity is removed by the substitution $x = 1 - t^2$.

5.76   Write the integral

$$\int_0^2 \frac{e^x}{(|x-1|)^{1/2}} \, dx$$

as the sum of two integrals whose integrands have an endpoint singularity. Remove the singularities by appropriate substitutions. Hint: Consider the method in exercise 5.75.

5.77   Estimate $\int_0^{\pi/4} \sin x / \sqrt{x}$ by the method used in the text for I_5.

5.78   Apply the method in the text for I_3 to the integral

$$\int_0^{\pi/2} \frac{\cos x}{(\pi - 2x)^{1/2}} \, dx$$

5.79   Try the substitution $x = 1/t$ on the integral

$$\int_1^\infty \frac{1}{1+x^2} \, dx$$

Project 6

Imagine that you are employed in the Applied Mathematics Group of the consulting engineering firm HYTECH Inc. Your boss gives you a table of values of some function f and asks you to approximate $I(f)$ as best you can. The value of $I(f)$ is critical in the design of a particular new product on which $1 million of research and development money has been invested. The data are as follows:

x	$f(x)$	x	$f(x)$	x	$f(x)$
0.6	0.12285	9.0	0.70456	17.4	−0.26973
1.2	0.23894	9.6	0.65704	18.0	−0.32258
1.8	0.34669	10.2	0.60097	18.6	−0.36825
2.4	0.44471	10.8	0.53757	19.2	−0.40629
3.0	0.53185	11.4	0.46814	19.8	−0.43639
3.6	0.60714	12.0	0.39400	20.4	−0.45810
4.2	0.66985	12.6	0.31654	21.0	−0.47228
4.8	0.71949	13.2	0.23709	21.6	−0.47813
5.4	0.75577	13.8	0.15699	22.2	−0.47617
6.0	0.77866	14.4	0.077554	2.8	−0.46674
6.6	0.78830	15.0	0.38080×10^{-6}	23.4	−0.45029
7.2	0.78508	15.6	−0.074513	24.0	−0.42734
7.8	0.76953	16.2	−0.14492		
8.4	0.74240	16.8	−0.21028		

A digital device is used for measurement with a relative error of $\pm 10^{-5}$. The function f is, by physical principles, known to be smooth.

After careful deliberation you and your boss devise two possible approaches to the problem:

Approach A

For a suitable quadrature formula $I_N(f)$, compute I_5, I_{10}, I_{15}, ..., I_{40}. Observe the convergence (or lack thereof) to a limiting value.

Approach B

For several different quadrature formulas (spline quadrature, Romberg, and so on), say I_{40}^1, ..., I_{40}^k, compute the approximate value of $I(f)$ and compare for agreement in the first few decimal digits.

Write a report to your boss that gives your findings. Clearly state the approach used, the final approximation to $I(f) = \int_{0.6}^{24} f(x)\, dx$, and the rationale for your work, including choice of method(s), error estimation used, how data accuracy influenced choice of method, and so forth.

When submitting your report, with which of the following statements would you concur?

1 "Here is my estimate for $I(f)$. I'm not very confident about the value, but isn't it true that some number is better than none?"

2 "I have used method _____, but without further information my estimate has little justification."

3 "The value of the integral is _____ to _____ correct digits. I'd stake my reputation and job on it."

4 None of the above. Supply your own statement.

If the smoothness of f was not known, how would your approach to the problem change? Could you estimate the derivatives of f that appear in the discretization error formulas for the various methods? Hint: See example 4.8.

Remark: This project has no exact answer. It is designed to test your understanding of the relative advantages and disadvantages of the methods in the text. The choice of a method is influenced by many factors—ease of programming, accuracy, efficiency, reliable error estimates, data imprecision, and so on.

5.6 NUMERICAL DIFFERENTIATION

Most derivative calculations in the calculus are relatively easy—one simply applies the appropriate formula, be it the product rule, chain rule, or whatever. Functions are known exactly by formula, and the derivative is found exactly provided the appropriate differentiation formula is properly used. On the other hand integration problems are considerably more difficult. Typically a certain amount of trial and error is involved (trying various substitutions, integration by parts, partial fractions) before it is decided that a particular approach will solve the problem.

In numerical computation the opposite is true: integration is "easy" and differentiation is "hard." This somewhat paradoxical statement is easily explained. In numerical work functions are approximately known only at a discrete set of points and the integral or derivative is almost never found exactly. One of the central questions is, Will the (presumably small) error in the data cause a large error in the result? For numerical differentiation the answer is more than likely yes, whereas the answer is more than likely no for numerical integration (see Theorem 5.5 and exercise 5.61).

The essential difficulty with numerical differentiation is its **inherent insta-bility** or ill-conditioning. One normally tries to avoid differentiating a function numerically. When unavoidable there are at least three methods that can be used. One method consists in differentiating the cubic spline interpolate of the data (Section 4.3). Another approach is to differentiate a least squares poly-nomial or piecewise polynomial fit of the data. This approach is discussed in Chapter 7. The third method for numerical differentiation is that of difference quotients and is the subject of this section.

Recall from calculus the definition of the derivative of f at point x:

$$\lim_{h \to 0} \frac{f(x + h) - f(x)}{h} = f'(x)$$

The definition immediately suggests the following difference quotient approx-imation:

$$f'(x) \simeq \frac{f(x + h) - f(x)}{h} \qquad |h| \text{ small}$$

In order to find the discretization error in this approximation, we employ Taylor's theorem. Assuming that f is three times continuously differentiable, we have

$$f(x + h) = f(x) + f'(x)h + f''(x)\frac{h^2}{2} + f'''(\xi)\frac{h^3}{3!} \tag{5.41}$$

for some point ξ between x and $x + h$. It follows that

$$\frac{f(x + h) - f(x)}{h} - f'(x) = f''(x)\frac{h}{2} + f'''(\xi)\frac{h^2}{3!}$$

and hence the approximation is first-order accurate.

A more accurate approximation is obtained as follows. Replace h by $-h$ in (5.41) to get

$$f(x - h) = f(x) - f'(x)h + f''(x)\frac{h^2}{2} - f'''(\gamma)\frac{h^3}{3!} \tag{5.42}$$

Then subtract (5.42) from (5.41) to obtain

$$f(x + h) - f(x - h) = 2f'(x)h + [f'''(\xi) + f'''(\gamma)]\frac{h^3}{3!}$$

from which it follows that

$$\frac{f(x + h) - f(x - h)}{2h} - f'(x) = \frac{[f'''(\xi) + f'''(\gamma)]}{2} \frac{h^2}{3!} \tag{5.43}$$

Therefore the centered difference quotient $[f(x + h) - f(x - h)]/2h$ is a second-order-accurate approximation to $f'(x)$.

Typically one is given a table of equally spaced values for f and the value of f' is sought at one or more of the tabular points.

EXAMPLE 5.15

The following data are taken from a five-place table for $f(x) = \cos x$:

x	$\cos x$
0.59	0.83904
0.60	0.82534
0.61	0.81965
0.62	0.81388
0.63	0.80803
0.64	0.80210
0.65	0.79608

Using the centered difference quotient, we find

$$f'(0.61) \simeq \frac{0.81388 - 0.82534}{0.02} = -0.573$$

and

$$f'(0.63) \simeq \frac{0.80210 - 0.81388}{0.02} = -0.588$$

The correct (to five places) values are -0.57287 and -0.58914.

Thus far we have considered only the discretization error. Next we consider the effect of data error and roundoff. Suppose $\tilde{f}(x_i) = f(x_i) + \varepsilon_i$, where $\tilde{f}$ is the known value and ε_i is the error due to measurement and/or roundoff. Then if $x_{i+1} = x_i + h$ and $x_{i-1} = x_i - h$, we find

$$\frac{\tilde{f}(x_{i+1}) - \tilde{f}(x_{i-1})}{2h} = \frac{f(x_{i+1}) - f(x_{i-1})}{2h} + \frac{\varepsilon_{i+1} - \varepsilon_{i-1}}{2h}$$

and by (5.43) it follows that

$$\frac{\tilde{f}(x_{i+1}) - \tilde{f}(x_{i-1})}{2h} - f'(x_i) = \left[\frac{f'''(\xi_i) + f'''(\gamma_i)}{2}\right] \frac{h^2}{3!} + \frac{\varepsilon_{i+1} - \varepsilon_{i-1}}{2h} \tag{5.44}$$

If M_3 is a bound for f''' and $\varepsilon = \max_j |\varepsilon_j|$, then

$$\left| \frac{\tilde{f}(x_{i+1}) - \tilde{f}(x_{i-1})}{2h} - f'(x) \right| \leq M_3 \frac{h^2}{6} + \frac{\varepsilon}{h}$$

As h approaches zero, the discretization error tends to zero as h^2, whereas the second component of the error approaches $+\infty$. Clearly this process is unstable. On the one hand we want h small for accuracy (small discretization error), but on the other hand we want h large for stability (small data error magnification). It is precisely this dilemma that makes numerical differentiation so difficult.

The behavior of the overall error in the centered difference approximation depends on which component in (5.44) is dominant. Assuming the discretization error is dominant for some range of point spacing, say between h and $4h$, one can estimate the error as follows. By assumption we have from (5.44) that

$$D_h \tilde{f}_i - f'_i \simeq Ch^2 \tag{5.45}$$

where $D_h \tilde{f}_i = [\tilde{f}(x_i + h) - \tilde{f}(x_i - h)]/2h$ and C depends on f''' but not h. Similarly by replacing h with $2h$, we have

$$D_{2h} \tilde{f}_i - f'_i \simeq 4Ch^2$$

and hence by subtraction we get

$$D_{2h}\tilde{f}_i - D_h \tilde{f}_i \simeq 3Ch^2 \tag{5.46}$$

We solve for Ch^2 in (5.46) and substitute into (5.45) to obtain the computable error estimate,

$$D_h\tilde{f}_i - f'_i \simeq \frac{D_{2h} \tilde{f}_i - D_h\tilde{f}_i}{3}$$

This estimate is valid only if the roundoff and data errors are negligible.

EXAMPLE 5.16

The following is a five-place table of natural logarithms

x	$\log x$
0.25	-1.38629
0.27	-1.30933
0.29	-1.23787
0.31	-1.17118
0.33	-1.10866
0.35	-1.04982
0.37	-0.99425

Then for $x_i = 0.31$ we compute

$$D_h \tilde{f}_i = \frac{-1.10866 - (-1.23787)}{0.04} = 3.23025$$

$$D_{2h} \tilde{f}_i = \frac{-1.04982 - (-1.30933)}{0.08} = 3.243875$$

The correct value of $f'(0.31)$ is $1/0.31 = 3.22580645$, and hence the actual error in $D_h \tilde{f_i}$ is

$$D_h \tilde{f_i} - f'(0.31) = 0.44436 \times 10^{-2}$$

The computed error estimate is

$$\frac{D_{2h} \tilde{f_i} - D_h \tilde{f_i}}{3} = 0.45416 \times 10^{-2}$$

Of the three methods for numerical differentiation there is no consensus among numerical analysts on a "best" method. The method of difference quotients is by far the easiest to implement; however, it is generally limited to equally spaced points. Moreover this method approximates f' only at a discrete set of points. Differentiation of the cubic spline interpolate or of a least squares polynomial furnishes an approximate derivative throughout an interval. All of these methods possess a certain amount of instability.

EXERCISES

5.80 From the following table of values of $f(x) = \tan x$:

x	$f(x)$
1.36	4.67344
1.38	5.17744
1.40	5.79788
1.42	6.58112
1.44	7.60183

compute an approximation to $f'(1.4)$ by a centered difference quotient and estimate the error. Compare with the exact value and exact error.

5.81 Use a pocket calculator to compute $(\cos h - 1)/h$ for $h = 0.01, 0.005, 0.0025, 0.00125, \ldots$. Observe that the computed values approach zero and then something "strange" happens. Explain this phenomenon.

5.82 Reconsider exercise 1.42.

5.83 Reconsider exercise 4.30.

5.84 Suppose f is six times continuously differentiable. In the derivation of (5.43) carry additional terms in the Taylor expansion and show that

$$\frac{f(x+h) - f(x-h)}{2h} - f'(x) = f'''(x) \cdot \frac{h^2}{3!} + f^{(5)}(x) \frac{h^4}{5!} + O(h^5)$$

5.85 Use the result of exercise 5.84 to show that

$$\frac{4D_h f_i - D_{2h} f_i}{3}$$

is a fourth-order-accurate approximation to $f'(x_i)$.

5.86 For $x_i = 1.4$, apply the result of exercise 5.85 to the data in exercise 5.80. Is this more accurate than the approximation computed in exercise 5.80?

5.87 Show that $f''(x_i)$ is approximated with second-order accuracy by

$$\frac{f(x_i + h) - 2f(x_i) + f(x_i - h)}{h^2} \equiv D_h^2 f_i$$

provided f is four times continuously differentiable.

5.88 If $\tilde{f}(x_i) = f(x_i) + \varepsilon_i$, show that

$$D_h^2 \tilde{f}_i - D_h^2 f_i = \frac{\varepsilon_{i+1} - 2\varepsilon_i + \varepsilon_{i-1}}{h^2}$$

NOTES AND COMMENTS

Section 5.1

The rules $T_N(f)$ and $S_N(f)$ are referred to as the composite trapezoidal and composite Simpson rules in many numerical analysis books. These results are presented in most books on the subject.

Section 5.2

Our derivation of the Euler-Maclaurin formula follows that of Dahlquist and Björck (1974). Given the polynomials $\{q_j\}$, this approach uses only elementary ideas—integration by parts and the chain rule.

   The usual derivation of the error formula (5.8) for Simpson's rule involves Hermite interpolation or an equivalent trick. See, for example, Conte and de Boor (1980).

   Some books present Romberg integration in a form somewhat different from that given in the text. The idea is to successively compute $T_1(f)$, $T_2(f)$, $T_4(f)$, and so on and the corresponding T table rows until it is ascertained that a particular diagonal entry is acceptable (by using the ratios R_j). See, for example, the book by Burden et al. (1981).

   Romberg integration is a special case of a more general idea called Richardson extrapolation.

Section 5.3

There are several types of adaptive quadrature methods. Our development of adaptive Simpson's rule is similar in spirit to that of Ralston and Rabinowitz (1978). An adaptive routine based on Romberg integration, called CADRE, is given in Rice (1971). The routine CADRE was developed by de Boor and has been found to be both efficient and reliable. The somewhat encyclopedic book by Davis and Rabinowitz (1975) contains additional methods and programs. The area of adaptive quadrature is of recent origin and a subject of current research.

Section 5.4

A more complete discussion of Gaussian quadrature is given in Atkinson

(1978) and Johnson and Riess (1982). The book by Stroud and Secrest (1966) gives an extensive tabulation of nodes and weights for the Gaussian rules.

Section 5.6

For a more detailed discussion of numerical differentiation, especially in connection with Richardson extrapolation (to the limit), see Conte and de Boor (1980).

DISCRETE VARIABLE METHODS FOR INITIAL VALUE PROBLEMS

6

A **differential equation** is an equation that involves an unknown function $y(t)$ and one or more of its derivatives. The equations

$$y'(t) + y(t) = e^{-t}$$

$$y'''(t) = \log|y(t)| + 2t^2 y''(t)$$

are both examples of differential equations. The former is called a **first-order** differential equation because it involves only the first derivative of the unknown function y. The latter is a third-order equation. By a **solution** of the differential equation we mean a function $y(t)$ that possesses the requisite number of derivatives and satisfies the equation for all t in some interval I. For example, the function

$$y(t) = te^{-t}$$

is a solution of the first equation on $I = (-\infty, +\infty)$ since

$$\frac{d}{dt}(te^{-t}) + te^{-t} = (e^{-t} - te^{-t}) + te^{-t}$$

$$= e^{-t} \qquad -\infty < t < +\infty$$

6.1   INTRODUCTION

We are primarily concerned with first-order equations, which can be written in the general form

$$y'(t) = f(t, y(t)) \tag{6.1}$$

for some known function f of two variables. For example, the first equation can be put in the form (6.1) with $f(t, y) = e^{-t} - y$. The simplest equation of this type occurs when f is independent of y:

$$y'(t) = f(t)$$

Then by calculus a solution is simply an antiderivative of $f(t)$:

$$y(t) = \int f(t)\, dt$$

If $f(t) = \cos 2t$, then solutions of $y'(t) = f(t)$ are given by

$$y(t) = \tfrac{1}{2} \sin 2t + C \tag{6.2}$$

where C is an arbitrary constant. Thus $y'(t) = \cos 2t$ has infinitely many solutions. However, if we also require that $y(0) = 1$, then the problem $y'(t) = \cos 2t$, $y(0) = 1$, has the unique solution $y(t) = 0.5 \sin 2t + 1$; in other words, the constant C in (6.2) must be unity. This is typical of the situation for the general equation (6.1). That is, equation (6.1) has infinitely many solutions (for reasonable f) but there is only one solution that satisfies the initial condition $y(t_0) = y_0$. This leads us to consider the following **initial value problem**:

$$y'(t) = f(t, y(t))$$
$$y(t_0) = y_0 \tag{6.3}$$

where the initial data (t_0, y_0) are specified. Initial value problems arise naturally in a variety of disciplines as models of the dynamic behavior of some physical phenomena. The independent variable t normally represents time, and the dependent variable y represents any one of a number of quantities—position of a projectile, voltage, population of a species, amount of a radioactive material, and so on.

EXAMPLE 6.1

RADIOCARBON DATING   Carbon 14 (^{14}C) is present in most living organisms, and when an organism dies the concentration of ^{14}C decreases through radioactive decay. It is known that the rate of radioactive decay of a substance is directly proportional to the amount of the substance present. Thus if $y(t)$ denotes the amount of ^{14}C present in a fossil at time t, then $y'(t)$ is the amount that decays per unit time and is proportional to $y(t)$, that is,

$$y'(t) = -ky(t)$$

The positive constant k is the decay constant of the substance, and for ^{14}C it is known that $k = 0.12449 \times 10^{-3}$ when t is measured in years. Thus if y is known at some initial time t_0, then the amount of ^{14}C present in a fossil at any time is given by the solution of the initial value problem:

$$y'(t) = -ky(t)$$
$$y(t_0) = y_0$$

The chemist Walter Libby discovered, in 1949, a procedure whereby this initial value problem can be used to determine the age of certain fossils. This important discovery resulted in his reception of a Nobel prize in 1960.

EXAMPLE 6.2

The logistic law of population growth is the first-order differential equation

$$y'(t) = \alpha y(t) - \beta y^2(t)$$

where $y(t)$ represents the population of a given species at time t. The constant α denotes the difference between the birth and death rates, and β is a constant that takes into account crowding effects and limitations of food supply. In many cases the values of these positive constants can be determined from experimental data. If the population is known at some time t_0, then the population at some future time $t > t_0$ can be predicted by solving the initial value problem

$$y'(t) = \alpha y(t) - \beta y^2(t)$$

$$y(t_0) = y_0$$

By rewriting the equation as

$$y'(t) = \beta y(t) \left[\frac{\alpha}{\beta} - y(t) \right]$$

we notice that $y'(t) > 0$ if $0 < y(t) < \alpha/\beta$ and $y'(t) < 0$ if $y(t) > \alpha/\beta$. Thus the population must be decreasing if $y(t)$ exceeds α/β [since $y'(t) < 0$] and it must be increasing if $y(t) < \alpha/\beta$. The ratio α/β is called the limiting population.

The following theorem gives sufficient conditions to make the initial value problem (6.3) uniquely solvable.

THEOREM 6.1

Suppose $f(t, y)$ and $\partial f/\partial y$ are continuous on the rectangle $R: a \leq t_0 \leq b$, $c \leq y_0 \leq d$ (Figure 6.1). Then for any initial data (t_0, y_0) in R, the problem (6.3) has a unique continuously differentiable solution $y(t)$ on some interval $I_r = [t_0 - r, t_0 + r], r > 0$.

We remark that if f and $\partial f/\partial y$ are continuous and bounded for $a \leq x \leq b$ and all y, then I_r can be replaced by $[a, b]$ in Theorem 6.1. Theorem 6.1 says that equation (6.3) is uniquely solvable for each initial data point (t_0, y_0) in R. For fixed $t_0 \in (a, b)$, consider several initial values $y_0^1, y_0^2, \ldots, y_0^m$ with (t_0, y_0^j) in R for each $j = 1, \ldots, m$. By the existence and uniqueness theorem there result m solutions of the differential equation $y' = f(t, y)$, say $y_1(t), \ldots, y_m(t)$, with $y_j(t_0) = y_0^j$. Thus there is a one-parameter family (viewing y_0 as the

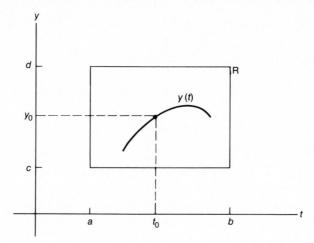

FIGURE 6.1

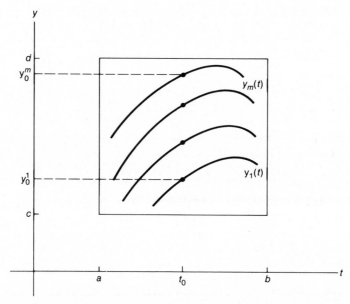

FIGURE 6.2

parameter) of solutions of $y' = f(t, y)$ that might look like what is depicted in Figure 6.2.

In addition to existence and uniqueness of solutions it is also important to know the effect of perturbations in y_0 on the solution. Will small perturbations result in a small change in the solution? Consider the **perturbed problem**

$$y'(t) = f(t, y(t))$$

$$y(t_0) = y_0 + \delta$$

with solution $y_\delta(t)$. Is it true that $|y(t) - y_\delta(t)|$ is small provided $|\delta|$ is sufficiently small? Fortunately the answer is yes if f satisfies the hypotheses of Theorem 6.1. We state this fact more precisely in the following theorem.

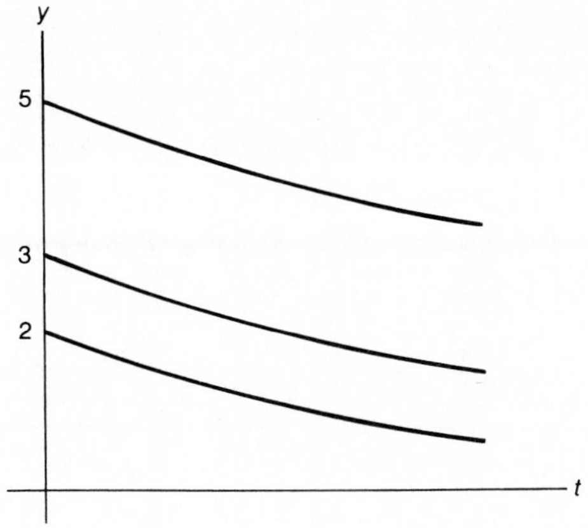

FIGURE 6.3

THEOREM 6.2

Suppose f satisfies the hypotheses of Theorem 6.1, $y(t)$ is the solution of (6.3), and $y_\delta(t)$ is the solution of the perturbed problem. Then if $(t, y_\delta(t)) \in R$, there exists a constant K, independent of δ, such that

$$|y(t) - y_\delta(t)| \leq K|\delta|$$

This theorem says that solutions of the initial value problem (6.3) **depend continuously on the initial data**. This is important since a numerical approximation to the solution likely introduces errors (due to roundoff or discretization) whose net effect is to solve a perturbed problem. These two theorems say that the initial value problem is well-posed (recall the discussion in Section 1.4). It is also true that solutions of (6.3) depend continuously on the function f, but Theorem 6.2 is sufficient for our purposes.

EXAMPLE 6.3

Consider the problem $y' = e^{-t} - y$, $y(0) = y_0$. Then $f(t, y) = e^{-t} - y$ and $\partial f/\partial y = -1$ are continuous on any rectangle. Solutions of this initial value problem are given by $y(t) = (t + y_0)e^{-t}$. For $y_0 = 2, 3, 5$, the solution curves are sketched as in Figure 6.3. Notice that each solution tends to zero as $t \to +\infty$. In the next example we see a considerably different behavior.

EXAMPLE 6.4

The differential equation $y'(t) - y(t) = -100e^{-100t}$ has solutions given by

$$y(t) = Ce^t + \tfrac{100}{101} e^{-100t}$$

where C is an arbitrary constant. If the initial condition $y(0) = \tfrac{100}{101}$ is prescribed, then the solution is $y_1 = \tfrac{100}{101}e^{-100t}$, and

$$\lim_{t \to +\infty} y_1(t) = 0$$

However, if the initial condition is $y(0) = y_0 > \frac{100}{101}$, then $C = y_0 - \frac{100}{101}$ and the solution $y_2(t) = (y_0 - \frac{100}{101})e^t + \frac{100}{101}e^{-100t}$ tends to $+\infty$ as $t \to +\infty$. The long-term behavior of the solution depends, in a critical way, on the prescribed initial data. This does not contradict Theorem 6.2 since the inequality therein is restricted to finite t.

Examples 6.3 and 6.4 illustrate a dichotomy in the behavior of solutions of $y' = f(t, y)$ with respect to perturbations in the initial condition. In Example 6.3 the solutions corresponding to $y(0) = y_0$ and $y(0) = y_0 + \delta$ are

$$y(t) = (t + y_0)e^{-t} \quad \text{and} \quad y_\delta(t) = (t + y_0 + \delta)e^{-t}$$

respectively. It follows that

$$|y(t) - y_\delta(t)| = |\delta|e^{-t} \le |\delta| \qquad t \ge 0$$

and hence a small perturbation in the initial condition results in a small change in the solution that is "damped out" as t increases. In contrast it is seen that solutions of Example 6.4 corresponding to $y(0) = y_0$ and $y(0) = y_0 + \delta$ satisfy

$$|y(t) - y_\delta(t)| = |\delta|e^t \qquad t \ge 0$$

For $0 \le t \le T$, the inequality of Theorem 6.2 is valid with $K = e^T$, and hence the problem is well-posed. However, the perturbation δ is magnified by e^t, and the two solutions deviate by an increasingly larger amount as t increases. Both of these problems are well-posed, but Example 6.4 is more sensitive to perturbations in the initial condition. Example 6.3 is well-conditioned and Example 6.4 is ill-conditioned with respect to the initial condition. **Conditioning** of an initial value problem is important when attempting to find a numerical solution.

The key to the conditioning of problem (6.3) is the sign of $\partial f / \partial y$. If $\partial f / \partial y < 0$ on a rectangle R that contains the solution curves, then the change $y(t) - y_\delta(t)$, due to a perturbation δ in the initial condition, is damped out as t increases. On the other hand the solution curve $y_\delta(t)$ deviates from $y(t)$ as t increases if $\partial f / \partial y > 0$. This is exactly the phenomenon we observed in Examples 6.3 and 6.4, where $\partial f / \partial y$ was -1 and 1, respectively.

Unfortunately the determination of the sign of $\partial f / \partial y$ (on an appropriate rectangle) is not always so simple. As an illustration consider, for $t > 0$, the Riccati problem:

$$y'(t) = y^2(t) - \frac{y(t)}{t} - \frac{1}{t^2} \qquad y(t_0) = y_0 \qquad (6.4)$$

We have $f(t, y) = y^2 - y/t - 1/t^2$ and $\partial f / \partial y = 2y - 1/t$. It follows that

$$\frac{\partial f}{\partial y}(t, y(t)) \begin{cases} > 0 & \text{if } y(t) > \dfrac{1}{2t} \\[2mm] < 0 & \text{if } y(t) < \dfrac{1}{2t} \end{cases}$$

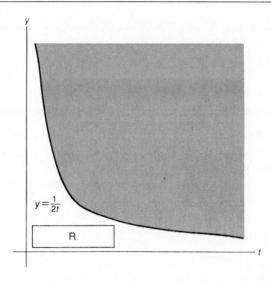

FIGURE 6.4

and the shaded area to the right of the hyperbola $y = 1/2t$ in Figure 6.4 gives the region where (6.4) is ill-conditioned, that is, where it is sensitive to perturbations in the initial condition. Moreover f and $\partial f/\partial y$ are undefined for $t = 0$, and hence problem (6.4) is well-conditioned in rectangles like R, as shown in Figure 6.4. In general $\partial f/\partial y$ may change sign, and hence the problem may be ill-conditioned in some places and well-conditioned in others. Conditioning can be very difficult to determine a priori.

As a model for conditioning we shall frequently refer to the problem

$$y'(t) = \lambda y(t) \qquad t > t_0$$

$$y(t_0) = y_0 \tag{6.5}$$

where λ is a given constant. It is easy to see that $u(t) = e^{\lambda t}$ is a solution of the differential equation, that is, $u' = \lambda u$. We claim that $y(t) = Cu(t)$ for some constant C. Indeed if we let $g(t) = y(t)/u(t) = e^{-\lambda t}y(t)$, then

$$g'(t) = e^{-\lambda t}y'(t) - \lambda e^{-\lambda t}y(t)$$

$$= e^{-\lambda t}[y'(t) - \lambda y(t)] = 0$$

This implies, however, that $g(t) = $ constant, that is, $y(t) = Ce^{\lambda t}$. The initial condition requires that $y(t_0) = Ce^{\lambda t_0} = y_0$, and consequently $C = y_0 e^{-\lambda t_0}$. Thus the unique solution of (6.5) is

$$y(t) = y_0 e^{\lambda(t - t_0)}$$

Note that $f(t, y) = \lambda y$ and $\partial f/\partial y = \lambda$. Hence problem (6.5) is well-conditioned for $\lambda < 0$ and ill-conditioned for $\lambda > 0$. The solutions are decaying exponentials for $\lambda < 0$, whereas for $\lambda > 0$ the exponentials grow without bound.

When assessing the performance of a numerical method for (6.3), it is important to know how the method is able to handle both well-conditioned and ill-conditioned problems. Clearly one cannot test a method on every such problem. The model problem (6.5) exhibits for $\lambda < 0$ the characteristic behavior of well-conditioned problems and for $\lambda > 0$ that of ill-conditioned problems. Thus problem (6.5) serves as a benchmark for comparing the performance of numerical methods.

The reader who has taken a first course in differential equations knows that problem (6.3) can be solved analytically only for certain special cases (linear, separable, exact, and so forth). In these special cases the solution is usually given by formula or expressed in terms of a power series. If the solution is given by a complicated formula or by a power series, it may be more convenient to solve the problem numerically. In most cases the initial value problem is not amenable to solution by analytical means and one must resort to a numerical method.

Before giving any specific numerical methods for problem (6.3), we briefly describe how one might seek an approximate solution. One approach would be to find a function of the form

$$u(t) = \sum_{n=1}^{N} a_n u_n(t) \tag{6.6}$$

[where $\{u_n(t)\}$ is a sequence of known functions] that satisfies the initial condition and approximately satisfies the differential equation, that is,

$$u'(t) \simeq f(t, u(t))$$

$$u(t_0) = y_0$$

For instance the choice $u_n(t) = \sin(n\pi t)$ in (6.6) may be appropriate if it is known that solutions of (6.3) are periodic. The approximation $u' \simeq f(t, u)$ could mean that

$$\max_{t_0 \leq t \leq T} |u'(t) - f(t, u(t))| \qquad \text{is small, some } T > t_0$$

or that

$$\int_{t_0}^{T} |u'(t) - f(t, u(t))|^2 \, dt \qquad \text{is small, some } T > t_0$$

or that

$$u'(t_i) = f(t_i, u(t_i)) \qquad t_i > t_0, i = 1, 2, \ldots, N$$

With this approach one determines an approximate solution $u(t)$ that is defined for all t (or at least for t in some interval). We shall not consider methods of this type.

The methods we do consider for approximating the solution of (6.3) are called **difference methods** or **discrete variable methods**. In this approach the solution is approximated at a set of discrete points called **grid** (or mesh) points. In much of our introductory discussion the grid points are equally spaced, say $t_i = t_0 + ih, i \geq 0$. Variable step methods are discussed in Sections 6.3 and 6.5. We denote the approximation to $y(t_i)$ by Y_i. Normally it is required to approximate y on a finite interval $[t_0, T]$, in which case $h = (T - t_0)/N$ for some integer N. The solution y of (6.3) and its approximation Y are sketched

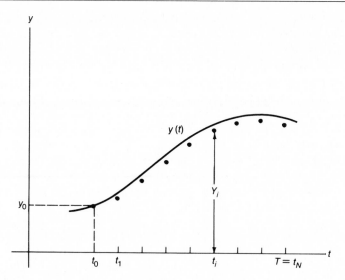

FIGURE 6.5

in Figure 6.5. A discrete variable method consists in a rule for computing the approximation Y_i at t_i in terms of one or more of the preceding values, that is, in terms of Y_j, $j \leq i - 1$. In this way we march across the interval from left to right and finally obtain Y_N. If Y_{n+1} can be computed knowing only Y_n, then the method is called a **single-step method**. If knowledge of $Y_n, Y_{n-1}, \ldots, Y_{n-k+1}$ is required for the computation of Y_{n+1}, the method is referred to as a **k-step method**.

A single-step discrete variable method takes the form

$$Y_{n+1} = Y_n + h\Phi(t_n, Y_n, h) \tag{6.7a}$$

or $\quad Y_{n+1} = Y_n + h\Phi(t_n, Y_n, Y_{n+1}, h) \tag{6.7b}$

for some function Φ, called an **increment function**. The increment function specifies the rule by which Y_{n+1} is determined knowing Y_n. A single-step method like (6.7a) is called **explicit** since Y_{n+1} can be computed explicitly given Y_n and the increment function. In (6.7b) the increment function depends on Y_{n+1}, and such a method is called **implicit**.

We begin the next section by considering a particularly simple single-step explicit method called Euler's method.

EXERCISES

6.1 Determine the order of each of the following differential equations:

(a) $y'''(t) = t^2 \cos t + [y'(t)]^4$
(b) $3y''(t) - 5y'(t) + 7y(t) = 0$
(c) $[y(t)]^2 + [y'(t)]^2 = 4$

6.2 Write each of the following in the form (6.1) and identify $f(t, y)$:

(a) $(t + e^{y(t)})y'(t) - 1 = 0$

(b) $\quad ty'(t) + 3y(t) = \dfrac{\sin t}{t}$

(c) $\quad e^{-t}y'(t) = e^{y(t)}$

Compute $\partial f/\partial y$ in each case.

6.3 Verify that y is a solution of the differential equation

(a) $\quad y'' + y = 0 \qquad y(t) = \cos t$

(b) $\quad \dfrac{dy}{dx} = e^{2x} + 3y \qquad y(x) = 4e^{3x} - e^{2x}$

(c) $\quad y' = \lambda y \qquad y(t) = 7e^{\lambda t}$

(d) $\quad y' = y^2 - \dfrac{y}{t} - \dfrac{1}{t^2} \qquad y(t) = \dfrac{1}{t}$

(e) $\quad \dfrac{dy}{dt} = \dfrac{-t}{3t + 2y} \qquad y(t) = \dfrac{2}{t^2} - t$

(f) $\quad y' - 2xy = x \qquad y(x) = e^{x^2} - \frac{1}{2}$

6.4 The unique solution of $y' = -ky$, $y(t_0) = y_0$ is given by

$$y(t) = y_0 \, e^{-k(t - t_0)}$$

(a) The half-life of a radioactive substance is the time required for half of a given amount of the substance to decay. Find the half-life of ^{14}C.

(b) A fossil is determined by experiment to have a ^{14}C concentration that is 30 percent of its original concentration. Determine the age of the fossil.

6.5 In 1845 Verhulst postulated that the population of the United States satisfies the logistic law with $\alpha = 0.03134$ and $\beta = 1.5887 \times 10^{-10}$. The unique solution of $y' = \alpha y - \beta y^2$, $y(t_0) = y_0$ is given by

$$y(t) = \frac{\alpha}{\beta} \left[\frac{y_0}{y_0 + (\alpha/\beta - y_0)e^{-(t - t_0)\alpha}} \right]$$

(a) Use the census figure $y(1790) = 3.9$ million and evaluate $y(1920)$, $y(1930)$, ..., $y(1970)$. Compare with the following:

Year	Population (millions)
1920	105.7
1930	122.8
1940	131.7
1950	150.7
1960	179.3
1970	203.2

(b) Show that the solution of the logistic law satisfies

$$\lim_{t \to \infty} y(t) = \frac{\alpha}{\beta}$$

(c) Based on parts (a) and (b), do you think that α and β need to be corrected to fit the population trend in recent years?

6.6 For each of the following equations find a rectangle in the ty plane where the existence and uniqueness of a solution through a specified initial point (t_0, y_0) is guaranteed:

(a) $y'(t) = t^2 - [y(t)]^2$

(b) $y'(t) = [y(t)]^2$

(c) $y'(t) = \dfrac{4ty(t)}{2 + y^2(t)}$

(d) $y'(t) = \dfrac{2t - 3y(t)}{t + y(t)}$

(e) $y'(t) + ty(t) = e^{-t}$

6.7 For each problem in exercise 6.6 determine a region where the differential equation is ill-conditioned with respect to perturbations of the initial data. Hint: Consider $\partial f / \partial y$.

6.8 Solve the equations

(a) $y'(t) = -3y(t)$ $y(2) = 4$

(b) $y'(t) = 6y(t)$ $y(0) = 3$

6.9 The solution of exercise 6.6(b) with initial condition $y(0) = 1$ is $y(t) = 1/(1 - t)$. Verify this. What is an interval I_r of existence and uniqueness for this problem? See Theorem 6.1.

*6.10 A function $f(t, y)$ is said to satisfy a **Lipschitz condition** on $a \leq t \leq b, c \leq y \leq d$ if there is a constant L such that

$$|f(t, y) - f(t, z)| \leq L|y - z|$$

for all $a \leq t \leq b$ and $c \leq y, z \leq d$. If $\partial f / \partial y$ is continuous, show that f satisfies a Lipschitz condition with

$$L = \max_{\substack{a \leq t \leq b \\ c \leq y \leq d}} \left| \frac{\partial f}{\partial y}(t, y) \right|$$

Hint: For fixed t apply the mean value theorem to f. Find the Lipschitz constant L for each problem in exercise 6.6 and specify the region where it is valid.

*6.11 It can be shown that the constant K in Theorem 6.2 is given by $K = e^{L(b-a)}$, where L is the Lipschitz constant given in exercise 6.10. Determine K for the rectangles you found in exercise 6.6. Based on the size of K, which of the problems of exercise 6.6 is least sensitive to perturbations in the initial condition? Most sensitive?

6.12 Suppose $g(t)$ is a given continuously differentiable function and consider the problem

$$y'(t) = \lambda[y(t) - g(t)] + g'(t)   y(0) = y_0$$

(a) Verify that the unique solution is given by $y(t) = [y_0 - g(0)]e^{\lambda t} + g(t)$.

(b) If $g(t) = t$ and $\lambda = -1$, sketch solution curves for $y_0 = 0, \pm 1, \pm 2$.

(c) Repeat part (b) with $\lambda = +1$.

(d) Characterize the conditioning of the problem in terms of λ.

6.13 Suppose λ, α are constants with $\lambda \neq -\alpha$ and consider the problem

$$y'(t) + \lambda y(t) = e^{\alpha t}   y(0) = y_0$$

(a) Verify that the unique solution is given by

$$y(t) = \frac{e^{\alpha t}}{\lambda + \alpha} + \left(y_0 - \frac{1}{\lambda + \alpha} \right) e^{-\lambda t}$$

(b) If $\lambda = \alpha = 1$, sketch solution curves for $y_0 = 0$, $\frac{1}{2}$, and 1.

(c) Characterize the conditioning for each of the following cases: (i) $\alpha = 1000$, $\lambda = 1$, (ii) $\alpha = 1$, $\lambda = 1000$, (iii) $\alpha = 1$, $\lambda = -1000$.

6.2 EULER'S METHOD

In this section we study Euler's method for approximating the solution of an initial value problem. Although this method is rarely used, it is conceptually simple and relatively easy to analyze. The error and stability analyses presented for Euler's method illustrate many of the central ideas involved in more sophisticated methods.

Suppose it is required to approximate the solution of

$$y'(t) = f(t, y(t)) \tag{6.8a}$$

$$y(t_0) = y_0 \tag{6.8b}$$

on some interval $[t_0, T]$. Let N be a positive integer and define the mesh points $t_i = t_0 + ih$ with $h = (T - t_0)/N$. Our aim is to construct $\{Y_i\}_{i=0}^{N}$ such that $Y_i \simeq y(t_i)$. Assume that y is twice continuously differentiable on $[t_0, T]$. By Taylor's theorem we have for each $t \in [t_0, T - h]$

$$y(t + h) = y(t) + y'(t)h + y''(\xi) \frac{h^2}{2}$$

for some point ξ, $t < \xi < t + h$. Using the fact that y satisfies differential equation (6.8a), this equation can be rewritten as

$$y(t + h) = y(t) + hf(t, y(t)) + y''(\xi) \frac{h^2}{2}$$

By neglecting the second-order (in h) term, we obtain the approximation

$$y(t + h) \simeq y(t) + hf(t, y(t)) \tag{6.9}$$

For $t = t_0$, the right-hand side of (6.9) is known and provides an approximation to $y_1 \equiv y(t_1)$. Therefore we define $Y_0 = y_0$ and

$$Y_1 = Y_0 + hf(t_0, Y_0)$$

It is not difficult to show that Y_1 determines the point where the vertical line $t = t_1$ intersects the tangent line to the graph of $y(t)$ at (t_0, y_0). The situation is depicted in Figure 6.6.

The point (t_1, Y_1) is likely not on the graph of the solution $y(t)$, but for small h we expect that $Y_1 \simeq y_1$. The approximation (6.9) suggests that Y_2 be defined by

$$Y_2 = Y_1 + hf(t_1, Y_1)$$

Since $Y_1 \neq y_1$, we are no longer following a tangent line to the solution curve in order to obtain Y_2 and can interpret Y_2 as follows. Consider a new initial value problem with the same differential equation (6.8a) but with the initial

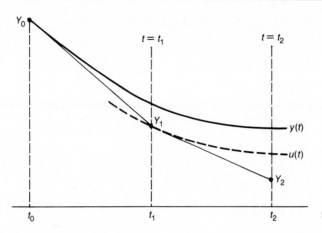

FIGURE 6.6

condition $y(t_1) = Y_1$. Assuming that (t_1, Y_1) is in a rectangle on which Theorem 6.1 applies, this new problem has a unique solution, call it $u(t)$:

$$u'(t) = f(t, u(t))$$

$$u(t_1) = Y_1$$

Then Y_2 determines the point where the vertical line $t = t_2$ intersects the graph of the tangent line to $u(t)$ at (t_1, Y_1). The hope is that the solution curve $u(t)$ is close to $y(t)$ and hence $Y_2 \simeq y_2 \equiv y(t_2)$. Continuing in this manner, we obtain Euler's method, which is given in the following algorithm.

ALGORITHM 6.1

Given the problem (6.8), define the Euler sequence $\{Y_i\}_{i=0}^{N}$ by

$$h \leftarrow \frac{T - t_0}{N}$$

$$Y_0 \leftarrow y_0$$

Do $n = 1, 2, \ldots, N - 1$:

$$\quad Y_{n+1} \leftarrow Y_n + hf(t_n, Y_n)$$

Euler's method is single-step explicit with increment function $\Phi(t_n, Y_n, h) = f(t_n, Y_n)$. In the following example we illustrate Euler's method applied to the problem of Example 6.3.

EXAMPLE 6.5

We apply Euler's method with the coarse mesh size $h = 0.05$ to the problem $y'(t) = e^{-t} - y(t)$, $y(0) = 1$. The numerical solution approximates $y(t) = (1 + t)e^{-t}$ reasonably well considering the choice of h.

```
C   A SIMPLE PROGRAM FOR EULER'S METHOD
        G(X)=(1.+X)*EXP(-X)
        T=1.0
        N=20
        H=T/FLOAT(N)
        Y=1.0
        WRITE(6,66)
   66   FORMAT(' ',8X,'EXACT',11X,'EULER',6X,'POINT'/)
        DO 2 J=1,N
          S=H*FLOAT(J)
          T=H*FLOAT(J-1)
          Y=Y+H*F(T,Y)
          YTRU=G(S)
          WRITE(6,88) YTRU,Y,S
    2   CONTINUE
   88   FORMAT(' ',2(2X,E14.7),3X,F5.2)
        STOP
        END
C
        REAL FUNCTION F(T,Y)
        F=EXP(-T)-Y
        RETURN
        END
```

EXACT	EULER	POINT
0.9987901E 00	0.1000000E 01	0.05
0.9953206E 00	0.9975615E 00	0.10
0.9898137E 00	0.9929253E 00	0.15
0.9824767E 00	0.9863144E 00	0.20
0.9735002E 00	0.9779352E 00	0.25
0.9630631E 00	0.9679785E 00	0.30
0.9513285E 00	0.9566205E 00	0.35
0.9384478E 00	0.9440238E 00	0.40
0.9245607E 00	0.9303386E 00	0.45
0.9097954E 00	0.9157030E 00	0.50
0.8942718E 00	0.9002444E 00	0.55
0.8780982E 00	0.8840796E 00	0.60
0.8613753E 00	0.8673161E 00	0.65
0.8441949E 00	0.8500526E 00	0.70
0.8266410E 00	0.8323792E 00	0.75
0.8087918E 00	0.8143786E 00	0.80
0.7907174E 00	0.7961261E 00	0.85
0.7724822E 00	0.7776905E 00	0.90
0.7541449E 00	0.7591344E 00	0.95
0.7357585E 00	0.7405147E 00	1.00

For the same problem we also compute the maximum error

$$\max_{0 \le n \le N} |y(t_n) - Y_n| = E_N$$

for $N = 20, 30, \ldots, 100$. The corresponding values of h are 0.05, 0.033, ...,

0.01, and the maximum error decreases steadily from 0.006 to 0.0012. Note that halving h reduces the error by a factor of approximately 0.5.

MAX ERROR	N
0.0060	20
0.0039	30
0.0029	40
0.0023	50
0.0019	60
0.0017	70
0.0015	80
0.0013	90
0.0012	100

EXAMPLE 6.6

We apply Euler's method to the problem $y'(t) = y(t) - 100e^{-100t}$, $y(0) = \frac{100}{101}$, whose unique solution is $y(t) = \frac{100}{101}e^{-100t}$. The results are as follows:

MAX ERROR	N
1.5342	100
0.7177	200
0.4671	300
0.3461	400
0.2748	500
0.2278	600
0.1946	700
0.1698	800
0.1506	900
0.1353	1000
0.1228	1100
0.1124	1200
0.1037	1300
0.0962	1400
0.0897	1500

This problem is not more complicated than that of Example 6.5 (the right-hand sides of the differential equations each contain exponentials and the unknown function), but Euler's method is much less accurate for this problem even with a considerably smaller mesh size. The source of the difficulty is the ill-conditioning of the problem itself. The Euler solution does not follow the solution curve $y(t)$ but rather jumps from one solution curve of the differential equation to another (recall the discussion of Figure 6.6) as it marches across the interval. For an ill-conditioned problem these solution curves tend to diverge from one another as t increases, and herein lies the difficulty. The mesh size must be very small in order that the solution curves, on which successive Euler values fall, be close to one another. All discrete variable methods experience this difficulty to some extent when applied to ill-conditioned initial value problems—such problems are difficult to solve numerically.

EXERCISES

6.14 Apply Euler's method to the problem $y' = -y$, $y(0) = 2$. Use $h = 0.25, 0.125$, and 0.0625 to compute the approximate value of $y(1) = 1/e$. Print Y_N, $y(1)$, and h in each case.

6.15 Would it be reasonable to apply Euler's method on $[0, 2]$ to the problem

$$y'(t) = \frac{ty(t)}{y(t) - 1} \qquad y(0) = 1$$

Explain your answer.

6.16 Suppose Euler's method is applied to the problem $y' = 10y$, $y(0) = 1$. If $h = 1/N$, show that $Y_N = (1 + 10/N)^N$. Note: It is known that, for any x, $\lim_{n \to \infty} (1 + x/n)^n = e^x$.

6.17 Repeat exercise 6.16 for the initial condition $y(0) = 1 + \delta$; that is, find an expression for the perturbed Euler solution Y_N^δ. Show that $Y_N^\delta - Y_N = \delta(1 + 10/N)^N$ and hence conclude that $Y_N^\delta - Y_N \simeq \delta e^{10}$.

6.18 Carry out the following experiment to test the initial value problem $y'(t) = -10[y(t) - t^2 - 1] + 2t$, $y(0) = 1$ for conditioning.

(a) Apply Euler's method with $h = 0.0025$ to compute the approximate value of $y(0.5)$.
(b) Repeat (a) with the perturbed initial condition $y(0) = 0.99$.
(c) Repeat (a) with $y(0) = 1.01$.
(d) On the basis of your results would you say that the initial value problem is well-conditioned? Explain. Note: It is not necessary to know the solution of the initial value problem in order to answer the question.

6.19 Use Euler's method to approximate the solution of the logistic law for the U.S. population given in exercise 6.5. With $t_0 = 1790$ and $h = 0.5$, compute the value of $y(1920)$. Compare with the census for 1920.

Error and stability analysis

When using any discrete variable method to solve an initial value problem, there are two fundamental sources of error: discretization and roundoff. The **discretization error** is the difference between the solution of the initial value problem and the exact solution of the particular discrete variable method, that is, $y - Y$. In what follows, we obtain a bound for the discretization error in Euler's method.

Let $e_n = y_n - Y_n$, where $y_n = y(t_n)$ and Y_n is the Euler solution. Then for $0 \leq n \leq N - 1$, we have the equations

$$y_{n+1} = y_n + hf(t_n, y_n) + \frac{h^2}{2} y''(\xi_n)$$

and $\quad Y_{n+1} = Y_n + hf(t_n, Y_n)$

By subtraction we find

$$e_{n+1} = e_n + h[f(t_n, y_n) - f(t_n, Y_n)] + \frac{h^2}{2} y''(\xi_n)$$

If $y''(t)$ is continuous on $[t_0, T]$ and M_2 is a bound for y'', then it follows that

$$|e_{n+1}| \leq |e_n| + h|f(t_n, y_n) - f(t_n, Y_n)| + \frac{h^2}{2} M_2$$

Assume that f and $\partial f/\partial y$ are continuous on a rectangle R, $t_0 \leq t \leq T$, $c \leq y \leq d$, that contains the solution curve. Then by the mean value theorem

$$f(t_n, y_n) - f(t_n, Y_n) = \frac{\partial f}{\partial y}(t_n, \gamma_n)(y_n - Y_n) \tag{6.10}$$

for some γ_n between y_n and Y_n. Thus if L is defined by

$$L = \max_{(t, y) \in R} \left| \frac{\partial f}{\partial y}(t, y) \right| \tag{6.11}$$

it follows from (6.10) and (6.11) that

$$|f(t_n, y_n) - f(t_n, Y_n)| \leq L|e_n|$$

Consequently we have the inequality

$$|e_{n+1}| \leq (1 + Lh)|e_n| + \frac{h^2}{2} M_2 \qquad 0 \leq n \leq N - 1 \tag{6.12}$$

which relates the error at the $(n + 1)$st step to the error at the nth step. By repeated use of (6.12) we find

$$|e_{n+1}| \leq (1 + Lh)^2 |e_{n-1}| + \frac{h^2}{2} M_2[1 + (1 + Lh)]$$

$$\cdots\cdots\cdots\cdots\cdots\cdots\cdots\cdots\cdots\cdots\cdots\cdots\cdots\cdots$$

$$|e_{n+1}| \leq (1 + Lh)^{n+1} |e_0| + \frac{h^2}{2} M_2 \sum_{j=0}^{n} (1 + Lh)^j$$

But $e_0 = 0$ and from knowledge of the sum of a geometric series it follows that

$$\sum_{j=0}^{n} (1 + Lh)^j = \frac{(1 + Lh)^{n+1} - 1}{Lh}$$

which implies the estimate

$$|e_{n+1}| \leq \frac{hM_2}{2L} [(1 + Lh)^{n+1} - 1] \qquad 0 \leq n \leq N - 1$$

This, together with exercise 6.21, establishes the following theorem. Also see exercise 6.22.

THEOREM 6.3

Suppose problem (6.8) satisfies the hypotheses of Theorem 6.1 and Y is defined by Algorithm 6.1. Then for h sufficiently small the discretization error $e = y - Y$ for Euler's method satisfies

$$\max_{0 \leq n \leq N} |e_n| \leq \frac{hM_2}{2L} (e^{L(T - t_0)} - 1) \tag{6.13}$$

The significance of this theorem is that, as $h \to 0$, the discretization error in Euler's method tends to zero as a multiple of h. We say that Euler's is a **first-order method** or is first-order accurate because the exponent of h in the bound (6.13), for the global discretization error, is 1.

EXAMPLE 6.7

Suppose we treat the inequality in Theorem 6.3 as an equality, that is, for some constant C independent of h,

$$E_N \equiv \max_{0 \leq n \leq N} |e_n| = Ch$$

Then if h is reduced to $h/2$, we have $E_{2N} = Ch/2$ and hence $E_N/E_{2N} = 2$ so that the error decreases linearly with h. From the output of Example 6.5 we computed this ratio for $N = 20$, 40, and 50 and obtained 2.07, 1.93, and 1.92, respectively. Similarly, from the output of Example 6.6 we find the ratios 2.03 and 2.02 for $N = 500$ and 700, respectively.

Next we consider the effect of roundoff error on the numerical solution. Suppose that an error ε_n is introduced at the nth step, $0 \leq n \leq N - 1$, of Euler's method. Instead of computing Y_n, we compute

$$\tilde{Y}_n = Y_{n-1} + hf(t_{n-1}, Y_{n-1}) + \varepsilon_n$$

Let us consider how this error propagates in subsequent computations, assuming that no additional errors are introduced. The computed values satisfy

$$\tilde{Y}_{k+1} = \tilde{Y}_k + hf(t_k, \tilde{Y}_k) \qquad n \leq k \leq N - 1$$

Ideally we want $|\tilde{Y}_{k+1} - Y_{k+1}|$, $n \leq k \leq N - 1$, to be insensitive to the initial perturbation $\tilde{Y}_n - Y_n = \varepsilon_n$. It would be desirable to have a bound like

$$|\tilde{Y}_k - Y_k| \leq C|\tilde{Y}_n - Y_n| \qquad C \text{ independent of } h \tag{6.14}$$

or $\quad |\tilde{Y}_k - Y_k| \leq \alpha^{k-n}|\tilde{Y}_n - Y_n| \qquad \text{some } \alpha, 0 < \alpha \leq 1 \tag{6.15}$

for $n \leq k \leq N$. Then the growth of the **propagated error** would be manageable. The bound (6.14) simply says that the propagated error is bounded independent of h, whereas for $\alpha < 1$ the bound (6.15) implies that the propagated error diminishes as k increases. For $\alpha = 1$ in (6.15), the error cannot grow in subsequent steps of the computation.

We demonstrate that (6.14) is valid for Euler's method applied to every problem in the form (6.8) that satisfies the hypotheses of Theorem 6.1. We have the equations

$$\tilde{Y}_{k+1} = \tilde{Y}_k + hf(t_k, \tilde{Y}_k) \qquad n \leq k \leq N - 1$$

$$Y_{k+1} = Y_k + hf(t_k, Y_k) \qquad 0 \leq k \leq N - 1$$

Thus $\gamma = \tilde{Y} - Y$ satisfies, for $k \geq n$,

$$\gamma_{k+1} = \gamma_k + h[f(t_k, \tilde{Y}_k) - f(t_k, Y_k)]$$

from which it follows that [see (6.10)]

$$|\gamma_{k+1}| \leq |\gamma_k|(1 + Lh) \qquad k \geq n$$

where L is given in (6.11). By iterating this inequality, we obtain

$$|\gamma_{k+1}| \leq |\gamma_n|(1 + Lh)^{k+1-n} \qquad n - 1 \leq k \leq N - 1$$

but $(1 + Lh)^{k+1-n} \leq (1 + Lh)^{N-n} \leq (1 + Lh)^N \leq e^{L(T-t_0)}$ and therefore we have, for each $k = n, n + 1, \ldots, N$,

$$|\tilde{Y}_k - Y_k| \leq e^{L(T-t_0)}|\tilde{Y}_n - Y_n|$$

Consequently (6.14) is valid with $C = e^{L(T-t_0)}$.

We say that a discrete variable method is **stable** if, for each problem satisfying the hypotheses of Theorem 6.1, there exists an $h_0 > 0$ such that a perturbation in the nth step produces a bounded change in the numerical solution for all subsequent steps and all $0 < h \leq h_0$. In other words, a method is stable if, for sufficiently small mesh size, the change in the numerical solution due to a perturbation at step n is bounded. The foregoing analysis shows that Euler's method is stable. The notion of a stable discrete variable method is akin to the concept of a well-conditioned algorithm.

In practice we compute with a given mesh size $h > 0$ and want to be assured that any errors introduced at a step do not have a large effect on the computed numerical solution. We say that a discrete variable method is **absolutely stable** for a given mesh size $h > 0$ if the change in Y_k due to a perturbation of size ε in Y_n, $n < k$, is no larger than ε. Unfortunately the analysis of absolute stability is dependent (in a nontrivial way) on the particular function $f(t, y)$ in the differential equation. We are content to analyze absolute stability only for the model equation $y' = \lambda y$, that is, for $f(t, y) = \lambda y$. Therefore we modify the definition of absolute stability as follows. We say that a discrete variable method is **absolutely stable** for a given $h > 0$ if the method applied to the model problem $y' = \lambda y$ has the property that a perturbation of size ε in step n, that is, in Y_n, results in a change in subsequent values that is no larger than ε.

To analyze absolute stability for Euler's method, we proceed as follows. Application of the method to the model equation $y' = \lambda y$ gives

$$Y_{k+1} = Y_k(1 + \lambda h) \qquad 0 \leq k \leq N - 1$$

and if Y_n is perturbed by ε_n then we compute

$$\tilde{Y}_{k+1} = \tilde{Y}_k(1 + \lambda h) \qquad n \leq k \leq N - 1$$

Therefore,

$$|\tilde{Y}_{k+1} - Y_{k+1}| = |1 + \lambda h| \, |\tilde{Y}_k - Y_k| \qquad n \le k \le N - 1$$

and it follows that

$$|\tilde{Y}_{k+1} - Y_{k+1}| = |1 + \lambda h|^2 |\tilde{Y}_{k-1} - Y_{k-1}|$$

$$\cdots\cdots\cdots\cdots\cdots\cdots\cdots\cdots\cdots\cdots\cdots\cdots$$

$$= |1 + \lambda h|^{k+1-n} |\tilde{Y}_n - Y_n| \qquad k \ge n - 1$$

Consequently if $-2 \le \lambda h \le 0$, then $|1 + \lambda h| \le 1$ and

$$|\tilde{Y}_{k+1} - Y_{k+1}| \le |\tilde{Y}_n - Y_n| = |\varepsilon_n|$$

Thus Euler's method is absolutely stable provided h and λ are such that λh is in the interval $I_A = [-2, 0]$, which is called the **interval of absolute stability**. If $\lambda > 0$, there is no choice of $h > 0$ for which $\lambda h \in I_A$. For $\lambda < 0$, we require $h \le -2/\lambda$ for absolute stability.

EXAMPLE 6.8

We illustrate the significance of absolute stability by applying Euler's method to the problem

$$y'(t) = e^t - 100y(t)$$

$$y(0) = 1$$

For this problem $f(t, y) = e^t - 100y$ and $\partial f/\partial y = -100$. This a well-conditioned differential equation whose solutions are of the form $Ce^{-100t} + e^t/101$. In particular the solution of the given initial value problem is

$$y(t) = \frac{e^t + 100e^{-100t}}{101}$$

We compute the maximum error on $[0, 1]$ for $h = 1/N$ with $N = 20, 70, 120, \ldots, 520$. The program and output are as follows:

```
C   MAXIMUM ERRORS FOR EULER'S METHOD
        G(X)=(EXP(X)+100.*EXP(-100.*X))/101.
        WRITE(6,55)
    55  FORMAT(' ',3X,'MAX ERROR',5X,'N'/)
        DO 2 N=20,520,50
          T=1.0
          Y=1.0
          H=T/FLOAT(N)
          ERR=0.0
          DO 1 J=1,N
            T=H*FLOAT(J-1)
            Y=Y+H*F(T,Y)
            S=H*FLOAT(J)
            YTRU=G(S)
            DIFF=ABS(Y-YTRU)
            IF(DIFF.GT.ERR) ERR=DIFF
```

```
1     CONTINUE
      WRITE(6,77) ERR,N
2     CONTINUE
77    FORMAT(' ',2X,E11.4,3X,I3)
      STOP
      END
C
      REAL FUNCTION F(T,Y)
      F=EXP(T)-100.*Y
      RETURN
      END
```

MAX ERROR	N
0.1089E 13	20
0.6616E 00	70
0.2653E 00	120
0.1421E 00	170
0.1043E 00	220
0.7953E-01	270
0.6599E-01	320
0.5536E-01	370
0.4836E-01	420
0.4246E-01	470
0.3818E-01	520

The maximum error is enormous for $h = 0.05$, significantly less for $h = \frac{1}{70}$, and subsequently decreases steadily. We expect the numerical solution for $h = 0.05$ to not be very accurate, but how do we explain the disastrous error that is observed? The exact Euler computation is

$$Y_{k+1} = Y_k + h(e^{t_k} - 100Y_k)$$

whereas the effect of roundoff is to compute the perturbed values

$$\tilde{Y}_{k+1} = \tilde{Y}_k + h(e^{t_k} - 100\tilde{Y}_k)$$

Thus $\tilde{Y}_{k+1} - Y_{k+1} = (1 - 100h)(\tilde{Y}_k - Y_k)$ and we can interpret $\gamma \equiv \tilde{Y} - Y$ as the exact solution of Euler's method applied to the model problem

$$u'(t) = -100u(t)$$

$$u(0) = 0$$

Absolute stability for $\gamma_{k+1} = (1 - 100h)\gamma_k$ requires h to satisfy $h \leq \frac{2}{100} = 0.02$. With $h = 0.05$ the perturbation γ is amplified by $1 - 100(0.05) = -4$ at each step, and with 20 steps the error is magnified by approximately $4^{20} \simeq 10^{12}$.

In applying Euler's or any other discrete variable method, it may be advantageous to decrease the mesh size and thereby improve the accuracy and/or stability of the method. However, because floating point arithmetic has

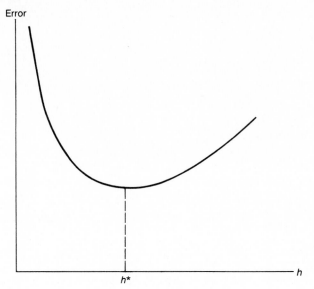

Error

h^*

h

FIGURE 6.7
Absolute error in
discrete variable
method.

a finite precision, there is a mesh size h^* beyond which no improvement is possible (Figure 6.7).

The overall error $y - \tilde{Y}$ in a discrete variable method can be written as

$$y_n - \tilde{Y}_n = y_n - Y_n + Y_n - \tilde{Y}_n$$

where the first component, the discretization error $y_n - Y_n$, gives the accuracy of the method in the absence of roundoff. The second component, $Y_n - \tilde{Y}_n$, is governed by stability. If the method is stable, then $|Y_n - \tilde{Y}_n| \leq C$ for some constant C; that is, the effect of perturbations due to roundoff is bounded. It may happen that C is large, especially if the initial value problem is ill-conditioned. For well-conditioned problems (that is, for $\partial f / \partial y < 0$) absolute stability of the method is of more importance. When a discrete variable method is applied to $y' = f(t, y)$ with $\partial f / \partial y$ negative and large in magnitude ($\partial f / \partial y \ll 0$), then absolute stability of the method is crucial; for $\partial f / \partial y > 0$ accuracy (that is, discretization error) is most important. The situation is illustrated graphically in Figure 6.8.

A problem in which $\partial f / \partial y \ll 0$ is referred to as a **stiff** problem. Stiff problems are the subject of considerable research and are not fully understood (especially for systems of ordinary differential equations). For $\lambda \ll 0$ the following problem is stiff:

$$y'(t) = \lambda[y(t) - g(t)] + g'(t) \qquad y(0) = y_0$$

where g is a given continuously differentiable function. The solution is $y(t) = [y_0 - g(0)]e^{\lambda t} + g(t)$. The first term on the right-hand side decays very rapidly, and $y(t) \simeq g(t)$. For instance if $\lambda = -1000$, $g(t) = t$, and $y_0 = 0.1$, then $y(t) \simeq t$ for $t \geq 0.005$ with an error of no more than 0.0067. Thus, for $t \geq 0.005$, $y(t)$ is essentially a linear function that can be approximated well by Euler's method with a relatively large h. However, absolute stability requires h to satisfy

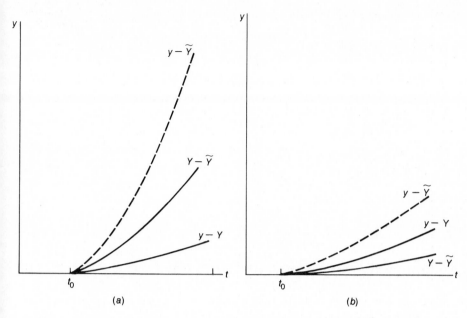

FIGURE 6.8
(a) Lack of absolute
stability. (b) Absolute
stability.

$h \leq \frac{2}{1000} = 0.002$. The effective treatment of stiff problems requires an implicit method.

The analysis we have discussed for Euler's method is indicative of the effects of accuracy and stability in applying discrete variable methods to initial value problems. In choosing a discrete variable method for a given problem, it is convenient to have an error bound (as in Theorem 6.3) that shows how the discretization error approaches zero as a power of h. This exponent of h is called the **order of accuracy** of the method and indicates how the error is reduced by a decrease in mesh size. Thus if the method is of order p and h is reduced to $h/2$, then the error should be reduced by a factor of 2^{-p}. The stability of a discrete variable method is essential, and knowledge of the interval of absolute stability may be useful in choosing a method and an appropriate mesh size.

EXERCISES

6.20 Use Example 1.6 to show that

$$\sum_{j=0}^{n} (1 + Lh)^j = \frac{(1 + Lh)^{n+1} - 1}{Lh}$$

6.21 For $x > 0$ it is known that $(1 + x/n)^n \leq e^x, n \geq 1$. Use this to show that

$$(1 + Lh)^{n+1} \leq e^{L(T-t_0)} \qquad 0 \leq n \leq N$$

*6.22 The derivation of Theorem 6.3 assumes $e_0 = 0$; however, if $y(0) = \pi$, then $e_0 \neq 0$. For $e_0 \neq 0$ show that the discretization error for Euler's method satisfies

$$\max_{1 \leq n \leq N} |e_n| \leq \frac{hM_2}{2L} (e^{L(T-t_0)} - 1) + |e_0| e^{L(t-t_0)}$$

6.23 The following problems are well-conditioned:

(a) $y'(t) = \dfrac{2y(t)}{1 + t}$, $y(0) = y_0$; solution is $y(t) = y_0(1 + t)^2$

(b) $y'(t) = e^{2t} - y(t)$, $y(0) = y_0$; solution is $y(t) = \dfrac{e^{2t}}{3} + (y_0 - \frac{1}{3})e^{-t}$

(c) $y'(t) = 2ty(t)[y(t) - 1]$, $y(0) = y_0 < 1$; solution is $y(t) = y_0/[y_0 + (1 - y_0)e^{t^2}]$

Test Euler's method on each problem with $y_0 = 0.5$, to approximate $y(1)$. Use $h = 0.005$ and print Y_N, $y(1) - Y_N$ in each case.

6.24 The following problems are ill-conditioned:

(a) $y'(t) = \dfrac{2y(t)}{100 + t}$, $y(0) = y_0$; solution is $y(t) = y_0(100 + t)^2$

(b) $y'(t) = e^{-t} + 100y(t)$, $y(0) = y_0$; solution is $y(t) = \dfrac{e^{-t}}{101} + (y_0 + \frac{1}{101})e^{100t}$

(c) $y'(t) = 1 + t^2 - 2ty(t) + [y(t)]^2$, $y(0) = y_0 > 0$; solution is $y(t) = t + \dfrac{y_0}{1 - ty_0}$

Test Euler's method on each problem with $y_0 = 5$ to approximate $y(0.18)$. Use $h = 0.005$ and print Y_N, $y(1) - Y_N$ in each case.

6.25 The following problems are stiff:

(a) $y'(t) = -10y(t) + 10 \sin t + \cos t$, $y(0) = y_0$; solution is $y(t) = y_0 e^{-10t} + \sin t$
(b) $y'(t) = -100[y(t) - t^3] + 3t^2$, $y(0) = y_0$; solution is $y(t) = y_0 e^{-100t} + t^3$

Determine a suitable value of h and apply Euler's method to approximate $y(1)$. Use $y_0 = 0$ in each case and print Y_N, $y(1) - Y_N$. What happens if Euler's method is applied with $h = 0.1$? Hint: Consider absolute stability.

6.26 The implicit method $Y_0 = y_0$, $Y_{n+1} = Y_n + hf(t_{n+1}, Y_{n+1})$ is called the **backwards Euler** method. Show that the interval of absolute stability for this method is $I_A = (-\infty, 0] \cup [2, +\infty)$.

6.27 Apply Euler's method to exercise 6.23(a) with $y_0 = 1$. For $h = 0.1$, 0.05, 0.025, and 0.0125 compute

$$\max_{0 \le n \le N} |y(1) - Y_N| = E_N$$

and print the ratios E_{2k}/E_k, $k = 10$, 20, and 40. Discuss your results in relation to Theorem 6.3.

6.28 Consider these two problems, each of which has the unique solution $y(t) = t^2$:

(a) $y' = 2t$, $y(0) = 0$
(b) $y' = -1000[y(t) - t^2] + 2t$, $y(0) = 0$

Apply Euler's method to each problem with $h = 0.01$ and $h = 0.001$ to approximate $y(1) = 1$. Report your findings, with particular attention given to accuracy and absolute stability.

6.29 Explain why accuracy (that is, discretization error) is more important than absolute stability for problems with $\partial f/\partial y > 0$.

6.3 RUNGE-KUTTA METHODS

The reason Euler's method is not used in practice is its low order of accuracy. By Theorem 6.3 it is seen that a change from mesh size h to mesh size $h/2$ reduces the error only by a factor of $\frac{1}{2}$. One method for increasing the order is to carry more terms in the Taylor expansion from which Euler's method is derived. Suppose the solution $y(t)$ of (6.8) is smooth; then by Taylor's theorem

$$y(t_{n+1}) = y(t_n) + hf(t_n, y_n) + \frac{h^2}{2!} y''(t_n) + \cdots + \frac{h^k}{k!} y^{(k)}(\xi_n)$$

where, as before, $t_n = t_0 + nh$ and $y_n = y(x_n)$. The derivatives of y can be computed in terms of partial derivatives of f using the chain rule and the differential equation. We have

$$y'(t) = f(t, y(t))$$

and hence

$$y'' = f_t + f_y y' = f_t + f_y f$$

where f_y and f_t denote the y and t partial derivatives, respectively. Thus if the Taylor expansion is truncated after three terms, there follows the approximation

$$y_{n+1} \simeq y_n + hf(t_n, y_n) + \frac{h^2}{2} [f_t(t_n, y_n) + f_y(t_n, y_n)f(t_n, y_n)]$$

which suggests the following discrete variable method (Taylor method of order 2):

$$Y_0 = y_0$$

$$Y_{n+1} = Y_n + hf(t_n, Y_n) + \frac{h^2}{2} [f_t(t_n, Y_n) + f_y(t_n, Y_n)f(t_n, Y_n)] \qquad n \geq 0$$

One obtains the Taylor method of order 3 by retaining one more term, y''', in the expansion. The chain rule applied to y'' gives

$$y''' = f_{tt} + 2f_{ty} f + f_{yy} f^2 + f_y^2 f + f_y f_t$$

Except for simple f this is not a practical approach because these partial derivatives must be determined analytically and repeatedly evaluated in the resultant method. Taylor methods are rarely used in a general-purpose library routine.

The German mathematicians Runge and Kutta discovered, in the late 1800s, a means of obtaining higher order methods without the necessity of differentiating f. The idea of the simplest Runge-Kutta methods is to obtain an approximation of the form

$$y_{n+1} \simeq y_n + h[a_1 f(t_n, y_n) + a_2 f(t_n + b_1 h, y_n + b_2 hf(t_n, y_n))] \qquad (6.16)$$

where the coefficients a_1, a_2, b_1, b_2 are determined such that the right-hand side differs from the left-hand side of (6.16) by Ch^p for p as large as possible. By using Taylor's theorem in two variables, it can be shown that $p = 3$ is the best possible value and that this occurs for a one-parameter family of coefficients given by $a_1 = 1 - \alpha$, $a_2 = \alpha$, $b_1 = b_2 = 1/2\alpha$. The derivation of this fact is somewhat tedious and not particularly instructive—we omit it. Instead we show that the one-parameter family of coefficients is at least plausible. By Taylor's theorem

$$y_{n+1} \simeq y_n + hf(t_n, y_n) + \frac{h^2}{2} y''(t_n)$$

and for any $\alpha \neq 0$

$$y''(t_n) \simeq \frac{y'(t_n + h/2\alpha) - y'(t_n)}{h/2\alpha} = \frac{f(t_n + h/2\alpha, \, y(t_n + h/2\alpha)) - f(t_n, \, y_n)}{h/2\alpha}$$

Combining these two approximations gives

$$y_{n+1} \simeq y_n + hf(t_n, \, y_n) + \alpha h \left[f\left(t_n + \frac{h}{2\alpha}, \, y\left(t_n + \frac{h}{2\alpha} \right) \right) - f(t_n, \, y_n) \right]$$

and since $y(t_n + h/2\alpha) \simeq y_n + (h/2\alpha)f(t_n, \, y_n)$ we obtain

$$y_{n+1} \simeq y_n + hf(t_n, \, y_n) + \alpha h \left[f\left(t_n + \frac{h}{2\alpha}, \, y_n + \frac{h}{2\alpha} f(t_n, \, y_n) \right) - f(t_n, \, y_n) \right]$$

which is just (6.16) with the aforementioned choice of coefficients.

Two common choices of the parameter are $\alpha = \frac{1}{2}$ (Heun's method) and $\alpha = 1$ (modified Euler method). The algorithms for these methods are as follows.

ALGORITHM 6.2

MODIFIED EULER METHOD   Given problem (6.8), generate the sequence $\{Y_n\}$ by

$$h \leftarrow \frac{T - t_0}{N}$$

$$Y_0 \leftarrow y_0$$

Do $n = 0, 1, \ldots, N - 1$:

$$\quad Y_{n+1} \leftarrow Y_n + hf\left(t_n + \frac{h}{2}, \, Y_n + \frac{h}{2} f(t_n, \, Y_n) \right)$$

ALGORITHM 6.3

HEUN'S METHOD

$$h \leftarrow \frac{T - t_0}{N}$$

$$Y_0 \leftarrow y_0$$

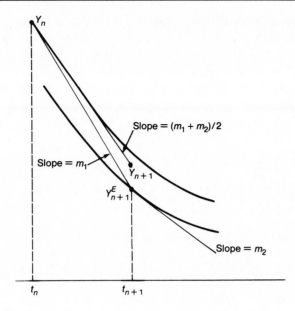

FIGURE 6.9
Heun's method.

Do $n = 0, 1, \ldots, N - 1$:

$$Y_{n+1} \leftarrow Y_n + \frac{h}{2}\left[f(t_n, Y_n) + f(t_{n+1}, Y_n + hf(t_n, Y_n))\right]$$

Heun's method has an appealing graphical interpretation (Figure 6.9) in terms of an average of two Euler steps. Let Y_{n+1}^E denote the result of applying one Euler step starting at Y_n; then the $(n + 1)$st step of Heun's method is

$$Y_{n+1} = \tfrac{1}{2}[Y_n + hf(t_n, Y_n)] + \tfrac{1}{2}[Y_n + hf(t_{n+1}, Y_n + hf(t_n, Y_n))]$$
$$= \tfrac{1}{2}Y_{n+1}^E + \tfrac{1}{2}[Y_n + hf(t_{n+1}, Y_{n+1}^E)]$$

The slope of the tangent line that determines Y_{n+1}^E is $m_1 = f(t_n, Y_n)$, and $m_2 = f(t_{n+1}, Y_{n+1}^E)$ is the slope of the tangent line that determines the next Euler step, Y_{n+2}^E. The line that determines Y_{n+1} has slope $(m_1 + m_2)/2$. Both the Heun and modified Euler methods are second-order accurate and stable. These second-order Runge-Kutta methods achieve a higher order of accuracy at a price—one more function evaluation per step than Euler's method. If N steps are carried out, then Euler's method requires N evaluations of f whereas $2N$ evaluations are needed in the Heun and modified Euler methods. However, it is likely that more Euler steps will be required to achieve the accuracy produced by either of the second-order methods.

EXAMPLE 6.9

For an illustration of the improvement in efficiency of Heun's method over Euler's method we consider the problem of Example 6.5. With $h = 0.01$ the maximum error is 0.15×10^{-4} for Heun's method:

```
C  A SIMPLE PROGRAM FOR HEUN'S METHOD
      G(X)=(1.+X)*EXP(-X)
      T=1.0
      N=100
      H=T/FLOAT(N)
      Y=1.0
      WRITE(6,66)
   66 FORMAT(' ',8X,'EXACT',11X,'HEUN ',7X,'POINT'/)
      ERR=0.0
      DO 2 J=1,N
        T=H*FLOAT(J-1)
        YE=Y+H*F(T,Y)
        YY=Y+H*F(T+H,YE)
        Y=.5*(YY+YE)
        S=H*FLOAT(J)
        YTRU=G(S)
        DIFF=ABS(Y-YTRU)
        IF(DIFF.GT.ERR) ERR=DIFF
        IF(J/10*10.NE.J) GO TO 2
        WRITE(6,88) YTRU,Y,S
    2 CONTINUE
   88 FORMAT(' ',2(2X,E14.7),3X,F5.2)
      WRITE(6,77) ERR
   77 FORMAT(' ',/2X,'THE MAXIMUM ERROR IS ',F9.6)
      STOP
      END
C
      REAL FUNCTION F(T,Y)
      F=EXP(-T)-Y
      RETURN
      END
```

EXACT	HEUN	POINT
0.9953206E 00	0.9953179E 00	0.10
0.9824767E 00	0.9824715E 00	0.20
0.9630631E 00	0.9630561E 00	0.30
0.9384478E 00	0.9384394E 00	0.40
0.9097955E 00	0.9097857E 00	0.50
0.8780983E 00	0.8780866E 00	0.60
0.8441950E 00	0.8441820E 00	0.70
0.8087919E 00	0.8087788E 00	0.80
0.7724823E 00	0.7724681E 00	0.90
0.7357587E 00	0.7357440E 00	1.00

THE MAXIMUM ERROR IS 0.000015

When Euler's method is applied with $h = 0.0025$, the maximum error is 0.285×10^{-3}:

EXACT	EULER	POINT
0.9953206E 00	0.9954280E 00	0.10
0.9824767E 00	0.9826596E 00	0.20
0.9630631E 00	0.9632975E 00	0.30
0.9384478E 00	0.9387132E 00	0.40
0.9097954E 00	0.9100765E 00	0.50
0.8780982E 00	0.8783824E 00	0.60
0.8441949E 00	0.8444726E 00	0.70
0.8087918E 00	0.8090564E 00	0.80
0.7724822E 00	0.7727280E 00	0.90
0.7357585E 00	0.7359826E 00	1.00

THE MAXIMUM ERROR IS 0.000285

For this problem Heun's method requires 200 evaluations of f (with $h = 0.01$) and Euler's method requires 400 function evaluations (with $h = 0.0025$). Moreover the maximum error for Euler's method is more than 15 times that of Heun's method.

These two second-order methods are currently used in some applications; however, one of the more popular Runge-Kutta methods is the fourth-order method given in the following algorithm.

ALGORITHM 6.4

CLASSICAL RUNGE-KUTTA METHOD Given the problem $y' = f(t, y)$, $y(t_0) = y_0$, generate values $\{Y_n\}_{n=0}^{N}$ as follows:

$Y_0 \leftarrow y_0$

$h \leftarrow \dfrac{T - t_0}{N}$

Do $n = 0, 1, \ldots, N - 1$:

$\quad k_1 \leftarrow f(t_n, Y_n)$

$\quad k_2 \leftarrow f\left(t_n + \dfrac{h}{2}, Y_n + \dfrac{h}{2} k_1\right)$

$\quad k_3 \leftarrow f\left(t_n + \dfrac{h}{2}, Y_n + \dfrac{h}{2} k_2\right)$

$\quad k_4 \leftarrow f(t_{n+1}, Y_n + hk_3)$

$\quad Y_{n+1} \leftarrow Y_n + \dfrac{h}{6}(k_1 + 2k_2 + 2k_3 + k_4)$

The method defined by this algorithm can be written as

$$Y_{n+1} = Y_n + h\Phi(t_n, Y_n, h) \qquad n = 0, 1, \dots \qquad (6.17)$$

where the increment function Φ is given by

$$\Phi(t, u, h) = \tfrac{1}{6}(k_1 + 2k_2 + 2k_3 + k_4)$$

$$k_1 = f(t, u)$$

$$k_2 = f\left(t + \frac{h}{2}, u + \frac{hk_1}{2}\right)$$

$$k_3 = f\left(t + \frac{h}{2}, u + \frac{hk_2}{2}\right)$$

$$k_4 = f(t + h, u + hk_3)$$

This method is constructed such that the exact solution of $y' = f(t, y)$ fails to satisfy (6.17) by an amount that is proportional to h^5. That is, the quantity

$$d_1^n(h) = y_{n+1} - [y_n + h\Phi(t_n, y_n, h)]$$

behaves like Ch^5 for some C that depends on y but not on h. The quantity $d_1^n(h)$ is called the **local discretization error** for method (6.17). The exponent of h in $d_1^n(h)$ exceeds by 1 the order of the method, and, as we shall see, this is typical in discrete variable methods.

In the next example we illustrate the classical Runge-Kutta method using a technique called **partial double precision**. When a single-step method of the form $Y_{n+1} = Y_n + h\Phi(t_n, Y_n, h)$ is used, the term $h\Phi(t_n, Y_n, h)$ is usually small relative to Y_n—it can be thought of as a correction to Y_n. If all computation is done in single precision, then the sum $Y_n + h\Phi(t_n, Y_n, h)$ is prone to roundoff error. On many computers the product of h and $\Phi(t_n, Y_n, h)$ is automatically done in double precision (when these two quantities are single-precision numbers), and therefore if the sum $Y_n + h\Phi(t_n, Y_n, h)$ is also computed in double precision the effect of roundoff can be significantly reduced. The technique of partial double precision can be implemented in two ways. Evaluate $\Phi(t_n, Y_n, h)$ in single precision with single-precision arguments and then either (a) perform the product and sum in double precision or (b) perform only the sum in double precision. In either case the result Y_{n+1} is saved for the next step in both its single- and double-precision forms. The single-precision form is used to evaluate Φ, and the double-precision form is used to compute the sum in the next step. The evaluation of Φ is most expensive and consequently is done in single precision. **Caution:** Some versions of FORTRAN do not permit option (b).

EXAMPLE 6.10

To illustrate the classical Runge-Kutta method, we solve the problem of Example 6.8, namely, $y'(t) = e^t - 100y(t)$, $y(0) = 1$. We use $h = 0.01$ and run both single-precision and partial double-precision versions of the method. Both implementations result in a maximum error of about 0.007, which is significantly better than the maximum error of about 0.038 produced by Euler's method for $h = 0.001923$ in Example 6.8. The partial double-precision program and output are as follows:

```
C   A SIMPLE PROGRAM FOR THE CLASSICAL 4TH
C   ORDER RUNGE-KUTTA METHOD USING
C   PARTIAL DOUBLE PRECISION
        DOUBLE PRECISION YY
        REAL KK,K1,K2,K3,K4
        G(X)=(EXP(X)+100.*EXP(-100.*X))/101.
        N=100
        A=0.0
        B=1.0
        H=(B-A)/FLOAT(N)
        YY=1.D0
        Y=YY
        ERR=0.0
        WRITE(6,66)
   66   FORMAT(' ',6X,'R-K',10X,'EXACT',6X,'POINT'/)
        DO 2 K=1,N
          T=H*FLOAT(K-1)
          TH=T+H/2.
          K1=F(T,Y)
          K2=F(TH,Y+H*K1/2.)
          K3=F(TH,Y+H*K2/2.)
          K4=F(T+H,Y+H*K3)
          KK=K1+2.*(K2+K3)+K4
          YY=YY+H*KK/6.
          Y=YY
          S=FLOAT(K)*H
          YTRU=G(S)
          DIFF=ABS(Y-YTRU)
          IF(DIFF.GT.ERR) ERR=DIFF
          IF(K/10*10.NE.K) GO TO 2
          WRITE(6,99) Y,YTRU,S
   99     FORMAT(' ',2X,F10.6,4X,F10.6,4X,F6.3)
    2   CONTINUE
        WRITE(6,88) ERR
   88   FORMAT(' ',/3X,'MAXIMUM ERROR= ',E14.7)
        STOP
        END
C
        REAL FUNCTION F(T,Y)
        F=EXP(T)-100.*Y
        RETURN
        END
```

R-K	EXACT	POINT
0.010997	0.010987	0.100
0.012093	0.012093	0.200

0.013365	0.013365	0.300
0.014771	0.014771	0.400
0.016324	0.016324	0.500
0.018041	0.018041	0.600
0.019938	0.019938	0.700
0.022035	0.022035	0.800
0.024353	0.024352	0.900
0.026914	0.026914	1.000

MAXIMUM ERROR= 0.7050445E-02

Runge-Kutta methods have been extensively studied, and methods of greater than fourth order are known. These higher order methods are more accurate than lower order ones; however, any fifth-order Runge-Kutta method requires six function evaluations, sixth-order requires seven evaluations, and for order $p \geq 7$ at least $p + 2$ function evaluations per step are necessary. Four is the highest order for which the order and number of function evaluations per step are equal—hence the popularity of fourth-order Runge-Kutta methods.

We conclude this section by giving a discretization error bound that applies to a large class of single-step methods. In particular this result applies to all of the single-step explicit methods discussed in the text.

Consider an explicit single-step method of the form

$$Y_0 = y_0$$

$$Y_{n+1} = Y_n + h\Phi(t_n, Y_n, h) \qquad n \geq 0, h = (T - t_0)/N$$

Define the **local discretization error** of the method $d_1^n(h)$ to be the amount by which the exact solution of $y' = f(t, y)$ fails to satisfy the method:

$$d_1^n(h) = y_{n+1} - [y_n + h\Phi(t_n, y_n, h)]$$

where the subscript 1 in d_1^n identifies a one-step method. This quantity depends on n, h, and derivatives of the unknown y. We have the following bound for the global discretization error $\max_{0 \leq n \leq N} |y_n - Y_n|$.

THEOREM 6.4

Assume that the increment function $\Phi(t, u, h)$ is continuous for $t_0 \leq t \leq T$, $-\infty < u < +\infty, 0 < h \leq h_0$ and satisfies

$$|\Phi(t, u, h) - \Phi(t, w, h)| \leq K|u - w|$$

for all (t, u, h), (t, w, h) in the region defined above. Then if $\Phi(t, y, 0) = f(t, y)$ and

$$|d_1^n(h)| \leq Ch^{p+1} \qquad 0 < h \leq h_0$$

the global discretization error of the method applied to problem (6.8) satisfies

$$\max_{0 \leq n \leq N} |y_n - Y_n| \leq \frac{Ch^p}{K} (e^{K(T - t_0)} - 1) \qquad (6.18)$$

is called the **order of accuracy** for
-order accurate. Notice that the
local error. The requirement
dition of the method and essen-
the right problem—the method
proof of this theorem is similar
apply the theorem, it is neces-

d hence the consistency condi-
osen to equal L, given by [see

$$(6.19)$$

owed, using Taylor's theorem,

r's method.

]

). The continuity of Φ follows

$t, u)) - f(t + h, w + hf(t, w))]$

$u - w + h(f(t, u) - f(t, w))|$

$|u - w|$

t. Since we did not derive the
utta methods), we can only
C depends, in a complicated
r analysis for the modified
hoice of K is sufficient.
other higher order methods
ffice it to say that Theorem
rithm 6.4 the order is $p = 4$
y (6.19).
or estimate because C gen-
of the unknown solution of

$y' = f(t, y)$. The proof of this result is interesting from a mathematical point of view but does not furnish the practititoner with any cogent insight regarding the application of single-step methods. What then is the significance of this theorem? It provides a sound theoretical basis on which the relative accuracy of single-step methods can be judged. Without consistency a method is doomed to failure, and without knowledge of the order the user has little rationale for a choice of method and mesh size.

We mention also that any method that satisfies Theorem 6.4 must necessarily be stable.

EXERCISES

6.30 Write a simple program for the modified Euler method and apply it to the problems of exercise 6.23. Use the same h and compare the results to Euler's method.

6.31 Apply the modified Euler method to the problems of exercise 6.24. Use the same h and compare the results to Euler's method.

6.32 Repeat exercise 6.25 for the modified Euler method. Hint: $I_A = [-2, 0]$ for the modified Euler method. Compare the results to Euler's method.

6.33* Show that $I_A = [-2, 0]$ for both the Heun and modified Euler methods.

6.34 Repeat exercise 6.23 for the classical Runge-Kutta method.

6.35 Repeat exercise 6.24 for the classical Runge-Kutta method.

6.36 Repeat exercise 6.25 for the classical Runge-Kutta method. Hint: $I_A = [-2.78, 0]$.

6.37 On the basis of exercises 6.23, 6.30, and 6.34 discuss the performance of the Euler, modified Euler, and classical Runge-Kutta methods for well-conditioned problems. Compare absolute error and number of function evaluations.

6.38 On the basis of exercises 6.24, 6.31, and 6.35 discuss the performance of the three methods for ill-conditioned problems. Compare relative error and number of function evaluations.

6.39 On the basis of exercises 6.25, 6.32, and 6.36 discuss the performance of the three methods for stiff problems. Compare relative and absolute error.

6.40 Repeat exercise 6.27 for Heun's method. Discuss the results in relation to Theorem 6.4.

6.41 Repeat exercise 6.27 for the classical Runge-Kutta method. Discuss the results in relation to Theorem 6.4.

6.42 For the modified Euler method show that $K = L(1 + Lh_0/2)$ is sufficient in Theorem 6.4.

6.43 The problem $y'(t) = f(t)$, $y(0) = y_0$ is particularly simple. Write down the equation for the classical Runge-Kutta method applied to this problem. Interpret the equation in terms of Simpson's rule.

Adaptive Runge-Kutta methods

In practice Runge-Kutta methods are not used in the straightforward manner described thus far in the text. Normally the user requires the numerical solution to deviate from the exact solution of $y' = f(t, y)$ by no more than some prescribed tolerance, say TOL. This tolerance then dictates the choice of mesh size to be used in the method. Since y is unknown, how can h be chosen given TOL? What is needed is a means of estimating the local error. This local estimator can then be used to determine an appropriate h. Moreover the mesh size should not be fixed because a relatively small choice of h may be required in some portions of the interval whereas a relatively large choice may suffice elsewhere. Thus a reasonable code for Runge-Kutta methods should incorporate a local error estimator and provide for mesh size changes in order to control the error.

To illustrate the ideas and techniques in this regard, we use the modified Euler method, that is, a second-order Runge-Kutta method. To estimate the local error, we use a third-order Runge-Kutta method. Specifically we consider the third-order method

$$U_{n+1} = U_n + h\tilde{\Phi}(t_n, U_n, h) \tag{6.20}$$

where $\tilde{\Phi}(t, u, h) = (2k_1 + 3k_2 + 4k_3)/9$ and

$$k_1 = f(t, u)$$

$$k_2 = f\left(t + \frac{h}{2}, u + \frac{hk_1}{2}\right)$$

$$k_3 = f\left(t + \frac{3h}{4}, u + \frac{3hk_2}{4}\right)$$

The local discretization error for (6.20)

$$\tilde{d}_1^n(h) = y_{n+1} - [y_n + h\tilde{\Phi}(t_n, y_n, h)]$$

is of the form $\tilde{d}_1^n(h) = \tilde{C}h^4$.

The modified Euler method can be written as

$$Y_{n+1} = Y_n + h\Phi(t_n, Y_n, h) \tag{6.21}$$

where $\Phi(t, u, h) = k_2$ and where its local discretization error $d_1^n(h) = y_{n+1} - [y_n + h\Phi(t_n, y_n, h)]$ is of the form Ch^3. To estimate $y_{n+1} - Y_{n+1}$, the error at step $n + 1$, we assume that $U_n = y_n = Y_n$. Then

$$y_{n+1} - Y_{n+1} = y_{n+1} - [y_n + h\Phi(t_n, y_n, h)] = d_1^n(h)$$

$$y_{n+1} - U_{n+1} = y_{n+1} - [y_n + h\tilde{\Phi}(t_n, y_n, h)] = \tilde{d}_1^n(h)$$

However,

$$y_{n+1} - Y_{n+1} = y_{n+1} - U_{n+1} + U_{n+1} - Y_{n+1}$$

and since $\tilde{d}_1^n(h)$ is of higher order than $d_1^n(h)$ we have the computable error estimate

$$y_{n+1} - Y_{n+1} \simeq U_{n+1} - Y_{n+1} \tag{6.22}$$

Next suppose that a local error tolerance TOL is given and we require the error per unit step to be no more than TOL. That is, we require

$$|y_{n+1} - Y_{n+1}| \leq \text{TOL} \cdot h$$

By controlling the local error per unit step, we actually control the global error. In view of (6.22) this suggests that h be chosen such that

$$\frac{|U_{n+1} - Y_{n+1}|}{h} \leq \text{TOL}$$

Note that, for $U_n = Y_n$, subtraction of (6.21) from (6.20) gives

$$\frac{U_{n+1} - Y_{n+1}}{h} = \frac{2k_1 + 4k_3 - 6k_2}{9} \tag{6.23}$$

where k_1, k_2, and k_3 are all evaluated at t_n, Y_n, and h. For simplicity we restrict mesh size changes to doubling or halving. Suppose we have found Y_n using mesh size h and want to compute the next step. We use h in (6.20) $-$ (6.21) with $U_n = Y_n$ and compute $(U_{n+1} - Y_{n+1})/h$ from (6.23). Then

If $\dfrac{|U_{n+1} - Y_{n+1}|}{h} \leq \text{TOL}$ accept Y_{n+1} and use h for next step

If $\dfrac{|U_{n+1} - Y_{n+1}|}{h} > \text{TOL}$ reject Y_{n+1} and repeat step with mesh size $h/2$

Actually we want to make sure that this procedure does not result in an endless loop in which the mesh size is repeatedly cut in half. Thus we place a lower bound on the permissible mesh size, say $\text{HMIN} \leq h$. If the mesh size strategy results in $h < \text{HMIN}$, then computation ceases and a signal is given for premature termination. Also, it is reasonable to increase the mesh size if the error estimate is significantly less than TOL. Since we allow only mesh size doubling, we use $2h$ for the next step if Y_{n+1} is accepted and the estimate is less than TOL/8, that is, if

$$\frac{|U_{n+1} - Y_{n+1}|}{h} < \frac{\text{TOL}}{8}$$

The rationale for choosing TOL/8 is as follows. Since the modified Euler method is locally third order, halving the mesh size reduces the error by about $2^{-3} = \frac{1}{8}$. In the following subroutine RK2ADP, we implement this adaptive (ADP) second-order Runge-Kutta method (RK2). Ample comments are supplied, and so the routine should be self-explanatory.

```
      SUBROUTINE RK2ADP(F,TO,T,YO,TOL,HMIN,H,Y,IFLG)
C   F:EXTERNAL FUNCTION FOR Y'=F(T,Y)
C   TO:INITIAL POINT
C   T:FINAL POINT
C   YO:INITIAL DATA, Y(TO)=YO
C   TOL:LOCAL ERROR TOLERANCE
C   HMIN:MINIMUM ALLOWABLE MESHSIZE
C   H:INITIAL MESHSIZE
C   Y:FINAL VALUE OF APPROXIMATE SOLUTION
C   IFLG:SIGNALS MODE OF RETURN,1 IS NORMAL, -1 IS ABNORMAL
      REAL K1,K2,K3
      DOUBLE PRECISION YY,DH,DK
```

```
C   INITIALIZATION
        IFLG=0
        TH=TO
        TOL1=.125*TOL
        Y=YO
        YY=YO
    1   TLEFT=T-TH
        IF(H.LE.TLEFT) GO TO 2
C   MAKE CERTAIN THAT FINAL POINT IS T
        H=TLEFT
        IFLG=1
C   MAIN COMPUTATION OF MOD-EULER APPROXIMATION
    2   K1=F(TH,Y)
        K2=F(TH+.5*H,Y+.5*H*K1)
        K3=F(TH+.75*H,Y+.75*H*K2)
        ESTERR=(2.*K1+4.*K3-6.*K2)/9.
        EST=ABS(ESTERR)
C   TEST FOR ACCEPTABLE VALUE
        IF(EST.GT.TOL) GO TO 3
C   ACCEPT YY,TAKE NEXT STEP STARTING AT TH+H
        DH=H
        DK=DH*K2
        YY=YY+DK
        Y=YY
        TH=TH+H
        WRITE(6,99)Y,TH,ESTERR
   99   FORMAT(' ',2X,3(E14.7,4X))
C   IFLG=1 SIGNALS THAT T HAS BEEN REACHED
        IF(IFLG.EQ.1) RETURN
C   INCREASE MESHSIZE IF APPROPRIATE
        IF(EST.LT.TOL1) H=2.*H
        GO TO 1
C   STEP REJECTED,RESTART WITH H/2 IF NOT TOO SMALL
    3   H=.5*H
        IFLG=0
        IF(H.GT.HMIN) GO TO 1
C   ADAPTIVE PROCEDURE FAILED
        IFLG=-1
        RETURN
        END
```

EXAMPLE 6.11

As an illustration of the use of RK2ADP and for comparison purposes, we apply the adaptive method to the problem of Example 6.9.

In that example Heun's method, which is comparable to the standard modified Euler method, was used with $h = 0.01$ and had a maximum error of 0.15×10^{-4}. The computation required 200 function evaluations. The calling program for RK2ADP and output are as follows:

```
      EXTERNAL F
      T0=0.0
      Y0=1.0
      H=.5
      HMIN=.005
      TOL=5.E-4
      T=1.0
      WRITE(6,66)
  66  FORMAT(' ',7X,'VALUE',13X,'POINT',12X,'ERREST'//)
      CALL RK2ADP(F,T0,T,Y0,TOL,HMIN,H,Y,IFLG)
      IF(IFLG.GE.0) STOP
      WRITE(6,77)
  77  FORMAT(' ',/4X,'PREMATURE TERMINATION:H TOO SMALL'/)
      WRITE(6,88) H,Y
  88  FORMAT(' ',/2X,'FINAL H= ',F7.5,2X,'FINAL Y= ',E12.5)
      STOP
      END
C
      REAL FUNCTION F(T,Y)
      F=EXP(-T)-Y
      RETURN
      END
```

VALUE	POINT	ERREST
0.9995155E 00	0.3125000E-01	0.2016624E-03
0.9981143E 00	0.6250000E-01	0.1905097E-03
0.9958535E 00	0.9375000E-01	0.1798471E-03
0.9927871E 00	0.1250000E 00	0.1696878E-03
0.9889668E 00	0.1562500E 00	0.1600583E-03
0.9844419E 00	0.1875000E 00	0.1507070E-03
0.9792596E 00	0.2187500E 00	0.1418723E-03
0.9734643E 00	0.2500000E 00	0.1334084E-03
0.9670988E 00	0.2812500E 00	0.1253552E-03
0.9602036E 00	0.3125000E 00	0.1177258E-03
0.9528174E 00	0.3437500E 00	0.1103083E-03
0.9449769E 00	0.3750000E 00	0.1033147E-03
0.9367170E 00	0.4062500E 00	0.9663899E-04
0.9280709E 00	0.4375000E 00	0.9028116E-04
0.9190703E 00	0.4687500E 00	0.8434719E-04
0.9097452E 00	0.5000000E 00	0.7841321E-04
0.9001239E 00	0.5312500E 00	0.7290310E-04
0.8902336E 00	0.5625000E 00	0.6771088E-04
0.8801000E 00	0.5937500E 00	0.6262460E-04
0.8697472E 00	0.6250000E 00	0.5796220E-04
0.8484659E 00	0.6875000E 00	0.2102322E-03
0.8265719E 00	0.7500000E 00	0.1770655E-03
0.8042176E 00	0.8125000E 00	0.1470778E-03
0.7815389E 00	0.8750000E 00	0.1200570E-03
0.7586573E 00	0.9375000E 00	0.9600320E-04
0.7356804E 00	0.1000000E 01	0.7417466E-04

Examination of the output reveals that the smallest mesh size used is 0.03125. The error at $T = 1$ is 0.783×10^{-4}, whereas in Example 6.9 the error at $T = 1$ is 0.147×10^{-4}.

The essential idea in this adaptive procedure is the use of a higher order method as an error monitor for the modified Euler method. This third-order method requires three function evaluations per step—one for each k_i. However, k_1 and k_2 are needed in the modified Euler method, and hence the error monitor necessitates only one additional evaluation. Had the third-order method required three additional evaluations, its use would have been difficult to justify.

A higher order adaptive Runge-Kutta method of recent (1969) origin is due to Fehlberg. The Runge-Kutta-Fehlberg (RKF) method uses a fourth-order Runge-Kutta method (not the classical one) and a fifth-order Runge-Kutta error monitor. A total of six function evaluations are needed per step. Shampine and Watts have written a thoroughly tested FORTRAN implementation of the RKF method, called RKF45. This subroutine provides for both absolute and relative error tolerances, is not restricted to mesh size doubling or halving, and is applicable to systems of differential equations. RKF45 is generally considered to be one of the better Runge-Kutta codes currently available. Another excellent Runge-Kutta code is subroutine DVERK in the IMSL package.

EXERCISES

6.44 Apply RK2ADP to the problems of exercise 6.23. Modify the program to count and print the total number of function evaluations. Use $\text{TOL} = 5 \times 10^{-5}$ and $\text{HMIN} = 0.0005$. Discuss your results, with particular attention to accuracy and efficiency. Compare with the results of exercise 6.37.

6.45 In RK2ADP the parameter TOL is an absolute error tolerance. For problems with large solutions TOL may not be appropriate. To illustrate this, apply the subroutine to the problem $y' = 1 + y^2$, $y(0) = 0$, with $T = 1.48$, $\text{TOL} = 10^{-5}$, and $\text{HMIN} = 0.0001$. Note: The solution is $y(t) = \tan t$.

6.46 In view of exercise 6.45 it is reasonable to modify the error tolerance specification. In RK2ADP replace TOL by TOLA and add a new parameter TOLR to the list. Change the definition of TOL1 to TOL1 = .125*TOLA and change the criterion for acceptance as follows. Insert the statement TOL2 = TOLA + TOLR*ABS(Y) immediately after the definition of EST and change the next statement to IF(EST.GT.TOL2)GO TO 3. The effect of these changes is to accept Y_{n+1} if $|U_{n+1} - Y_{n+1}|/h < \text{TOLA} + \text{TOLR}|Y_n|$. Repeat exercise 6.45 with $\text{TOLA} = 10^{-5}$, $\text{TOLR} = 10^{-3}$, and $\text{HMIN} = 0.0001$.

6.47 Apply RK2ADP, as modified in exercise 6.46, to part (c) of exercise 6.23. Choose suitable values of TOLA, TOLR, and HMIN so that the approximate solution has approximately four correct decimal digits. With your choices what is the number of correct decimal digits in the approximation to $y(1) = 0.2689414$?

6.48 Apply the modified version of RK2ADP to part (a) of exercise 6.24 and approximate $y(0.18) = 50,180.162$. Can you choose appropriate parameters to give three correct decimal digits? Is TOLA important in this problem? Justify your answer.

6.49 Apply the modified version of RK2ADP to part (b) of exercise 6.25. Determine suitable parameters to approximate $y(1) = 1$ with four correct decimal digits. Is TOLR important in this problem? Justify your answer.

6.50 Can you think of any reason for putting an upper bound HMAX on the mesh size? Hint: If arbitrary mesh size changes are permitted, a region in which the solution is changing rapidly may be skipped.

Project 7

In this project you are required to test the performance of RK2ADP as modified in exercises 6.44 and 6.46 on several types of problems. The test problems are as follows:

(a) Well-conditioned

$$y'(t) = \frac{2y(t)}{1 + t}, \ y(0) = 1; \text{ solution: } y(t) = (1 + t)^2$$

(b) Ill-conditioned

$$y'(t) = e^{-t} + 100y(t), \ y(0) = 0; \text{ solution: } y(t) = \frac{e^{100t} - e^{-t}}{101}$$

(c) Stiff $(\lambda < 0)$

$$y'(t) = \lambda[y(t) - \sin t] + \cos t, \ y(0) = 0; \text{ solution: } y(t) = \sin t$$

(d) Singular solution

$$y'(t) = |y(t)|^{3/2}, \ y(0) = 1; \text{ solution: } y(t) = \frac{4}{(2 - t)^2}$$

Perform the following computations:

1 For (a) choose suitable parameters to approximate the solution on $[0, 1]$ with

(i) relative error of no more than 10^{-2}, 10^{-4}, and 10^{-6}

(ii) absolute error of no more than 10^{-1}, 10^{-3}, and 10^{-5}

2 Repeat (1) for problem (b) on $[0, 0.1]$.

3 Repeat part (ii) of (1) for problem (c) on $[0, \pi/2]$ with $\lambda = -1, -10$, and -50.

4 Repeat part (i) of (1) for problem (d) on $[0, 1.9]$, $[0, 1.95]$, and $[0, 1.975]$.

For each program run, print the value, point, error estimate, and total number of function evaluations.

Based on your findings, write a report that summarizes the performance of RK2ADP. Discuss the efficiency of the routine versus the error specification and HMIN for each problem. For which problem is the performance most satisfactory? Least satisfactory? Explain. How does the routine handle the stiff problem for decreasing λ? Is the output for problem (d) predictable? Compare the accuracy obtained to that requested—do only for the final point in each run.

Determine whether an adaptive single-step routine is available in the library of your computer facility. If so, what features does your library program incorporate that are not in RK2ADP? Apply your library routine to one of the previous problems.

6.4 MULTISTEP METHODS

Single-step methods require information about the numerical solution only at t_n in order to compute the approximation Y_{n+1} at the next grid point $t = t_{n+1}$. In multistep methods it is necessary to have information at one or more additional preceding points, say at $t_{n-1}, \ldots, t_{n-k+1}$ for some $k \geq 2$. A widely used class of multistep methods can be derived by applying quadrature formulas to an appropriate integral form of the differential equation.

The idea is the following. Let $h = (T - t_0)/N$ and $t_n = t_0 + nh$ as before and suppose that the exact solution is known at t_k, $k \leq n$. Then by integrating $y' = f(t, y)$ from t_n to t_{n+1} we obtain

$$\int_{t_n}^{t_{n+1}} y'(t)\, dt = y_{n+1} - y_n = \int_{t_n}^{t_{n+1}} f(t, y(t))\, dt$$

or, equivalently,

$$y_{n+1} = y_n + \int_{t_n}^{t_{n+1}} f(t, y(t))\, dt \tag{6.24}$$

To obtain the quadrature formula for (6.24), we let $p_{k-1}(x)$ denote the polynomial of degree $\leq k - 1$ that interpolates $f(t, y(t))$ at $t_n, t_{n-1}, \ldots, t_{n-k+1}$. Then we use the approximation

$$\int_{t_n}^{t_{n+1}} f(t, y(t))\, dt \simeq \int_{t_n}^{t_{n+1}} p_{k-1}(t)\, dt \tag{6.25}$$

and integrate $p_{k-1}(t)$ to obtain a quadrature of the form

$$\int_{t_n}^{t_{n+1}} p_{k-1}(t)\, dt = h\{A_0 f_n + A_1 f_{n-1} + \cdots + A_{k-1} f_{n-k+1}\}$$

where f_j denotes $f(t_j, y(t_j))$. The weights $A_0, \ldots, A_k$ have been tabulated for many choices of k and so has the corresponding **local discretization error**

$$d_k^n(h) = \int_{t_n}^{t_{n+1}} f(t, y(t))\, dt - \int_{t_n}^{t_{n+1}} p_{k-1}(t)\, dt$$

$$= y_{n+1} - \left\{ y_n + h \sum_{i=0}^{k-1} A_i f_{n-i} \right\}$$

Thus the exact solution of $y' = f(t, y)$ satisfies

$$y_{n+1} = y_n + h \sum_{i=0}^{k-1} A_i f_{n-i} + d_k^n(h)$$

where $d_k^n(h)$ depends on n, k, and y. A k-step discrete variable method results when $d_k^n(h)$ is ignored and y is replaced by Y:

$$Y_{n+1} = Y_n + h \sum_{i=0}^{k-1} A_i f(t_{n-i}, Y_{n-i}) \qquad n \geq k - 1 \tag{6.26}$$

The philosophy behind this approach is as follows. Since the solution of $y' = f(t, y)$ satisfies (6.26) approximately [assuming $d_k^n(h)$ is small], one hopes that the exact solution of (6.26) is approximately equal to y at the grid points.

The simplest case is $k = 1$, which, as we next demonstrate, results in Euler's method. For $k = 1$ the constant polynomial that interpolates f at t_n is $p_0(t) = f_n = y_n'$. By the error formula for polynomial interpolation we have

$$y'(t_n) - p_0(t) = y'[t_n, t](t - t_n)$$

and hence

$$y_{n+1} = y_n + \int_{t_n}^{t_{n+1}} p_0(t) \, dt + d_1^n(h) = y_n + hf(t_n, y_n) + d_1^n(h) \tag{6.27}$$

where

$$d_1^n(h) = \int_{t_n}^{t_{n+1}} y'[t_n, t](t - t_n) \, dt$$

Since $t - t_n \geq 0$ on $[t_n, t_{n+1}]$, an application of the integral mean value theorem gives

$$d_1^n(h) = y'[t_n, c_n] \int_{t_n}^{t_{n+1}} (t - t_n) \, dt$$

$$= \frac{h^2}{2} y''(\xi_n) \qquad \text{some } \xi_n \in (t_n, t_{n+1})$$

Therefore Euler's method results when $d_1^n(h)$ is dropped in (6.27) and y is replaced by Y. For $d_1^n(h)$ small (that is, small h) the exact solution of $y' = f(t, y)$ satisfies Euler's method approximately: $y_{n+1} \simeq y_n + hf(t_n, y_n)$.

For $k = 2$ the linear interpolate of $y' = f$ at t_n, t_{n-1} is $p_1(t) = f_{n-1} + \Delta f_{n-1}(t - t_{n-1})$, and by the error formula for polynomial interpolation we have

$$y'(t) - p_1(t) = f(t, y(t)) - p_1(t) = y'[t_{n-1}, t_n, t](t - t_{n-1})(t - t_n)$$

By straightforward calculation it follows that

$$\int_{t_n}^{t_{n+1}} p_1(t) \, dt = h(\tfrac{3}{2} f_n - \tfrac{1}{2} f_{n-1})$$

and there results the two-step method

$$Y_{n+1} = Y_n + h[\tfrac{3}{2} f(t_n, Y_n) - \tfrac{1}{2} f(t_{n-1}, Y_{n-1})] \qquad n \geq 1$$

TABLE 6.1   Adams-Bashforth
coefficients

k	A_0	A_1	A_2	A_3
1	1			
2	$\frac{3}{2}$	$-\frac{1}{2}$		
3	$\frac{23}{12}$	$-\frac{16}{12}$	$\frac{5}{12}$	
4	$\frac{55}{24}$	$-\frac{59}{24}$	$\frac{37}{24}$	$-\frac{9}{24}$

Notice that two starting values, Y_0 and Y_1, are needed in this two-step method. In practice one chooses $Y_0 = y_0$ and Y_1 is determined by an appropriate single-step method. We postpone the discussion of starting procedures until later in the section.

The local discretization error for this two-step method is

$$d_2^n(h) = \int_{t_n}^{t_{n+1}} y'[t_{n-1}, t_n, t](t - t_{n-1})(t - t_n)\, dt$$

and, as before, the integral mean value theorem gives

$$d_2^n(h) = y'[t_{n-1}, t_n, c_n] \int_{t_n}^{t_{n+1}} (t - t_{n-1})(t - t_n)\, dt$$

$$= \tfrac{5}{12}h^3 y'''(\xi_n) \qquad \text{some } \xi_n \in (t_{n-1}, t_{n+1})$$

By now the procedure for the derivation of these multistep methods should be clear: Replace the integral in (6.24) by the integral of $p_{k-1}(t)$, ignore the local discretization error $d_k^n(h)$, and in the resultant expression use Y in place of y. We summarize the coefficients $A_0, \ldots, A_{k-1}$ in Table 6.1 for several values of k. These methods are all explicit and are commonly called Adams-Bashforth methods.

The local discretization error for Adams-Bashforth methods takes the form

$$d_k^n(h) = \gamma_k h^{k+1} y^{(k+1)}(\xi_n) \qquad \text{some } \xi_n$$

where the error coefficient γ_k is given in Table 6.2. The expression for $d_k^n(h)$ is obtained from the formula for polynomial interpolation error.

A popular Adams-Bashforth method is the four-step method given in the following algorithm.

TABLE 6.2   Error coefficients
for Adams-Bashforth methods

k	1	2	3	4
γ_k	$\frac{1}{2}$	$\frac{5}{12}$	$\frac{3}{8}$	$\frac{251}{720}$

ALGORITHM 6.5

Given problem (6.8), determine $\{Y_i\}_{i=0}^N$ by

$$h \leftarrow \frac{T - t_0}{N}$$

$$Y_0 \leftarrow y_0$$

Y_1, Y_2, Y_3 are determined by a single-step method.

Do $n = 3, \ldots, N - 1$:

$$Y_{n+1} \leftarrow Y_n + \frac{h}{24} \left[55f(t_n, Y_n) - 59f(t_{n-1}, Y_{n-1}) \right.$$
$$\left. + 37f(t_{n-2}, Y_{n-2}) - 9f(t_{n-3}, Y_{n-3}) \right]$$

When using a multistep method, it is particularly important to have accurate starting values. A common practice is to use a single-step method whose local discretization error is of the same order of accuracy as that of the multistep method at hand. Thus in the next example we generated the starting values by the classical Runge-Kutta method.

EXAMPLE 6.12

The four-step Adams-Bashforth method is applied to the problem of Example 6.5: $y'(t) = e^{-t} - y(t)$, $y(0) = 1$. For $h = 0.01$ we used partial double precision and obtained the following results:

A–B	EXACT	POINT
0.9953212	0.9953206	0.100
0.9824769	0.9824767	0.200
0.9630637	0.9630631	0.300
0.9384480	0.9384478	0.400
0.9097959	0.9097955	0.500
0.8780985	0.8780983	0.600
0.8441949	0.8441950	0.700
0.8087920	0.8087919	0.800
0.7724822	0.7724823	0.900
0.7357588	0.7357587	1.000

The computed values agree with the exact solution to nearly six decimal digits.

In Example 6.12 we used the classical Runge-Kutta method, whose local discretization is fifth-order accurate, to start the four-step Adams-Bashforth method. Since both methods have the same local order of accuracy, why should we invest the additional effort to code a single-step starting method and a multistep "main" method? Why not simply use the single-step method?

A partial answer lies in the relative efficiency of the methods. Normally the bulk of computational effort is involved in the evaluation of f. Each step of the classical Runge-Kutta method requires four function evaluations, whereas, for $n \geq 3$, the determination of Y_{n+1} requires only one evaluation in Algorithm 6.5. Consequently the multistep method can be considerably faster than the single-step Runge-Kutta method of the same order. A more significant consideration is error estimation and control through mesh size changes. In Section 6.5 we see that local error estimation is relatively easy and inexpensive when multistep methods are used in a predictor-corrector scheme.

Multistep implicit methods are found the same way explicit methods are except that now the polynomial also interpolates f at t_{n+1}. In this context let $q_k(t)$ denote the polynomial of degree $\leq k$ that interpolates $f(t, y(t))$ at t_{n+1}, t_n, $\ldots, t_{n-k+1}$. Then using the quadrature approximation

$$\int_{t_n}^{t_{n+1}} f(t, y(t)) \, dt \simeq \int_{t_n}^{t_{n+1}} q_k(t) \, dt$$

and proceeding as before, we obtain an implicit k-step method of the form

$$Y_{n+1} = Y_n + h\{a_0 f(t_{n+1}, Y_{n+1}) + a_1 f(t_n, Y_n) + \cdots$$
$$+ a_k f(t_{n-k+1}, Y_{n-k+1})\}$$

For $k = 1$ the quadrature formula is the familiar trapezoidal rule and the resulting single-step method is

$$Y_{n+1} = Y_n + \frac{h}{2} [f(t_{n+1}, Y_{n+1}) + f(t_n, Y_n)] \qquad n \geq 0$$

Other choices of k and the corresponding coefficients are given in Table 6.3. These implicit multistep methods are commonly called Adams-Moulton methods.

The **local discretization error**

$$d_k^n(h) = y_{n+1} - \left\{ y_n + h \sum_{i=0}^{k} a_i f_{n-i+1} \right\}$$

for the k-step Adams-Moulton method can be shown to be of the form

$$d_k^n(h) = \gamma_{k+1}^* h^{k+2} y^{(k+2)}(\xi_n)$$

for some constant γ_{k+1}^* and some $\xi_n \in (t_{n-k+1}, t_{n+1})$. Several values for γ_{k+1}^*

TABLE 6.3   Adams-Moulton coefficients

k	a_0	a_1	a_2	a_3	a_4
1	$\frac{1}{2}$	$\frac{1}{2}$			
2	$\frac{5}{12}$	$\frac{8}{12}$	$-\frac{1}{12}$		
3	$\frac{9}{24}$	$\frac{19}{24}$	$-\frac{5}{24}$	$\frac{1}{24}$	
4	$\frac{251}{720}$	$\frac{646}{720}$	$-\frac{264}{720}$	$\frac{106}{720}$	$-\frac{19}{720}$

TABLE 6.4 Error coefficients for Adams-Moulton

k	1	2	3	4
γ^*_{k+1}	$-\frac{1}{12}$	$-\frac{1}{24}$	$-\frac{19}{720}$	$-\frac{3}{160}$

are given in Table 6.4. Note that the local error of a k-step Adams-Moulton method is one order higher than that of a k-step Adams-Bashforth method.

Adams-Moulton methods have the disadvantage, which is common to all multistep methods, that they are not self-starting. The values Y_1, Y_2, ..., Y_{k-1} must be determined by other means for a k-step method. The major disadvantage of implicit multistep methods like Adams-Moulton is that Y_{n+1} cannot be determined exactly—even in the absence of roundoff. It is not even apparent that Y_{n+1} exists. To illustrate, consider the one-step Adams-Moulton (that is, trapezoidal rule) method

$$Y_{n+1} = Y_n + \frac{h}{2} [f(t_{n+1}, Y_{n+1}) + f(t_n, Y_n)] \qquad (6.28)$$

How do we know that there is a number Y_{n+1} that satisfies (6.28)? If Y_{n+1} does exist, how do we find it? Assuming for the moment that Y_{n+1} exists, we could try to find it by iteration:

$$Y^{(i+1)}_{n+1} = Y_n + \frac{h}{2} [f(t_{n+1}, Y^{(i)}_{n+1}) + f(t_n, Y_n)] \qquad i \geq 0$$

where $Y^{(0)}_{n+1}$ is some educated guess for Y_{n+1}. The hope is that $\lim_{i \to \infty} Y^{(i)}_{n+1} = Y_{n+1}$. It turns out that this iteration converges to the solution Y_{n+1} of (6.28) for h sufficiently small. The precise statement of this fact is given in the following general theorem for Adams-Moulton methods.

THEOREM 6.5

Suppose f and $\partial f/\partial y$ are continuous for $t_0 \leq t \leq T$ and $-\infty < y < +\infty$ and L is defined by $\max_{t_0 \leq t \leq T} \max_{-\infty < y < \infty} |\partial f(t, y)/\partial y|$. Then if $h \leq 1/(|a_0|L)$, the k-step Adams-Moulton method is uniquely solvable and Y_{n+1} can be determined by

$$Y_{n+1} = \lim_{i \to \infty} Y^{(i)}_{n+1}$$

where $\{Y^{(i)}_{n+1}\}$ is given by

$$Y^{(i+1)}_{n+1} = Y_n + h\{a_0 f(t_{n+1}, Y^{(i)}_{n+1}) + \cdots + a_k f(t_{n-k+1}, Y_{n-k+1})\}$$

where $Y^{(0)}_{n+1}$ is arbitrary.

It follows, for the one-step method, that the iteration converges to Y_{n+1} if $h \leq 2/L$ since $a_0 = \frac{1}{2}$. This theorem gives sufficient conditions for existence and uniqueness of a solution and also provides a means of determining it—at least approximately. Unfortunately this iteration may converge very slowly unless the initial guess $Y_{n+1}^{(0)}$ is sufficiently close to Y_{n+1}. This can be very costly because each iteration requires an evaluation of f. For this and other reasons (which we discuss later) implicit methods are almost never used by themselves. Rather they are used in conjunction with explicit multistep methods to form a predictor-corrector pair. The explicit method predicts a good initial guess $Y_{n+1}^{(0)}$ for the iteration, and Y_{n+1} can usually be determined with sufficient accuracy in one or two iterations. This is the subject of the next section.

EXERCISES

6.51 Fill in the details of the derivation of $d_2^n(h)$ for the two-step Adams-Bashforth formula.

6.52 Derive the two-step Adams-Bashforth formula.

6.53 Show that the local discretization error for Algorithm 6.5 is given by

$$
d_4^n(h) = \int_{t_n}^{t_{n+1}} y'[t_n, t_{n-1}, t_{n-2}, t_{n-3}, t] \prod_{i=0}^{3} (t - t_{n-i}) \, dt
$$

$$
= \frac{y^{(5)}(\xi_n)}{4!} \int_{t_n}^{t_{n+1}} \prod_{i=0}^{3} (t - t_{n-i}) \, dt \qquad \text{some } \xi_n \in (t_{n-3}, t_{n+1})
$$

Hint: See (4.13) and Theorem 4.3.

6.54 Write a simple program for the two-step Adams-Bashforth method and apply it to exercise 6.23(a). Use partial double precision and the modified Euler method for starting values.

6.55 Write a subroutine AB4 for Algorithm 6.5 that has a parameter list (F,TO,YO,T,N) and accepts an external function F for the right-hand side, the initial data in TO and YO, the final point T, and the number N of steps from TO to T. The subroutine should use the classical Runge-Kutta method for starting values and print the point t_n and value Y_n at each step n. Apply AB4 to one problem each of exercises 6.23, 6.24, and 6.25. Try several values of N and discuss your results.

*6.56 Instead of (6.24) one can integrate the differential equation over $[t_{n-1}, t_{n+1}]$ to obtain

$$
y_{n+1} = y_{n-1} + \int_{t_{n-1}}^{t_{n+1}} f(t, y(t)) \, dt
$$

Replace the integrand by the polynomial of degree ≤ 2 that interpolates f at t_{n-1}, t_n, and t_{n+1}. Use this to derive the implicit formula (Simpson's rule)

$$
Y_{n+1} = Y_{n-1} + \frac{h}{3} [f(t_{n+1}, Y_{n+1}) + 4f(t_n, Y_n) + f(t_{n-1}, Y_{n-1})]
$$

Show that I_A for this method is the single point $\{0\}$.

6.57 Derive the one-step Adams-Moulton formula and its local discretization error $d_1^n(h)$.

6.58 Write a simple program to carry out the iteration

$$Y_{n+1}^{(i)} = Y_n + \frac{h}{2} \left[f(t_{n+1}, Y_{n+1}^{(i-1)}) + f(t_n, Y_n) \right] \qquad i \geq 1$$

Terminate the iteration when $|Y_{n+1}^{(i)} - Y_{n+1}^{(i-1)}| \leq \varepsilon |Y_{n+1}^{(i)}|$. Apply the program to the problem $y'(t) = 2y(t)/(10 + t)$, $y(0) = 1$, and compute Y_2 for an appropriate h and ε. Hint: See Theorem 6.5 and exercise 2.27. The solution of the initial value problem is $y(t) = (10 + t)^2$.

6.59 Show that $I_A = (-\infty, 0]$ for the one-step Adams-Moulton formula (trapezoidal rule).

6.60 Show how Newton's method could be applied to find Y_2 in exercise 6.58.

6.5 PREDICTOR-CORRECTOR METHODS

For clarity and ease of exposition we describe a second-order predictor-corrector scheme. The techniques carry out to higher order methods with relatively minor modifications. Thus we use the second-order implicit method

$$\text{C:} \quad Y_{n+1} = Y_n + \frac{h}{2} \left[f(t_{n+1}, Y_{n+1}) + f(t_n, Y_n) \right]$$

and generate the initial guess $Y_{n+1}^{(0)}$ for the iteration described in the previous section by the explicit second-order method

$$\text{P:} \quad Y_{n+1} = Y_n + \frac{h}{2} \left[3f(t_n, Y_n) - f(t_{n-1}, Y_{n-1}) \right]$$

The latter is called the **predictor** (P) formula, and the former is the **corrector** (C). In practice the corrector is not iterated to convergence as described in Theorem 6.5; rather the corrector is applied a fixed number of times (usually one or two), say k, and then $Y_{n+1}^{(k)}$ is accepted as the approximation to Y_{n+1}. To illustrate the calculations involved, we take $k = 1$ and then with the notation $f_n^{(i)} \equiv f(t_n, Y_n^{(i)})$ we compute

$$\text{P:} \quad Y_{n+1}^{(0)} = Y_n^{(1)} + \frac{h}{2} (3f_n^{(1)} - f_{n-1}^{(1)})$$

$$\text{E:} \quad f_{n+1}^{(0)} = f(t_{n+1}, Y_{n+1}^{(0)})$$

$$\text{C:} \quad Y_{n+1}^{(1)} = Y_n^{(1)} + \frac{h}{2} (f_n^{(1)} + f_{n+1}^{(0)})$$

$$\text{E:} \quad f_{n+1}^{(1)} = f(t_{n+1}, Y_{n+1}^{(1)})$$

Then $Y_{n+1}^{(1)}$ is accepted as an approximation to y_{n+1} and used in the next step. The process so described is referred to as the PECE mode, where E stands for evaluate. A less expensive mode, again with $k = 1$, is the PEC mode, given by

P: $\quad Y_{n+1}^{(0)} = Y_n^{(1)} + \dfrac{h}{2}(3f_n^{(0)} - f_{n-1}^{(0)})$

E: $\quad f_{n+1}^{(0)} = f(t_{n+1}, Y_{n+1}^{(0)})$

C: $\quad Y_{n+1}^{(1)} = Y_n^{(1)} + \dfrac{h}{2}(f_{n+1}^{(0)} + f_n^{(0)})$

which requires one fewer evaluation per step. Even though the PECE mode is more expensive, it is preferred because of its superior stability characteristics. If k iterations of the corrector are used, then the mode is either $P(EC)^kE$ or $P(EC)^k$, but we do not pursue this approach.

The following algorithm defines the PECE mode for the second-order predictor-corrector scheme.

ALGORITHM 6.6

Given the problem $y' = f(t, y)$, $y(t_0) = y_0$, and the final point T, generate the sequence $\{Y_n\}$ by

$$h \leftarrow \frac{T - t_0}{N}$$

$Y_0^{(1)} \leftarrow y_0$

$Y_1^{(1)} \leftarrow$ determined by a second-order one-step method

Do $n = 1, 2, \ldots, N - 1$:

$$Y_{n+1}^{(0)} \leftarrow Y_n^{(1)} + \frac{h}{2}[3f(t_n, Y_n^{(1)}) - f(t_{n-1}, Y_{n-1}^{(1)})]$$

$$Y_{n+1}^{(1)} \leftarrow Y_n^{(1)} + \frac{h}{2}[f(t_{n+1}, Y_{n+1}^{(0)}) + f(t_n, Y_n^{(1)})]$$

In this algorithm we do not explicitly identify the evaluation phase—its use is implicit.

In what follows, we describe three subroutines that implement an adaptive variable mesh size predictor-corrector scheme using Algorithm 6.6. This scheme is similar in spirit to the adaptive Runge-Kutta method of Section 6.3 in that a local error estimator is used to control the error per unit step and only mesh size halving or doubling is utilized. We begin by deriving the local error estimator. Assume that $Y_{n-1}^{(1)} = y_{n-1}$ and $Y_n^{(1)} = y_n$ in the predictor; then

$$y_{n+1} - Y_{n+1}^{(0)} = y_{n+1} - \left\{ y_n + \frac{h}{2}[3f(t_n, y_n) - f(t_{n-1}, y_{n-1})] \right\} \tag{6.29}$$

$$= \gamma_2 h^3 y'''(\xi_n)$$

where γ_2 is given in Table 6.2. Similarly from the corrector we find

$$y_{n+1} - Y^{(1)}_{n+1} = y_{n+1} - \left\{ y_n + \frac{h}{2} \left[f(t_{n+1}, Y^{(0)}_{n+1}) + f(t_n, y_n) \right] \right\}$$

$$= \gamma_2^* h^3 y'''(\mu_n) + \frac{h}{2} \left[f(t_{n+1}, y_{n+1}) - f(t_{n+1}, Y^{(0)}_{n+1}) \right] \qquad (6.30)$$

In equation (6.30) the mean value theorem gives, for some c_n between y_{n+1} and $Y^{(0)}_{n+1}$,

$$f(t_{n+1}, y_{n+1}) - f(t_{n+1}, Y^{(0)}_{n+1}) = \frac{\partial f}{\partial y}(t_n, c_n)(y_{n+1} - Y^{(0)}_{n+1})$$

and by (6.29) we find that

$$\frac{h}{2} \left[f(t_{n+1}, y_{n+1}) - f(t_{n+1}, Y^{(0)}_{n+1}) \right] = \gamma_2 \frac{h^4}{2} y'''(\xi_n) \frac{\partial f}{\partial y}(t_{n+1}, c_n)$$

Since this is a fourth-order term (in h), we neglect it in (6.30) and write

$$y_{n+1} - Y^{(1)}_{n+1} \simeq \gamma_2^* h^3 y'''(\mu_n) \qquad (6.31)$$

Assuming that y''' does not vary much on $[t_n, t_{n+1}]$, we subtract (6.29) from (6.31) and obtain

$$Y^{(0)}_{n+1} - Y^{(1)}_{n+1} \simeq (\gamma_2^* - \gamma_2)h^3 y'''$$

$$\simeq \frac{\gamma_2^* - \gamma_2}{\gamma_2^*} \gamma_2^* h^3 y'''$$

Hence by (6.31) there results

$$Y^{(0)}_{n+1} - Y^{(1)}_{n+1} \simeq \frac{\gamma_2^* - \gamma_2}{\gamma_2^*} (y_{n+1} - Y^{(1)}_{n+1})$$

or, equivalently,

$$y_{n+1} - Y^{(1)}_{n+1} \simeq \frac{-\gamma_2^*}{\gamma_2 - \gamma_2^*} (Y^{(0)}_{n+1} - Y^{(1)}_{n+1}) = \frac{Y^{(0)}_{n+1} - Y^{(1)}_{n+1}}{6} \qquad (6.32)$$

since $\gamma_2^* = -\frac{1}{12}$ and $\gamma_2 = \frac{5}{12}$. Equation (6.32) furnishes a computable estimate for $y_{n+1} - Y^{(1)}_{n+1}$ that is used in exactly the same way as in RK2ADP to control mesh size. Thus if $|Y^{(0)}_{n+1} - Y^{(1)}_{n+1}|/6h > \text{TOL}$, we reject $Y^{(1)}_{n+1}$ and repeat the step with mesh size $h/2$. If $\text{TOL}/8 \leq |Y^{(0)}_{n+1} - Y^{(1)}_{n+1}|/6h \leq \text{TOL}$, we accept $Y^{(1)}_{n+1}$ and use h for the next step. The mesh size is doubled if $\text{TOL}/8 > |Y^{(0)}_{n+1} - Y^{(1)}_{n+1}|/6h$. The following is the main subroutine for the second-order Adams-Bashforth-Moulton (ABM) adaptive (ADP) scheme:

```
      SUBROUTINE ABMADP(F,TO,YO,H,T,TOL,HMIN,YOUT,IFLG)
C  F:EXTERNAL FUNCTION FOR Y'=F(T,Y)
C  TO:INITIAL POINT
C  YO:INITIAL VALUE, Y(TO)=YO
C  T:FINAL POINT WHERE APPROXIMATION IS REQUIRED
C  H:INITIAL MESHSIZE, MUST BE > HMIN,MUST BE < T-TO
```

```
C   TOL:LOCAL ABSOLUTE ERROR TOLERANCE
C   HMIN:MINIMUM ALLOWABLE MESHSIZE
C   YOUT:FINAL VALUE OF APPROXIMATION
C   IFLG:SIGNALS MODE OF RETURN,-1 MEANS ACCURATE START NOT
C        FOUND,0 IS PREMATURE TERMINATION,1 IS NORMAL EXIT
C   FUN:WORKING ARRAY OF F VALUES
C   Y:WORKING ARRAY OF VALUES FOR APPROX. SOLUTION
          REAL FUN(4),Y(4)
C   INITIALIZATION
          TOL1=.125*TOL
          IFLG=-1
          Y(2)=YO
          FUN(2)=F(TO,YO)
          TH=TO
C   GENERATE STARTING VALUES
    1     CALL MODEUL(FUN,Y,TH,H,F)
          GO TO 3
C   MAKE SURE LAST POINT IS T
    2     TLEFT=T-TH
          IF(H.LT.TLEFT) GO TO 3
          IFLG=1
          HSTEP=H-TLEFT
          H=TLEFT
          Y(1)=Y(3)
          FUN(1)=FUN(3)
          CALL MODEUL(FUN,Y,TH,HSTEP,F)
          Y(2)=Y(3)
          Y(3)=Y(1)
          FUN(2)=FUN(3)
          FUN(3)=FUN(1)
C   MAKE PREDICTOR-CORRECTOR STEP
    3     CALL ABAM(FUN,Y,TH,H,F,ERR)
C   TEST FOR ACCEPTABLE VALUE
          IF(IFLG.EQ.1) GO TO 7
          TOLH=TOL*H
          IF(ERR.GT.TOLH) GO TO 6
C   ACCEPT VALUE
    7     TH=TH+H
          YOUT=Y(3)
          IF(IFLG.GE.0) YOUT=Y(4)
          WRITE(6,88) YOUT,TH,ERR
   88     FORMAT(' ',2X,E14.7,4X,F9.6,4X,E14.7)
C   IFLG=1 MEANS THAT T HAS BEEN REACHED ; NORMAL RETURN
          IF(IFLG.EQ.1) RETURN
          IFLG=0
C   IS H TOO SMALL?
          TOL1H=TOL1*H
          IF(ERR.LT.TOL1H) GO TO 5
C   CONTINUE USING H FOR NEXT STEP
          DO 4 J=1,3
            Y(J)=Y(J+1)
            FUN(J)=FUN(J+1)
```

```
      4   CONTINUE
          GO TO 2
C   CONTINUE USING 2H FOR NEXT STEP
      5   H=2.*H
          Y(3)=Y(4)
          FUN(3)=FUN(4)
          GO TO 2
C   REJECT APPROXIMATION,REDUCE H TO H/2 AND REPEAT STEP
      6   YOUT=Y(3)
          H=.5*H
C   MAKE SURE NEW H IS OK
          IF(HMIN.GT.H) RETURN
C   HAS AT LEAST ONE P-C STEP BEEN ACCEPTED?
          IF(IFLG.LT.0) GO TO 1
          Y(4)=Y(3)
          FUN(4)=FUN(3)
          Y(1)=Y(2)
          FUN(1)=FUN(2)
C   OBTAIN NECESSARY VALUES FOR REDUCED MESHSIZE
C   USING MODIFIED EULER METHOD
          CALL MODEUL(FUN,Y,TH,H,F)
          Y(2)=Y(3)
          Y(3)=Y(4)
          FUN(2)=FUN(3)
          FUN(3)=FUN(4)
          GO TO 2
          END
```

When an adaptive multistep scheme is being coded, there arises a difficulty not present in single-step adaptive methods. In halving the mesh size, it is necessary to generate new starting values for the next step. One way to accomplish this is to use a single-step method of the same order. Our code uses this approach with subroutine MODEUL, which computes a single step of the modified Euler method. This subroutine is also used to generate sufficiently accurate starting values at the initiation of the scheme.

```
          SUBROUTINE MODEUL(FUN,Y,T,H,F)
C   GIVEN VALUES AT T IN FUN(2),Y(2) COMPUTE ONE STEP
C   AND STORE VALUES FOR STEP IN FUN(3),Y(3)
          REAL FUN(4),Y(4)
          DOUBLE PRECISION DH,YY
          DH=H
          YY=Y(2)
          PHI=F(T+.5*H,Y(2)+.5*H*FUN(2))
          YY=YY+DH*PHI
          Y(3)=YY
          FUN(3)=F(T+H,Y(3))
          RETURN
          END
```

The subroutine ABAM performs one predictor-corrector step in PECE mode and returns the estimated error in ERR:

```
      SUBROUTINE ABAM(FUN,Y,T,H,F,ERR)
C  EXPLANATION OF VALUES IN WORKING ARRAYS Y,FUN
C  Y:Y(2)=Y(T-H),Y(3)=Y(T),Y(4)IS VALUE FOR THIS STEP
C    T IS POINT WHERE MOST RECENT
C    APPROXIMATION WAS ACCEPTED
C    Y(1) IS DUMMY STORAGE USED WHEN H IS HALVED
C  FUN:FUN(J)=VALUE OF F AT Y(J) J=2,3,4
C    FUN(1) IS DUMMY STORAGE
      REAL FUN(4),Y(4)
      DOUBLE PRECISION DH,DY0,DY1
      DH=H
      DY1=Y(3)
C  PREDICT
      DY0=DY1+.5D0*DH*(3.*FUN(3)-FUN(2))
      Y0=DY0
C  EVALUATE
      FUN(4)=F(T+H,Y0)
C  CORRECT
      DY1=DY1+.5D0*DH*(FUN(4)+FUN(3))
      Y(4)=DY1
C  EVALUATE
      FUN(4)=F(T+H,Y(4))
C  ESTIMATE LOCAL ERROR
      ERR=ABS(Y(4)-Y0)/6.0
      RETURN
      END
```

The following is a calling program for subroutine ABMADP and a sampling (to save space) of the output applied to the problem as described in Examples 6.9 and 6.11.

EXAMPLE 6.13

For the initial value problem $y'(t) = e^{-t} - y(t)$, $y(0) = 1$, we used ABMADP with the same TOL specification as in Example 6.11 for RK2ADP. The error at $T = 1$ is 0.777×10^{-4}, whereas for RK2ADP the error is 0.783×10^{-4}.

```
      EXTERNAL F
      T0=0.0
      Y0=1.0
      T=1.0
      H=.1
      HMIN=.001
      TOL=5.E-4
      WRITE(6,77)
77    FORMAT(' ',6X,'VALUE',12X,'POINT',9X,'ERREST'/)
      CALL ABMADP(F,T0,Y0,H,T,TOL,HMIN,YOUT,IFLG)
```

```
        IF(IFLG.GT.0) STOP
        WRITE(6,66) H,YOUT
  66    FORMAT(' ',/2X,'FINAL H= ',F8.5,2X,'LAST Y= ',E14.7)
        STOP
        END
C
        REAL FUNCTION F(T,Y)
        F=EXP(-T)-Y
        RETURN
        END
```

ABRIDGED OUTPUT:

VALUE	POINT	ERREST
0.9999951E 00	0.003125	0.8145968E-06
0.9999466E 00	0.009375	0.9934105E-08
0.9996026E 00	0.028125	0.3178914E-06
0.9970988E 00	0.078125	0.2433856E-05
0.9924615E 00	0.128125	0.2255042E-05
0.9859048E 00	0.178125	0.2086163E-05
0.9819691E 00	0.203125	0.2006689E-05
0.9776268E 00	0.228125	0.1937151E-05
0.9678112E 00	0.278125	0.1788139E-05
0.9566275E 00	0.328125	0.1658996E-05
0.9442326E 00	0.378125	0.1519918E-05
0.9163949E 00	0.478125	0.1094739E-04
0.8853033E 00	0.578125	0.9278456E-05
0.8518291E 00	0.678125	0.7828076E-05
0.8167076E 00	0.778125	0.6566444E-05
0.7805554E 00	0.878125	0.5473693E-05
0.7438869E 00	0.978125	0.4529953E-05
0.7358366E 00	1.000000	0.3320971E-04

A popular higher order predictor-corrector method uses the four-step Adams-Bashforth formula as the predictor and the three-step Adams-Moulton formula as the corrector. The resultant method is fourth-order accurate and, once started, requires only two function evaluations per step. If implemented in PECE mode, this method has the error estimator

$$y_{n+1} - Y_{n+1}^{(1)} \simeq \frac{-\gamma_4^*}{\gamma_4 - \gamma_4^*} (Y_{n+1}^{(0)} - Y_{n+1}^{(1)}) = \tfrac{19}{270} (Y_{n+1}^{(0)} - Y_{n+1}^{(1)}) \tag{6.33}$$

which can be used to control mesh size and error per unit step in an adaptive scheme. Note that no additional function evaluations are required for this error estimate—it is free and is built into the method. By way of contrast the Runge-Kutta-Fehlberg method requires six evaluations per step. Thus multistep methods can be very fast, and their efficiency explains why they are becoming increasingly popular. We should point out that programs for adaptive multistep methods are generally more complicated than single-step adaptive methods owing to the difficulty of starting values whenever mesh size is changed.

EXERCISES

6.61 Apply ABMADP to exercise 6.23. Modify the program to count and print the total number of function evaluations. Use TOL $= 5 \times 10^{-5}$ and HMIN $= 0.0005$. Discuss your results, with particular attention to accuracy and efficiency. Compare with the results of exercise 6.44.

6.62 Repeat exercise 6.45 for ABMADP.

6.63 Modify ABMADP to accept both an absolute (TOLA) and a relative (TOLR) error tolerance. We suggest the following changes: Rename TOL TOLA throughout the program and add TOLR to the parameter list. Change TOL1 to TOL1 $= 0.125 *$ TOLA, and change the statement defining TOLH to TOLH $=$ H $*$ (TOLA + TOLR $*$ ABS(Y)). Repeat exercise 6.62 with TOLA $= 10^{-5}$, TOLR $= 10^{-3}$, and HMIN $= 0.0001$.

6.64 Repeat exercise 6.47 for the modified version of ABMADP.

6.65 Repeat exercise 6.48 for the modified version of ABMADP.

6.66 Repeat exercise 6.49 for the modified version of ABMADP.

6.67 Mimic the derivation of (6.32) in order to establish the error estimate (6.33).

6.68 Write a subroutine ABAM(F,TO,YO,T,N) that applies the four-step Adams-Bashforth predictor and the three-step Adams-Moulton corrector formula in PECE mode. Use the classical Runge-Kutta method for starting values. The subroutine accepts the right-hand-side function F, the initial data in YO and TO, the final point in T, and the number of steps from TO to T in N. At each step print $Y_{n+1}^{(1)}$, t_{n+1}, and (6.33). Apply ABAM to one problem each of exercises 6.23, 6.24, and 6.25. Try several values of N and discuss your results.

Project 8

In this project you are required to write and test a program for a four-step, adaptive, variable mesh size scheme. Use the four-step Adams-Bashforth predictor formula

$$Y_{n+1}^{(0)} = Y_n^{(1)} + \frac{h}{24} \left[55f(t_n, Y_n^{(1)}) - 59f(t_{n-1}, Y_{n-1}^{(1)}) \right.$$

$$\left. + 37f(t_{n-2}, Y_{n-2}^{(1)}) - 9f(t_{n-3}, Y_{n-3}^{(1)}) \right]$$

and the three-step Adams-Moulton corrector formula

$$Y_{n+1}^{(1)} = Y_n^{(1)} + \frac{h}{24} \left[9f(t_{n+1}, Y_{n+1}^{(0)}) + 19f(t_n, Y_n^{(1)}) \right.$$

$$\left. - 5f(t_{n-1}, Y_{n-1}^{(1)}) + f(t_{n-2}, Y_{n-2}^{(1)}) \right]$$

in PECE mode as the basis for a subroutine ABAM4(FUN,Y,T,H,F,ERR) that performs one predictor-corrector step from T to T + H and returns the estimated error [see (6.33)] in ERR. The working arrays FUN and Y should each have DIMENSION equal to 8, and the jth entry should contain information at T + $(j - 7)$H for $2 \leq j \leq 8$, for example, FUN(6) = f(T − H,Y(6)), Y(6) $\simeq y$(T − H). Use FUN(1) and Y(1) as dummy storage. ABAM4 should use FUN(J),Y(J) for J = 4,5,6,7 to compute FUN(8) and Y(8).

In order to start the procedure or reduce the mesh size, use the classical fourth-order Runge-Kutta method. Write a subroutine CRK4(FUN,Y,T,H,F) that uses information at T, in FUN(4) and Y(4), to compute three Runge-Kutta steps and stores the corresponding information in locations 5, 6, and 7 of FUN and Y. When using CRK4 for mesh size halving, some data shifting in FUN and Y will be necessary.

Write a calling subroutine BM4ADP(F,TO,YO,H,T,TOL,HMIN, YOUT,IFLG) whose parameters have the same functions as in ABMADP. Since the corrector is locally fifth-order accurate, use TOL1 = TOL/ 2^5 = TOL/32. Begin by storing y_0 and $f(t_0, y_0)$ in Y(4) and FUN(4), respectively. Make certain that the last computed value is at T and use only mesh size doubling or halving. Carefully study subroutine ABMADP to make certain that you understand how the values in FUN and Y are used when the mesh size is changed. Use this as a guide for coding BM4ADP.

Test your programs on some of the well-conditioned problems given in the examples or on some of the exercises your instructor has assigned. Compare outputs with exact solutions. After you are assured that the programs are functioning properly, try BM4ADP on at least one problem each of exercises 6.23, 6.24, and 6.25. Specify TOL = 5×10^{-4} and an appropriate HMIN and report your results. Pay particular attention to efficiency, obtainable accuracy, and observed stability (or lack thereof).

Next modify BM4ADP to accept both an absolute (TOLA) and a relative (TOLR) error tolerance. Use these parameters in the same way as discussed for RK2ADP (exercise 6.46) and ABMADP (exercise 6.63). Repeat the computations with appropriately chosen parameters.

6.6 EXTENSION TO FIRST-ORDER SYSTEMS AND CONCLUDING REMARKS

Thus far we have restricted our attention to a single first-order equation with an initial condition. In this section we demonstrate how explicit discrete variable methods can be applied to systems of first-order equations.

In biological and physical applications there often arise differential equations that involve several unknown functions, say $y_1(t), \ldots, y_k(t)$, wherein the rate of change of the variable y_i depends not only on t but on some of the other variables as well. Equations of this type can be used to model very complicated physical systems. A system of first-order equations takes the form

$$y_1'(t) = f_1(t, y_1(t), \ldots, y_k(t))$$

$$y_2'(t) = f_2(t, y_1(t), \ldots, y_k(t)) \qquad \qquad (6.34)$$

$$\cdots\cdots\cdots\cdots\cdots\cdots\cdots\cdots$$

$$y_k'(t) = f_k(t, y_1(t), \ldots, y_k(t))$$

where $f_1, \ldots, f_k$ are given functions of $k + 1$ variables. For example,

$$x'(t) = -x(t) + e^{-2t}y(t)$$

$$y'(t) = y(t) - e^{2t}x(t)$$

(6.35)

is a system of two first-order equations. When the initial conditions $x(0) = 0$, $y(0) = 1$ are imposed, the solution is given by $x(t) = e^{-t} \sin t$, $y(t) = e^t \cos t$. This system can be written more compactly using vector notation. Let $\mathbf{X}(t)$ denote the column vector $[x(t), y(t)]^T$ and $\mathbf{F}$ be the vector function whose components are given by the right-hand sides of (6.35). Then the vector form of (6.35) with initial condition is

$$\mathbf{X}'(t) = \mathbf{F}(t, \mathbf{X}(t))$$

$$\mathbf{X}(0) = [0, 1]^T$$

More generally the system (6.34) can be written as $\mathbf{X}'(t) = \mathbf{F}(t, \mathbf{X}(t))$ with $\mathbf{X}(t) = [y_1(t), \ldots, y_k(t)]^T$ and $\mathbf{F}(t, \mathbf{X}(t)) = [f_1(t, \mathbf{X}(t)), \ldots, f_k(t, \mathbf{X}(t))]^T$. If $y_1(0) = y_1^0, \ldots, y_k(0) = y_k^0$ are specified, then the initial condition is $\mathbf{X}(0) = [y_1^0, \ldots, y_k^0]^T$.

All of the discrete variable methods presented in the text extend to first-order systems with initial conditions. We demonstrate how the classical Runge-Kutta algorithm can be applied to $\mathbf{X}'(t) = \mathbf{F}(t, \mathbf{X}(t))$, $\mathbf{X}(t_0) = \mathbf{X}_0$. We trust that, with this as a guide, the reader can make necessary modifications to apply other other explicit methods for first-order systems.

The subroutine SYSRK4(X0,T0,N,X,K,T) is written for a system of K first-order equations and performs N steps of the classical Runge-Kutta method starting with T0 and ending with T. The initial condition is X0, and array X contains the numerical solution. Subroutine FUN(X,T,K,A) accepts array X of dimension K and T and then returns the value of F(T,X) in array A.

```
      SUBROUTINE SYSRK4(X0,T0,N,X,K,T)
C   PARTIAL DOUBLE-PRECISION CLASSICAL RK METHOD
C   FOR SYSTEM OF UP TO 8 EQUATIONS
C   X0:INITIAL DATA
C   T0:INITIAL POINT; X(T0)=X0
C   K:NUMBER OF FIRST ORDER EQUATIONS
C   X:APPROXIMATE SOLUTION
C   N:DETERMINES MESH SIZE
C   T:FINAL POINT
      REAL X0(K),X(K),Y(8),K1(8),K2(8),K3(8),K4(8)
      DOUBLE PRECISION DH,DX(8)
      H=(T-T0)/FLOAT(N)
      DH=H
      TH=T0
      DO 1 J=1,K
        X(J)=X0(J)
        DX(J)=X0(J)
    1 CONTINUE
      DO 6 J=1,N
        HINC=.5*H
        THINC=TH+HINC
```

```
         CALL FUN(X,TH,K,K1)
         DO 2 I=1,K
           Y(I)=X(I)+HINC*K1(I)
2        CONTINUE
         CALL FUN(Y,THINC,K,K2)
         DO 3 I=1,K
           Y(I)=X(I)+HINC*K2(I)
3        CONTINUE
         CALL FUN(Y,THINC,K,K3)
         DO 4 I=1,K
           Y(I)=X(I)+H*K3(I)
4        CONTINUE
         CALL FUN(Y,TH+H,K,K4)
         DO 5 I=1,K
           DX(I)=DX(I)+DH*(K1(I)+2.*(K2(I)+K3(I))+K4(I))/6.0
           X(I)=DX(I)
5        CONTINUE
         TH=TO+H*FLOAT(J)
         WRITE(6,66) TH,(X(I),I=1,K)
6     CONTINUE
66    FORMAT(' ',2X,'T= ',E12.5,2X,'X:',4(2X,E14.7)/)
      RETURN
      END
```

EXAMPLE 6.14

We apply SYSRK4 to the system (6.35) with initial conditions $x(0) = 0$, $y(0) = 1$. For $T = \pi/4$ and $N = 20$ (that is, $h = \pi/80$), the calling program and output are as follows:

```
REAL X(2),X0(2)
X0(1)=0.0
X0(2)=1.0
TO=0.0
T=.7853982
N=20
CALL SYSRK4(X0,TO,N,X,2,T)
STOP
END


SUBROUTINE FUN(X,T,K,A)
REAL X(K),A(K)
A(1)=-X(1)+EXP(-2.*T)*X(2)
A(2)=X(2)-EXP(2.*T)*X(1)
RETURN
END


T=  0.39270E-01  X:   0.3774792E-01   0.1039248E 01
T=  0.78540E-01  X:   0.7253259E-01   0.1078371E 01
T=  0.11781E 00  X:   0.1044747E 00   0.1117231E 01
```

```
T=  0.15708E 00  X:  0.1336943E 00  0.1155683E 01
T=  0.19635E 00  X:  0.1603103E 00  0.1193568E 01
T=  0.23562E 00  X:  0.1844406E 00  0.1230721E 01
T=  0.27489E 00  X:  0.2062011E 00  0.1266961E 01
T=  0.31416E 00  X:  0.2257065E 00  0.1302098E 01
T=  0.35343E 00  X:  0.2430692E 00  0.1335929E 01
T=  0.39270E 00  X:  0.2583997E 00  0.1368239E 01
T=  0.43197E 00  X:  0.2718058E 00  0.1398801E 01
T=  0.47124E 00  X:  0.2833934E 00  0.1427372E 01
T=  0.51051E 00  X:  0.2932652E 00  0.1453698E 01
T=  0.54978E 00  X:  0.3015217E 00  0.1477513E 01
T=  0.58905E 00  X:  0.3082604E 00  0.1498534E 01
T=  0.62832E 00  X:  0.3135760E 00  0.1516466E 01
T=  0.66759E 00  X:  0.3175602E 00  0.1530998E 01
T=  0.70686E 00  X:  0.3203017E 00  0.1541807E 01
T=  0.74613E 00  X:  0.3218862E 00  0.1548553E 01
T=  0.78540E 00  X:  0.3223966E 00  0.1550882E 01
```

At $\pi/4$ the exact solution, to seven decimal digits, is given by $x(\pi/4) = 0.3223969$, $y(t) = 1.550883$.

First-order systems can also arise from a single higher order equation in one unknown. Every kth-order equation for the variable $y(t)$ can be converted to a system of k first-order equations involving the variables $x_1(t) = y(t)$, $x_2(t) = y'(t)$, $x_3(t) = y''(t)$, ..., $x_k(t) = y^{(k-1)}(t)$. The procedure should be clear with the aid of a few examples.

EXAMPLE 6.15

The third-order initial value problem

$$y'''(t) - y''(t) + [y'(t)]^2 + e^t y(t) = \cos t$$

$$y(0) = 0 \qquad y'(0) = 2 \qquad y''(0) = 1$$

is converted to a first-order system as follows. Define $x_1(t) = y(t)$, $x_2(t) = y'(t)$, and $x_3(t) = y''(t)$. Then

$$x_1'(t) = x_2(t)$$

$$x_2'(t) = x_3(t)$$

$$x_3'(t) = \cos t - e^t x_1(t) - [x_2(t)]^2 + x_3(t)$$

and, in terms of the new variables, the initial conditions are $x_1(0) = 0$, $x_2(0) = 2$, and $x_3(0) = 1$. Thus the third-order initial value problem has been converted to

$$\mathbf{X}'(t) = \mathbf{F}(t, \mathbf{X}(t))$$

$$\mathbf{X}(0) = [0, 2, 1]^T$$

where $\mathbf{F}(t, \mathbf{X}) = [x_2, x_3, \cos t - e^t x_1 - x_2^2 + x_3]^T$.

EXAMPLE 6.16

Consider the system of second-order equations

$$x''(t) = -a_1 x(t) + a_2 y(t) + f(t)$$

$$y''(t) = a_3 x(t) - a_4 y(t) + g(t)$$

$$x(t_0) = 2,\ x'(t_0) = 0,\ y(t_0) = 4,\ y'(t_0) = 1$$

where the a_i's are constants and f, g are given functions. This second-order initial value problem can be converted to an initial value problem for a first-order system of four equations as follows. Define $x_1(t) = x(t)$, $x_2(t) = x'(t)$, $x_3(t) = y(t)$, $x_4(t) = y'(t)$. Then

$$x_1'(t) = x_2(t)$$

$$x_2'(t) = -a_1 x_1(t) + a_2 x_3(t) + f(t)$$

$$x_3'(t) = x_4(t)$$

$$x_4'(t) = a_3 x_1(t) - a_4 x_3(t) + g(t)$$

$$x_1(t_0) = 2,\ x_2(t_0) = 0,\ x_3(t_0) = 4,\ x_4(t_0) = 1$$

Consequently the initial value problem for the second-order system can be written as

$$\mathbf{X}'(t) = \mathbf{F}(t, \mathbf{X}(t)) \qquad \mathbf{X}(t_0) = [2, 0, 4, 1]^T$$

where

$$\mathbf{F}(t, \mathbf{X}) = [x_2,\ -a_1 x_1 + a_2 x_3 + f(t),\ x_4,\ a_3 x_1 - a_4 x_3 + g(t)]^T$$

Second-order equations like these arise in electrical network analysis and mechanical vibration.

When attempting to solve a higher order equation or a system of higher order equations, it is frequently useful to convert the equations to a first-order system. Then any of the discrete variable methods of the text can be applied. We mention that there are special discrete variable methods that apply directly to higher order (most notably second-order) equations.

We conclude the chapter by discussing the relative advantages and disadvantages of the methods presented. Unfortunately there is no "best" method, and in fact for any given method there is an initial value problem for which the method does not perform well. Thus there is no known method that outperforms every other method on all initial value problems. For general-purpose differential equation solving, there are essentially two types of methods in common use. These are the Runge-Kutta methods and the Adams family of predictor-corrector methods. In a good modern library subroutine these two types are coded for systems of first-order equations with error estimators and variable mesh size. Our subroutines RK2ADP and ABMADP, although not incorporating sophisticated mesh size changes and roundoff error controls, serve as reasonable introductory models of such library routines.

Runge-Kutta methods have many desirable attributes. They are self-starting and stable, provide good accuracy, and require little overhead cost (in core storage). Their major disadvantage is the large number of function evaluations required by the method and error estimator per step. If $f(t, y)$ is relatively inexpensive to evaluate, then Runge-Kutta methods can be very effective.

The Adams predictor-corrector methods require only two function evaluations per step in PECE mode and provide an automatic (and free) error estimate. However, predictor-corrector subroutines are somewhat complicated, require more substantial overhead costs, and need special techniques for mesh size changes. All of the Adams-Bashforth and Adams-Moulton methods are stable, but we should mention that other multistep methods (not discussed in the text) have less desirable stability properties.

Some initial value problems require special treatment. The essential feature of stiff problems is that one component of the solution is decaying at a significantly faster rate than the others. For a single equation we indicated that $\partial f / \partial y \ll 0$ is the key to stiffness. For a system of first-order equations the analysis is more complicated—it involves the eigenvalues of the jacobian of f. In any case the difficulty in numerically solving stiff equations is that the mesh size h must be sufficiently small—that is, $h \, \partial f / \partial y \in I_A$—for absolute stability, whereas a larger value of h may be more than adequate for accuracy. The numerical method must cope with rapidly decaying transients even when these transients contribute very little to the solution of the initial value problem. In an adaptive scheme for stiff problems it is desirable to use formulas with a large interval of absolute stability, thereby permitting a large range of acceptable mesh sizes. The effective treatment of stiff equations requires an implicit method.

For reference Table 6.5 summarizes several important features of the discrete variable methods given in the text.

TABLE 6.5   Summary of Discrete Variable Methods

Method	Order	I_A	k-Step	Type
Euler	1	$[-2, 0]$	1	E
Backward Euler	1	$(-\infty, 0] \cup [2, +\infty)$	1	I
Runge-Kutta	2	$[-2, 0]$	1	E
	3	$[-2.51, 0]$	1	E
	4	$[-2.78, 0]$	1	E
Adams-Bashforth	2	$[-1, 0]$	2	E
	3	$[-0.54, 0]$	3	E
	4	$[-0.3, 0]$	4	E
Adams-Moulton	2	$(-\infty, 0]$	1	I
	3	$(-6, 0]$	2	I
	4	$(-3, 0]$	3	I
ABAM (PECE mode)	4	$(-1.25, 0]$	4	P—C
ABAM (PEC mode)	4	$(-0.16, 0]$	4	P—C
Simpson	4	$\{0\}$	2	I

When a discrete variable method is being chosen for an initial value problem, there are many considerations: the accuracy of the initial condition and the right-hand-side function, the conditioning of the problem, the amount of human effort required for program preparation, the amount of computer effort required, and the desired accuracy of the numerical solution. The successful solution of the numerical problem depends on many factors, the most important of which are the insight gained from the physical phenomena the initial value problem is modeling, the stability and accuracy guidelines from the theory of the chosen method, previous computational experience with the method, and the computer being used. We hope that the exposition in the text and the computation required in the exercises furnish the reader with many of the requisite tools in this regard.

EXERCISES

6.69 Write the following system in the form $X'(t) = F(t, X(t))$:

$$x'(t) + z(t) = t^2 x(t) + [y(t)]^2 - e^t$$

$$y'(t) - x(t) = 2tz(t) + e^{-2t}$$

$$z'(t) + ty(t) = x(t) \cos t - 3$$

6.70 Convert the following problem to a form suitable for SYSRK4:

$$y'''(t) + y''(t) - y'(t) = y(t) + \cos t$$

$$y(0) = 0 \qquad y'(0) = 1 \qquad y''(0) = 0$$

6.71 Consider the problem

$$x_1'(t) = x_1(t) - e^{-t}$$

$$x_2'(t) = -x_2(t) + e^t$$

$$x_1(0) = 1 \qquad x_2(0) = 2$$

Would it be appropriate to use SYSRK4 to solve this problem? Is there a better approach?

6.72 Explain how you would solve the following problem numerically:

$$x''(t) = -4x(t) + t$$

$$y''(t) = 9y(t) - 1$$

$$x(0) = 1 \qquad x'(0) = 0 \qquad y(2) = 0 \qquad y'(2) = 1$$

6.73 The following problem arises as a model for a two-species population in which one species preys on the other:

$$u'(t) = u(t)[a_{11} - a_{12} v(t)]$$

$$v'(t) = v(t)[-a_{21} + a_{22} u(t)]$$

$$u(0) = u_0 \qquad v(0) = v_0$$

These are commonly called the **predator-prey equations** and were proposed in the 1920s by Lotka and Volterra. Write a subroutine FUN for solving these equations by SYSRK4.

6.74 The following problem arises as a model for a two-species population sharing a common food supply:

$$u'(t) = u(t)[b_1 - a_{11} u(t) - a_{12} v(t)]$$

$$v'(t) = v(t)[b_2 - a_{21} u(t) - a_{22} v(t)]$$

$$u(0) = u_0 \qquad v(0) = v_0$$

Write a subroutine FUN for solving these equations by SYSRK4.

6.75 Use SYSRK4 to solve the predator-prey equations with $a_{11} = 4\pi$, $a_{12} = \pi$, $a_{21} = \pi$, $a_{22} = \pi/2$, $u_0 = 3$, and $v_0 = 5.732051$. Use T = 1, N = 50, and plot the numerical solution for the prey population u and the predator population v. The solution of this problem is given approximately by

$$u(t) \simeq 2[1 + \cos (2\pi t + \pi/3)]$$

$$v(t) \simeq 4[1 + 0.5 \sin (2\pi t + \pi/3)]$$

Plot the numerical solution in the vu plane.

6.76 In the problem of exercise 6.74 suppose $b_1 = 10$, $a_{11} = 0$, $a_{12} = 1$, $b_2 = 2$, $a_{21} = 1$, $a_{22} = 1$, $u_0 = 50$, and $v_0 = 40$. For these data it is known that $\lim_{t \to \infty} v(t) = 0$. Use SYSRK4 with several values of T and N to see if v is decaying. Plot the numerical solution in the vu plane.

6.77 Use SYSRK4 to solve the system on [0, 2]

$$x'(t) = x(t) + 3y(t) + \sin t$$

$$y'(t) = x(t) - y(t) - \cos t$$

$$x(0) = 0 \qquad y(0) = \tfrac{1}{3}$$

Solution

$$x(t) = \frac{e^{2t}}{10} - e^{-2t} - \frac{\sin t}{5} + 2\frac{\cos t}{5}$$

$$y(t) = \frac{e^{2t}}{30} + e^{-2t} - 2\frac{\sin t}{15} - \frac{\cos t}{5}$$

6.78 Solve $y''(t) + y(t) = 0$, $y(0) = 1$, $y'(0) = 0$, on $[0, \pi]$ by any convenient numerical method. What functions do you know which satisfy $y'' = -y$? Can you guess the solution of the initial value problem?

NOTES AND COMMENTS

Section 6.1

We address only the two aspects of initial value problems that are most important for numerical computation: well-posedness and conditioning. In the jargon of differential equations a problem whose solution depends continuously on the initial condition is called stable. We have not used this terminology in order to avoid confusion with stability of discrete variable methods. In theoretical texts on ordinary differential equations, other types of stability—asymptotic and relative—are analyzed.

Of the many introductory texts on differential equations we mention only the two with which we are most familiar. These are Boyce and DiPrima (1977) and Braun (1975). Both books are written for post-calculus students, and the latter gives detailed applications to a variety of real-world problems, including detection of art forgeries, diagnosis of diabetes, insecticide spraying, and spread of epidemics. More theoretical treatments are given in the classic works by Birkhoff and Rota (1969) and Coddington and Levinson (1955).

Section 6.3

A detailed development of Runge-Kutta methods can be found in the advanced texts by Gear (1971), Henrici (1962), and Lapidus and Seinfeld (1971). A complete listing of RKF45 and a guide to its use can be found in Forsythe et al. (1977).

Section 6.4

We present the derivation of only the Adams family of multistep methods. This was done for two reasons: (1) They are the most effective and widely used multistep methods. (2) Their derivation is a straightforward application of polynomial interpolation. In fact, these formulas and their corresponding error terms constitute the most important application of polynomial interpolation in modern numerical analysis.

Section 6.5

There are several excellent FORTRAN codes for adaptive predictor-corrector methods based on the Adams family. Examples of such subroutines are DIFSUB by Gear (1971), STEP and DE by Shampine and Gordon (1975), and DVOGER, which is available in the IMSL package of library programs. The Shampine and Gordon text is devoted exclusively to methods of this type.

We have omitted a discussion of stability for multistep methods. This important topic requires a more substantial mathematical background than is assumed of the reader. Most of the fundamental results in this area are due to Dahlquist and can be found in Henrici (1962) or Gear (1971).

Section 6.6

Systems of differential equations arise most frequently in applications; however, a thorough discussion of methods for their numerical solution is beyond the scope of the text. See the advanced books by Lambert (1973), Shampine and Gordon (1975), and Gear (1971). The effective treatment of systems, especially stiff systems, is the subject of extensive current research.

The general numerical analysis texts by Conte and de Boor (1980), Johnson and Riess (1982), and Atkinson (1978) present a more theoretical discussion of discrete variable methods. These well-written books are intended for advanced undergraduate students.

LEAST SQUARES APPROXIMATION

One of the fundamental procedures in numerical methods is the replacement of a function f by another function that approximates f. The approximating function should be simple for computational purposes and should also be a sufficiently accurate approximation to f. In Chapter 4 the approach was to obtain an approximation by interpolation, and implicit in the process was the assumption that values of f are known with high accuracy. We found that polynomial and piecewise polynomial interpolates can successfully be used in numerical integration and discrete variable methods for initial value problems. In this chapter we consider other methods for approximating f when its values are contaminated by noise. More precisely we consider the situation where data $\{(x_i, y_i)\}_{i=1}^m$ are given

$$y_i = f(x_i) + \varepsilon_i$$

with ε_i bounded, say $|\varepsilon_i| \leq \varepsilon$, but otherwise of unknown character. The noise ε_i could be the result of measurement error and may be random or rapidly varying in some unknown manner. The bound ε need not be small relative to the magnitude of f; however, in many physical applications it is the case that $y_i = (1 - \rho_i)f(x_i)$ with $|\rho_i| \leq \rho$ and ρ the known relative error of the measurement device.

The problem is to recover f or, more realistically, an approximation to f from the data. Since an interpolate of the data would inherit the measurement error (noise), we require another approach—one in which the noise is somehow damped out. The method of least squares is designed so that, on the

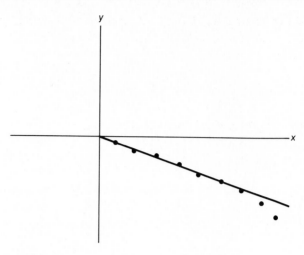

FIGURE 7.1

average, the data noise is suppressed, thereby enabling the determination of a reasonable approximation called the least squares fit.

7.1 INTRODUCTION TO DATA FITTING

Consider a simple physics experiment in which a spring is stretched from its equilibrium position by a known external force. If x denotes the spring displacement and y the force exerted by the spring (which is equal in magnitude to the external force but opposite in direction), then a plot of the data $\{(x_i, y_i)\}_{i=1}^{m}$ looks like Figure 7.1. For x not too large it appears that the force and displacement are related by the linear relationship $y = -kx$, $k > 0$, which is known as Hooke's law. How can the spring constant k be determined from the experimental data? We postulate that the force exerted by the spring is $f(x) = -kx$ and determine k by requiring that the residuals

$$r(x_i) = f(x_i) - y_i \qquad 1 \le i \le m$$

be small. This could be accomplished by choosing k so as to minimize one of the following:

$$\max_{1 \le i \le m} |r(x_i)| = E_\infty(k)$$

$$\sum_{i=1}^{m} |r(x_i)| = E_1(k)$$

$$\sum_{i=1}^{m} [r(x_i)]^2 = E_2(k)$$

where E_1, E_2, and E_∞ furnish different measures of how well the straight line $y = -kx$ fits the data. By minimizing one of these quantities, we select the particular line (that is, k) that, in some sense, best fits the data. From statistical considerations the choice of E_2 is appropriate and, as we shall see, easy to minimize. The minimization of $E_\infty(k)$ or $E_1(k)$ is possible, though much more difficult than that of $E_2(k)$.

The problem of minimizing $E_2(k)$ is a special case of a slightly more general problem: Given data $\{(x_i, y_i)\}_{i=1}^m$, find a_0 and a_1 in $f(x) = a_1 x + a_0$ that minimize

$$E_2(a_0, a_1) = \sum_{i=1}^m [f(x_i) - y_i]^2 = \sum_{i=1}^m (a_1 x_i + a_0 - y_i)^2$$

If $\hat{a}_0, \hat{a}_1$ minimize $E_2(a_0, a_1)$, then by calculus it follows that

$$\frac{\partial E_2}{\partial a_k}(\hat{a}_0, \hat{a}_1) = 0 \qquad k = 0, 1 \tag{7.1}$$

A simple calculation gives

$$\frac{\partial E_2}{\partial a_0}(\hat{a}_0, \hat{a}_1) = 2 \sum_{i=1}^m (\hat{a}_1 x_i + \hat{a}_0 - y_i)$$

$$\frac{\partial E_2}{\partial a_1}(\hat{a}_0, \hat{a}_1) = 2 \sum_{i=1}^m (\hat{a}_1 x_i + \hat{a}_0 - y_i)x_i$$

Therefore by (7.1) the solution $\hat{a}_0, \hat{a}_1$ must satisfy

$$m\hat{a}_0 + \left(\sum_{i=1}^m x_i\right)\hat{a}_1 = \sum_{i=1}^m y_i$$

$$\left(\sum_{i=1}^m x_i\right)\hat{a}_0 + \left(\sum_{i=1}^m x_i^2\right)\hat{a}_1 = \sum_{i=1}^m x_i y_i \tag{7.2}$$

which are called the **normal** equations. The foregoing analysis shows that a necessary condition for $\hat{a}_0, \hat{a}_1$ to minimize $E_2(a_0, a_1)$ is that $\hat{a}_0, \hat{a}_1$ must satisfy the normal equations. It can be shown that minimizing $E_2(a_0, a_1)$ is equivalent to solving the normal equations. The problem of minimizing $E_2(a_0, a_1)$ is called the **least squares problem for linear functions,** and the function $\hat{f}(x) = \hat{a}_1 x + \hat{a}_0$ that minimizes $E_2(a_0, a_1)$ is called the **linear least squares fit** of the data $\{(x_i, y_i)\}_{i=1}^m$. Least squares data fitting is a fundamental tool in statistical analysis and, in this context, is called linear regression.

EXAMPLE 7.1

A plot of the data $\{(x_i, y_i)\}_{i=1}^7$ indicates that, for some constants a_0 and a_1, $y = f(x) = a_0 + a_1 x$. Verify this.

x_i	y_i
0.6	7.05
1.3	12.2
1.64	14.4
1.8	15.2
2.1	17.4
2.3	19.6
2.44	20.2

To find the linear least squares fit, we compute

$$\sum_{i=1}^{7} x_i = 12.18 \qquad \sum_{i=1}^{7} x_i^2 = 23.6332$$

$$\sum_{i=1}^{7} x_i y_i = 201.974 \qquad \sum_{i=1}^{7} y_i = 106.05$$

and solve the normal equations (7.2) given by

$$7\hat{a}_0 + 12.18\hat{a}_1 = 106.05$$

$$12.18\hat{a}_0 + 23.6332\hat{a}_1 = 201.974$$

The solution is easily found to be $\hat{a}_0 = 2.708287$ and $\hat{a}_1 = 7.150410$, rounded to seven digits. Thus the straight line that best fits the data in a least squares sense is $\hat{f}(x) = 2.708287 + 7.15041x$ and $E_2(\hat{a}_0, \hat{a}_1) = 0.4917996$.

The method of least squares data fitting is not restricted to linear functions (polynomial of degree 1); in principle the method applies to any functional form for f. To illustrate this, we suppose it is required to fit data $\{(x_i, y_i)\}_{i=1}^{m}$ by a function of the form

$$f(x) = a + bx^2 + ce^{-x}$$

Then we define $E_2(a, b, c) = \sum_{i=1}^{m} [f(x_i) - y_i]^2$ and conclude, as before, that E_2 is minimized by $\hat{a}, \hat{b}, \hat{c}$ if

$$\frac{\partial E_2}{\partial a} (\hat{a}, \hat{b}, \hat{c}) = \frac{\partial E_2}{\partial b} (\hat{a}, \hat{b}, \hat{c}) = \frac{\partial E_2}{\partial c} (\hat{a}, \hat{b}, \hat{c}) = 0$$

Direct computation of the partial derivatives gives the following normal equations for $\hat{a}, \hat{b},$ and $\hat{c}$:

$$m a + \left(\sum_{i=1}^{m} x_i^2 \right) \hat{b} + \left(\sum_{i=1}^{m} e^{-x_i} \right) \hat{c} = \sum_{i=1}^{m} y_i$$

$$\left(\sum_{i=1}^{m} x_i^2 \right) \hat{a} + \left(\sum_{i=1}^{m} x_i^4 \right) \hat{b} + \left(\sum_{i=1}^{m} e^{-x_i} x_i^2 \right) \hat{c} = \sum_{i=1}^{m} y_i x_i^2 \qquad (7.3)$$

$$\left(\sum_{i=1}^{m} e^{-x_i} \right) \hat{a} + \left(\sum_{i=1}^{m} e^{-x_i} x_i^2 \right) \hat{b} + \left(\sum_{i=1}^{m} e^{-2x_i} \right) \hat{c} = \sum_{i=1}^{m} y_i e^{-x_i}$$

These normal equations can be written as a symmetric 3×3 matrix problem in the unknowns $\hat{a}, \hat{b},$ and $\hat{c}$. The equations (7.3) may appear more formidable than those of (7.2); however, the only essential difference is that (7.3) gives a 3×3 linear system of equations whereas (7.2) is a 2×2 linear system. In the next example the normal equations are significantly different from (7.2) or (7.3).

Suppose a least squares fit of the form $f(x) = ae^{bx}$ is required for data $\{(x_i, y_i)\}_{i=1}^{m}$. Then

$$E_2(a, b) = \sum_{i=1}^{m} (ae^{bx_i} - y_i)^2$$

and the normal equations $\partial E_2/\partial a = \partial E_2/\partial b = 0$ are

$$\hat{a} \sum_{i=1}^{m} \exp(2\hat{b}x_i) = \sum_{i=1}^{m} y_i \exp(\hat{b}x_i)$$

$$\hat{a}^2 \sum_{i=1}^{m} x_i \exp(2\hat{b}x_i) = \hat{a} \sum_{i=1}^{m} y_i x_i \exp(\hat{b}x_i)$$

(7.4)

These are nonlinear equations in $\hat{a}$ and $\hat{b}$, and it is not clear that solutions exist or how to find them. In the functional form $f(x) = ae^{bx}$ the unknown coefficients a and b appear in a nonlinear way, whereas for $f(x) = a_1 x + a_0$ or $f(x) = a + bx^2 + ce^{-x}$ the coefficients appear linearly. Whenever the coefficients in f appear linearly, the minimization of E_2 is called a **linear least squares problem**. This is not to be confused with a least squares fit by a linear function (that is, a polynomial of degree 1). We shall consider only linear least squares problems and suggest in the exercises how one handles simple nonlinear problems.

In the next section we consider linear least squares problems more systematically in a general context. We introduce an appropriate inner product and norm structure on the problem and show that the normal equations always take a particular form. This is an excellent example, in which a more abstract point of view actually simplifies the statement of the problem and, more important, reveals a principle that applies to several (seemingly) unrelated least squares problems. We urge the reader not to be impatient with the theoretical discussion in the next section. A careful reading of this material will enable the student to recognize the methods of Sections 7.3 and 7.4 as simply special cases of the linear least squares problem.

EXERCISES

7.1 Show that the least squares estimate of the spring constant is

$$\hat{k} = -\frac{\sum_{i=1}^{m} x_i y_i}{\sum_{i=1}^{m} x_i^2}$$

7.2 Define the **data means** by

$$\bar{x} = \frac{1}{m} \sum_{i=1}^{m} x_i \quad \text{and} \quad \bar{y} = \frac{1}{m} \sum_{i=1}^{m} y_i$$

If $\hat{a}_0, \hat{a}_1$ solve (7.2), show that $\bar{y} = \hat{a}_0 + \hat{a}_1 \bar{x}$—that is, $(\bar{x}, \bar{y})$—is on the least squares line.

*7.3 Let $\hat{f}(x) = \hat{a}_0 + \hat{a}_1 x$ be the least squares fit of data $\{(x_i, y_i)\}_{i=1}^{m}$. Use the normal equations to show that for any $f(x) = a_0 + a_1 x$

$$\sum_{i=1}^{m} [f(x_i) - y_i]^2 = \sum_{i=1}^{m} [f(x_i) - \hat{f}(x_i)]^2 + \sum_{i=1}^{m} [\hat{f}(x_i) - y_i]^2$$

or, equivalently,

$$E_2(a_0, a_1) = \sum_{i=1}^{m} [f(x_i) - \hat{f}(x_i)]^2 + E_2(\hat{a}_0, \hat{a}_1)$$

Conclude that $E_2(\hat{a}_0, \hat{a}_1) \leq E_2(a_0, a_1)$ for all a_0, a_1. This demonstrates that any solution of the normal equations (7.2) minimizes E_2.

7.4 Let $\hat{f}(x) = \hat{a}_0 + \hat{a}_1 x$ minimize $E_2(a_0, a_1)$. Show that

$$E_2(\hat{a}_0, \hat{a}_1) = \sum_{i=1}^{m} y_i^2 - \sum_{i=1}^{m} [\hat{f}(x_i)]^2$$

Hint: Expand $E_2(\hat{a}_0, \hat{a}_1)$ and use the normal equations.

7.5 If $\bar{x} = 0$, solve equations (7.2) and interpret the results.

7.6 If $\bar{x} = 0 = \bar{y}$, solve equations (7.2) and interpret the results.

7.7 Suppose the least squares fit $\hat{f}(x) = \hat{a}_0 + \hat{a}_1 x$ satisfies $E_2(\hat{a}_0, \hat{a}_1) = 0$. Conclude that $\hat{f}(x_i) = y_i$, $1 \leq i \leq m$.

7.8 If $m = 2$, show that the least squares fit $\hat{f}(x) = \hat{a}_0 + \hat{a}_1 x$ is the linear Lagrange interpolate of the data.

7.9 Compute the least squares fit of these data by a linear function:

x_i	y_i
−0.4	0.774597
−0.2	0.894427
0	1.000000
0.2	1.095445
0.4	1.183216

7.10 Compute the least squares fit of the form $f(x) = a_0 + a_1 x^2$ for these data:

x_i	y_i
−0.4	1.0770
−0.2	1.0198
0	1.0000
0.2	1.0198
0.4	1.0770

7.11 For the given data make a careful plot of x versus $\log|y|$:

x_i	y_i
−0.5	0.552
0	1.507
0.5	4.083
1.0	11.06
1.5	31.00

On the basis of your plot suggest a suitable form for a least squares fit of the data.

7.12 Consider the nonlinear least squares problem of fitting data $\{(x_i, y_i)\}_{i=1}^{m}$ by a function of the form $f(x) = (ax + b)^{-1}$. One can linearize this problem by finding the least squares fit of data $\{(x_i, 1/y_i)\}_{i=1}^m$ by a function of the form $g(x) = ax + b$. Carry out this procedure for these data:

x_i	y_i
1.0	0.931
1.4	0.473
1.8	0.297
2.2	0.224
2.6	0.168

7.13 In the text we demonstrate that a fit of the form $f(x) = ae^{bx}$ leads to nonlinear normal equations. Let $A = \log|a|$ and define $g(x) = \log|f(x)| = A + bx$. The fact that the coefficients A and b appear linearly in g suggests that a reasonable estimate of a and b can be found from the least squares fit of data $\{(x_i, \log|y_i|)\}_{i=1}^m$ by a function of the form $A + bx$. Estimate a and b using this procedure for the data of exercise 7.11.

7.14 One means of solving (7.4) is as follows. If $\hat{a} \neq 0$, solve each equation for $\hat{a}$ in terms of $\hat{b}$ and equate the two to yield

$$\frac{\sum\limits_{i=1}^m y_i \exp(\hat{b}x_i)}{\sum\limits_{i=1}^m \exp(2\hat{b}x_i)} = \frac{\sum\limits_{i=1}^m x_i y_i \exp(\hat{b}x_i)}{\sum\limits_{i=1}^m x_i \exp(2\hat{b}x_i)}$$

In terms of $\hat{b}$ this is an equation of the form $F(\hat{b}) = G(\hat{b})$ or, equivalently, $F(\hat{b}) - G(\hat{b}) = 0$. Thus $\hat{b}$ is a root of the equation $F(b) - G(b) = 0$ that can be "found" by the methods of Chapter 2. Once $\hat{b}$ is found, then $\hat{a}$ is given by $\hat{a} = F(\hat{b})$ or $\hat{a} = G(\hat{b})$. Use the secant method to find $\hat{b}, \hat{a}$ for the data of exercise 7.11.

7.15 Suppose data $\{(i, y_i)\}_{i=1}^{10}$ are given (that is, $x_i = i$) and a least squares fit of the form $f(x) = a_0 + a_1 x + a_2 x^2 + a_3 x^3$ is required. The normal equations give a 4×4 linear system with coefficient matrix

$$A = \begin{bmatrix} 10 & 55 & 385 & 3025 \\ 55 & 385 & 3025 & 25333 \\ 385 & 3025 & 25333 & 220825 \\ 3025 & 25333 & 220825 & 1978405 \end{bmatrix}$$

Verify this. Determine $\|A\|$ and use ELIMIN and BAKSUB to solve $Ax = [1, 0, 0, 0]^T$. Estimate $\|A^{-1}\|$ by (3.22) and conclude that the condition number of A satisfies $K(A) \geq 10^7$.

7.2 THE LINEAR LEAST SQUARES PROBLEM

The general linear least squares problem is formulated as follows. Given data $\{(x_i, y_i)\}_{i=1}^m$ and a postulated functional form

$$f(x) = a_0 \phi_0(x) + \cdots + a_n \phi_n(x) \tag{7.5}$$

where $\{\phi_k\}_{k=0}^n$ are known functions, find coefficients $\hat{a}_0, \ldots, \hat{a}_n$ that minimize

$$E(a_0, \ldots, a_n) = \sum_{i=1}^m [f(x_i) - y_i]^2 = \sum_{i=1}^m \left[\sum_{k=0}^n a_k \phi_k(x_i) - y_i \right]^2 \tag{7.6}$$

In (7.6) and in the sequel we drop the subscript on $E_2(a_0, \ldots, a_n)$. The data points are assumed to be ordered such that $x_1 < x_2 < \cdots < x_m$, and normally m is much larger than n. We assume that the functions $\{\phi_k\}_{k=0}^n$ are **linearly independent**. This means that each ϕ_k cannot be written as a linear combination of $\{\phi_j\}_{j=1}^n$, $j \neq k$. That is, we assume it is not possible to write ϕ_k as

$$\phi_k(x) = \sum_{\substack{j=0 \\ j \neq k}}^n \alpha_j \phi_j(x) \tag{7.7}$$

If (7.7) were possible for some k, then it would be superfluous to include ϕ_k in (7.5). The functions $\{\phi_j\}_{j=0}^n$ are called a **basis** for the linear least squares problem.

EXAMPLE 7.2

If $n = 1$ and $\phi_0(x) = 1$, $\phi_1(x) = x$, then we are seeking the least squares fit by a linear function. More generally, the choice $\phi_k(x) = x^k$, $0 \le k \le n$, gives the least squares fit by a polynomial of degree n.

EXAMPLE 7.3

If the data are known to be periodic, it may be appropriate to choose $\phi_k(x) = \cos kx$, $0 \le k \le n$, or $\phi_k(x) = \sin kx$ for some suitable value of n.

EXAMPLE 7.4

In Chapter 4 we saw that cubic splines provide an excellent approximation to smooth functions. If we let $a = x_1$, $b = x_m$, and $h = (b - a)/(n - 2)$, then the functions $\{\phi_k(x)\}_{k=0}^n$ could be chosen by [see (4.42)]

$$\phi_k(x) = \Phi(h^{-1}(x - t_k))$$

where $t_k = a + (k - 1)h$.

The particular choice of basis functions $\phi_0, \ldots, \phi_n$ is generally influenced by the suspected character of the underlying function as evidenced by a plot of the data, by the principles of the discipline in which the data are found, by the use intended for the resultant least squares fit, and by the precision to which the data is known. Data obtained in the social sciences may be subjective in nature and must be treated differently from data obtained in an electronics laboratory using a digital measurement device. For nondeterministic data the least squares fit must be interpreted in terms of statistical hypotheses.

In order to formulate the minimization of (7.6) in a more abstract setting, we introduce an inner product and norm that are related to the data points.

For any two functions u and v defined on the data points $\{x_i\}_{i=1}^m$, we define the **inner product**

$$(u, v) = \sum_{i=1}^m u(x_i)v(x_i)$$

It is easy to see that the inner product has the following properties:

(i) $(u, u) \geq 0$

(ii) $(u, v) = (v, u)$

(iii) $(\alpha u + \beta v, w) = \alpha(u, w) + \beta(v, w)$

where α, β are real scalars and u, v, w are defined on the data points. Moreover, for certain functions it is also true that

(iv) $(u, u) = 0$ if and only if $u(x) = 0$ for all x

Clearly $(u, u) = 0$ means that $\sum_{i=1}^m [u(x_i)]^2 = 0$ or, equivalently, $u(x_i) = 0$, $1 \leq i \leq m$. For example, if u is a polynomial of degree $k < m$, then $(u, u) = 0$ means that u has m distinct roots. However, a polynomial of degree k can have at most k real roots. Therefore condition (iv) holds for polynomials of degree $< m$. We assume that each basis function ϕ_k satisfies (iv). It can be shown that every f in the form (7.5) satisfies (iv) if and only if each basis function satisfies (iv).

We define a **norm** on such functions by

$$\|u\| = \sqrt{(u, u)}$$

The norm is a measure of the "size" of functions, and $\|u - v\|$ is a measure of how "close" u is to v (and vice versa). By (iv) it is seen that $\|u - v\| = 0$ if and only if $u \equiv v$. The norm has the following properties:

(i′) $\|u\| \geq 0$

(ii′) $\|u + v\| \leq \|u\| + \|v\|$

(iii′) $\|\alpha u\| = |\alpha|\,\|u\|$

(iv′) $\|u\| = 0$ if and only if $u \equiv 0$

Of these properties only (ii′) does not follow directly from the definition of $\|\cdot\|$. The proof of (ii′), which is called the triangle inequality, though not very difficult, would be an unnecessary distraction at this point—we omit it.

Let us reformulate the least squares problem in terms of this norm. We have

$$E(a_0, a_1, \ldots, a_n) = \|f - y\|^2 = \left\| \sum_{k=0}^n a_k \phi_k - y \right\|^2$$

where we interpret $y = y(x)$ to be a function whose values are the measured data, that is, $y_i = y(x_i)$. Thus $E(a_0, \ldots, a_n)$ gives the square of the "distance" between $f(x) = \sum_{k=0}^{n} a_k \phi_k(x)$ and $y(x)$. Moreover the least squares problem consists in finding that f, call it $\hat{f} = \sum_{k=0}^{m} \hat{a}_k \phi_k$, that is closest to y. Thus

$$E(\hat{a}_0, \ldots, \hat{a}_n) = \|\hat{f} - y\|^2 = \min_{f} \|f - y\|^2$$

where the minimum is taken over all functions f of the form (7.5).

Next we characterize the least squares problem in terms of the normal equations. The normal equations can be derived by taking partial derivatives of E; however, we take another (but equivalent) approach. Suppose $\hat{f}$ minimizes $\|f - y\|^2$ and consider, for t real, the function

$$F(t) = \|\hat{f} + tf - y\|^2$$

where f is an arbitrary (but fixed) function of the form (7.5). F is a real-valued (nonnegative) function of the real variable t. We note that $\hat{f} + tf$ is also in the form (7.5), that is, $\hat{f} + tf = \sum_{k=0}^{n} (\hat{a}_k + ta_k)\phi_k$, and hence

$$F(t) \geq F(0) = \|\hat{f} - y\|^2 \qquad \text{all } t$$

Thus F has a minimum at $t = 0$, and by calculus we must have $F'(0) = 0$. To compute the derivative of F, we write

$$F(t) = (\hat{f} - y + tf, \hat{f} - y + tf)$$
$$= \|\hat{f} - y\|^2 + 2t(\hat{f} - y, f) + t^2\|f\|^2$$

where we have used properties (ii) and (iii). Now it is easy to see that

$$F'(t) = 2(\hat{f} - y, f) + 2t\|f\|^2$$

and hence $F'(0) = 0$ means that

$$(\hat{f} - y, f) = 0 \qquad \text{all } f \text{ in form (7.5)}$$

Each function $f = \phi_k$, however, is in the form (7.5), that is, $a_k = 1$ and $a_j = 0$ for $j \neq k$, and hence we must have

$$(\hat{f} - y, \phi_k) = 0 \qquad 0 \leq k \leq n$$

Since $\hat{f} = \sum_{j=0}^{n} \hat{a}_j \phi_j$, the above equation can be rewritten as (exercise 7.19)

$$(\hat{f}, \phi_k) = \sum_{j=0}^{n} (\phi_j, \phi_k)\hat{a}_j = (y, \phi_k) \qquad 0 \leq k \leq n$$

This is a system of $n + 1$ linear equations, called the **normal equations**, for the unknowns $\hat{a}_0, \hat{a}_1, \ldots, \hat{a}_n$. We have shown that the normal equations are necessary for $\hat{f}$ to minimize $\|f - y\|^2$; in fact the normal equations are also sufficient (exercise 7.26). In the following theorem we state the relevant facts about the linear least squares problem.

THEOREM 7.1

Suppose every function of form (7.5) satisfies property (iv'). Then the linear least squares problem has a unique solution $\hat{f} = \sum_{j=0}^{n} \hat{a}_j \phi_j$ that satisfies the normal equations

$$\sum_{j=0}^{n} (\phi_j, \phi_k) \hat{a}_j = (y, \phi_k) \qquad 0 \le k \le n$$

The normal equations determine the $(n + 1) \times (n + 1)$ matrix problem $\mathbf{G}\hat{\mathbf{a}} = \mathbf{c}$, where $\hat{\mathbf{a}} = [\hat{a}_0, \ldots, \hat{a}_n]^T$, $\mathbf{c} = [(y, \phi_0), \ldots, (y, \phi_n)]^T$, and the **Gram matrix G** is given by

$$g_{jk} = (\phi_j, \phi_k)$$

Notice that $\mathbf{G}$ is symmetric and, by Theorem 7.1, invertible. However, we caution the reader that $\mathbf{G}$ is ill-conditioned for many choices of n and $\{\phi_k\}_{k=0}^{n}$.

In the remainder of the chapter we apply the theory of this section to several specific cases. Sections 7.3 and 7.4 are concerned with particular choices of the basis consisting of polynomials and cubic splines, respectively. In Section 7.5 our theory is not directly applicable; however, the essential ideas developed herein are applied to least squares problems other than discrete data fitting.

EXERCISES

7.16 Suppose f, g are two functions in the form (7.5). Show that, for any scalars α and β, $\alpha f + \beta g$ is also in the form (7.5).

7.17 Show that the inner product $(\cdot, \cdot)$ satisfies conditions (i), (ii), and (iii).

7.18 Suppose every f in (7.5) satisfies (iv). Show that this implies that each basis function ϕ_k satisfies (iv).

7.19 Show that $(\sum_{j=0}^{n} a_j u_j, v) = \sum_{j=0}^{n} a_j(u_j, v)$. Hint: Use property (iii).

7.20 Show that $(\hat{f} - y, f) = 0$ for all f in (7.5) is equivalent to $(\hat{f} - y, \phi_k) = 0$, $0 \le k \le n$. Hint: Use properties (ii) and (iii).

7.21 Are $\{1, \sin x, \cos x\}$ linearly independent? Justify your answer.

7.22 Are $\{1, \sin^2 x, \cos^2 x\}$ linearly independent? Justify your answer.

*7.23 Let $\{\phi_k(x)\}_{k=0}^{n}$ be the basic cubic spline functions defined in Example 7.4. Show that $\{\phi_k(x)\}_{k=0}^{n}$ are linearly independent. Hint: Each $\phi_k(x)$ is nonzero only on a small interval; in other words, $\phi_k(x) = 0$ for $x \le t_{k-2}$ or $x \ge t_{k+2}$.

7.24 Show that (i'), (iii'), and (iv') are valid for the basic functions $\{\phi_k(x) = x^k\}_{k=0}^{n}$.

7.25 Let f, g be any two functions defined on $\{x_i\}_{i=1}^{m}$.

(a) Show that $\|f - g\|^2 = \|f\|^2 - 2(f, g) + \|g\|^2$.
(b) Conclude that $\|f - g\|^2 = \|f\|^2 + \|g\|^2$ if and only if $(f, g) = 0$.

7.26 Suppose $\hat{f}$ satisfies the normal equations. Let f be in the form (7.5) and show that $\|f - y\|^2 = \|f - \hat{f}\|^2 + \|\hat{f} - y\|^2$. Hint: Consider $\|(f - \hat{f}) - (y - \hat{f})\|^2$ and use the normal equations and exercise 7.25. Conclude that $\|\hat{f} - y\|^2 \le \|f - y\|^2$ for all f in (7.5). This shows that every solution of the normal equations minimizes $\|f - y\|^2$.

7.27 Show that $(y - \bar{y}, \bar{y}) = 0$ where $\bar{y} = (\sum_{i=1}^{m} y_i)/m$.

7.28 The basis functions are said to be orthogonal if, for $i \neq j$, $(\phi_i, \phi_j) = 0$. If the basis functions are orthogonal, show that the least squares data fit $\hat{f}$ is given by

$$\hat{f}(x) = \sum_{k=0}^{n} \frac{(y, \phi_k)}{\|\phi_k\|^2} \phi_k(x)$$

For orthogonal basis functions does the Gram matrix have a simple structure?

7.3 POLYNOMIAL APPROXIMATION

Polynomials are the functions most often used for least squares data fitting. In this section we examine this choice and discuss some of its computational aspects.

Given data $\{(x_i, y_i)\}_{i=1}^{m}$, the problem is to find $\hat{a}_0, \ldots, \hat{a}_n$ such that

$$E(a_0, \ldots, a_n) = \sum_{i=1}^{m} \left(\sum_{k=0}^{n} a_k x_i^k - y_i \right)^2$$

is minimized. Moreover in many problems of this type the choice of n is not clear. If $n = m - 1$, then by interpolating the data one can make $E = 0$, but, as previously observed, this is usually not recommended. Normally one wants $n \ll m$. Since high degree polynomials tend to oscillate, we restrict our consideration to $n \leq 6$.

Suppose that some $n \leq 6$ has been chosen and the corresponding least squares fit is required. The obvious choice of basis is $\phi_k(x) = x^k$, $0 \leq k \leq n$. Unfortunately this choice usually results in an ill-conditioned Gram matrix. For computational purposes we need a choice of $\{\phi_k(x)\}_{k=0}^{n}$ such that

(a) Every polynomial p of degree $\leq n$ can be written as a linear combination of $\phi_0, \ldots, \phi_n$; that is, for some scalars $\{c_k\}_{k=0}^{n}$

$$p(x) = \sum_{k=0}^{n} c_k \phi_k(x)$$

(b) The Gram matrix $\mathbf{G} = [(\phi_i, \phi_j)]$ is reasonably well-conditioned.

Experience has shown that the modified Chebyshev polynomials are a satisfactory choice for the ϕ_k's. Recall from Section 5.4 that the Chebyshev polynomials are given by

$$T_n(t) = \cos (n \arccos t) \qquad n \geq 0$$

or, equivalently, these polynomials can be generated by the recursive formula

$$T_n(t) = 2t T_{n-1}(t) - T_{n-2}(t) \qquad n \geq 2$$

where $T_0(t) \equiv 1$, $T_1(t) = t$.

For the convenience of the reader we list the first seven Chebyshev polynomials:

$$T_0(t) = 1$$

$$T_1(t) = t$$

$$T_2(t) = 2t^2 - 1$$

$$T_3(t) = 4t^3 - 3t$$

$$T_4(t) = 8t^4 - 8t^2 + 1$$

$$T_5(t) = 16t^5 - 20t^3 + 5t$$

$$T_6(t) = 32t^6 - 48t^4 + 18t^2 - 1$$

The modified Chebyshev polynomials for the interval $[a, b]$ are found by using the change of variables

$$t(x) = \frac{2x - (a + b)}{b - a}$$

That is, the **modified Chebyshev polynomials** for $[a, b]$ are

$$\tilde{T}_n(x) = T_n(t(x)) = T_n\left(\frac{2x - (a + b)}{b - a}\right)$$

Given data $\{(x_i, y_i)\}_{i=1}^m$ ordered such that $a = x_1 < x_2 < \cdots < x_m = b$, we choose $\phi_k(x) = \tilde{T}_k(x)$. Then property (a) is valid, and if the data points are distributed more or less uniformly in $[a, b]$ the Gram matrix is reasonably well-conditioned. With this choice of $\{\phi_k\}_{k=0}^n$ we minimize $E(c_0, \ldots, c_n) = \sum_{i=1}^m [\sum_{k=0}^n c_k \tilde{T}_k(x_i) - y_i]^2$ and the normal equations are

$$\sum_{k=0}^n (\tilde{T}_j, \tilde{T}_k)\hat{c}_k = (y, \tilde{T}_j) \qquad 0 \le j \le n$$

where $\tilde{f} = \sum_{k=0}^n \hat{c}_k \tilde{T}_k$ is the desired least squares polynomial fit of degree n.

EXAMPLE 7.5

Suppose data are give in the interval $[1, 5]$ (with $x_1 = 1$, $x_m = 5$) and a polynomial fit of degree 3 is required. Then we use

$$t(x) = \frac{2x - 6}{4}$$

and the modified Chebyshev polynomials are

$$\tilde{T}_0(x) = T_0(t(x)) = 1$$

$$\tilde{T}_1(x) = T_1(t(x)) = 0.5x - 1.5$$

$$\tilde{T}_2(x) = T_2(t(x)) = 0.5x^2 - 3x + 3.5$$

$$\tilde{T}_3(x) = T_3(t(x)) = 0.5x^3 - 4.5x^2 + 10.5x - 9$$

Every polynomial p of degree ≤ 3 can be written as

$$p(x) = c_0 \tilde{T}_0(x) + c_1 \tilde{T}_1(x) + c_2 \tilde{T}_2(x) + c_3 \tilde{T}_3(x)$$

for some constants $\{c_i\}_{i=0}^3$ (exercises 7.29 and 7.30). We emphasize that minimizing

$$E(c_0, \ldots, c_n) = \sum_{i=1}^m \left[\sum_{k=0}^n c_k \tilde{T}_k(x_i) - y_i \right]^2$$

is equivalent to minimizing

$$E(a_0, \ldots, a_n) = \sum_{i=1}^m \left(\sum_{k=0}^n a_k x_i^k - y_i \right)^2$$

in the absence of roundoff error.

In order to compute the inner products, it is necessary to evaluate each $\tilde{T}_k$ at the data points. For this purpose we have written subroutine CHEBY(A,B,K,Z,CVAL), which evaluates $\tilde{T}_k(Z)$ and returns the value in CVAL. In CHEBY the coefficients for $\{T_j(x)\}_{j=1}^6$ are stored in array COEF. The subroutine listing is as follows:

```
      SUBROUTINE CHEBY(A,B,N,Z,CVAL)
C  A,B:ENDPOINTS OF INTERVAL WHERE DATA IS GIVEN
C  N:DEGREE OF POLYNOMIAL; MUST BE < 7
C  Z:POINT WHERE MODIFIED CHEBYSHEV POLY IS EVALUATED
C  CVAL:VALUE OF POLY AT Z
C  COEF:ARRAY OF COEFFICIENTS FOR CHEBY POLYS
      REAL COEF(6,7)/0.,-1.,0.,1.,0.,-1.,1.,0.,-3.,0.,
     * 5.,0.,0.,2.,0.,-8.,0.,18.,0.,0.,4.,0.,-20.,0.,
     * 0.,0.,0.,8.,0.,-48.,0.,0.,0.,0.,16.,0.,0.,0.,0.,
     * 0.,0.,32./
      X=(2.*Z-A-B)/(B-A)
      IF(N.GT.0) GO TO 1
      CVAL=1.0
      RETURN
 1    N1=N+1
      SUM=COEF(N,N1)
      DO 2 J=1,N
        JJ=N1-J
        SUM=COEF(N,JJ)+SUM*X
 2    CONTINUE
      CVAL=SUM
      RETURN
      END
```

The normal equations are generated and solved in subroutine LSPOLY. The inner products are accumulated in double precision and then stored in the single-precision working array W. Subroutines ELIMIN and BAKSUB are used to solve the normal equations by Gaussian elimination.

```
      SUBROUTINE LSPOLY(X,Y,M,C,N1,N2,W,S)
C  LSPOLY DETERMINES THE LEAST SQUARES FIT FOR
C  A POLY OF DEGREE=N; N=N1-1
C  X,Y:ARRAYS OF DATA POINTS AND VALUES
C  M:NUMBER OF DATA POINTS
C  N1:NUMBER OF FUNCTIONS IN FORM OF LEAST SQ. FIT
C  N2:N1+1
C  C:COEFFICIENTS IN LEAST SQ. FIT
C  W,S:ARRAYS FOR ELIMIN AND BAKSUB
      REAL X(M),Y(M),C(N1),T(51,7),W(N1,N2),S(N1)
      DOUBLE PRECISION DD,DB,P,Q,R
C  EVALUATE MODIFIED CHEBYSHEV POLYS AT DATA POINTS
C  AND STORE IN ARRAY T
      DO 2 J=1,N1
        DO 2 I=1,M
          CALL CHEBY(X(1),X(M),J-1,X(I),T(I,J))
    2   CONTINUE
C  ACCUMULATE INNER PRODUCTS IN DOUBLE-PRECISION
C  AND GENERATE WORKING ARRAY W FOR GAUSS ELIM.
      DO 6 I=1,N1
        DB=0.D0
        DO 3 K=1,M
          R=T(K,I)
          DB=DB+R*Y(K)
    3   CONTINUE
        W(I,N2)=DB
        DO 5 J=1,N1
          DD=0.D0
          DO 4 K=1,M
            P=T(K,I)
            Q=T(K,J)
            DD=DD+P*Q
    4     CONTINUE
          W(I,J)=DD
    5   CONTINUE
    6 CONTINUE
      CALL ELIMIN(W,N1,N2,IFLG,S)
      IF(IFLG.LT.0) GO TO 9
      CALL BAKSUB(W,N1,N2,C)
      RETURN
    9 WRITE(6,99)
   99 FORMAT(' ',2X,'NORMAL MATRIX IS SINGULAR')
      RETURN
      END
```

The subroutine LSFIT can be used to evaluate the resultant least squares fit of the data at any point z.

```
      SUBROUTINE LSFIT(C,N1,A,B,Z,VALU)
C  LSFIT EVALUATES THE LEAST SQUARES FIT AT Z
C  C:COEFFICIENTS OF CHEBYSHEV POLYS IN LEAST SQ. FIT
```

```
C   N1:NUMBER OF POLYNOMIALS;N+1=N1 WHERE N IS THE
C       HIGHEST DEGREE CHEBYSHEV POLY
C   A,B:ENDPOINTS OF INTERVAL WHICH CONTAINS DATA POINTS
C       A IS LEAST AND B IS LARGEST DATA POINT
C   VALU:VALUE OF LEAST SQUARES POLY FIT AT Z
        REAL C(N1)
        N=N1-1
        VALU=C(1)
        IF(N.EQ.0) RETURN
        DO 2 J=1,N
           CALL CHEBY(A,B,J,Z,CVAL)
           VALU=VALU+CVAL*C(J+1)
   2    CONTINUE
        RETURN
        END
```

EXAMPLE 7.6

For $g(x) = (x - 2)^6$ on $[1, 3]$, we compute polluted data

$$y_i = g(x_i) + \varepsilon_i \qquad 1 \leq i \leq 51$$

where $x_i = 1 + 0.04(i - 1)$ and ε_i is a random number that satisfies $|\varepsilon_i| \leq 10^{-3}$. We compute the polynomial of degree 6 that best fits the data in a least squares sense. Let $\hat{f} = \sum_{k=0}^{6} \hat{c}_k \tilde{T}_k$ denote the least squares fit; then we compute the maximum error

$$\max_{1 \leq i \leq 201} |\hat{f}(z_i) - g(z_i)|$$

and root mean square (RMS) error

$$\left\{ \frac{1}{201} \sum_{i=1}^{201} [\hat{f}(z_i) - g(z_i)]^2 \right\}^{1/2}$$

where $z_i = 1 + 0.01(i - 1)$. Thus we sample the error at all data points and at three additional points between each adjacent pair of data points. The main program and output are as follows:

```
        REAL X(51),Y(51),C(7),W(7,8),S(7)
        DOUBLE PRECISION DT
        F(X)=(X-2.)**6
        EPS=1.E-3
        DO 1 J=1,51
           T=1.+.04*FLOAT(J-1)
           DT=T
           X(J)=T
           CALL NOISE(DT,RAND)
           Y(J)=F(T)+(2.*RAND-1.0)*EPS
   1    CONTINUE
        M=51
        N=6
```

```
      N1=N+1
      N2=N+2
      CALL LSPOLY(X,Y,M,C,N1,N2,W,S)
      ABSERR=0.0
      RMSERR=0.0
      DO 2 J=1,201
        Z=1.0+.01*FLOAT(J-1)
        CALL LSFIT(C,N1,X(1),X(M),Z,VALU)
        ERR=F(Z)-VALU
        TERM=ERR**2
        RMSERR=RMSERR+TERM
        DIFF=ABS(ERR)
        IF(DIFF.GT.ABSERR) ABSERR=DIFF
    2   CONTINUE
      WRITE(6,88) ABSERR
   88 FORMAT(' ',2X,'MAXIMUM ERROR IS ',E14.7/)
      RMSERR=SQRT(RMSERR/201.)
      WRITE(6,77) RMSERR
   77 FORMAT(' ',2X,'ROOT MEAN SQUARE ERROR= ',E14.7)
      STOP
      END
```

```
MAXIMUM ERROR IS   0.1000583E-02

ROOT MEAN SQUARE ERROR=   0.1000092E-02
```

Subroutine NOISE(U,RAND) accepts a double-precision number in U, called a seed, and returns a single-precision "random" number RAND in the interval $[0, 1]$. Thus $2.*RAND - 1.0$ is a random number in $[-1, 1]$. We do not furnish a listing of NOISE because it is machine-dependent.

In statistical analysis a measure of goodness of fit for a least squares polynomial is given by the quantities

$$R = \frac{\|\hat{f} - \bar{y}\|^2}{\|y - \bar{y}\|^2}$$

$$\sigma = \frac{\|\hat{f} - y\|^2}{m - n - 1}$$

where $\hat{f}$ is the least squares fit of degree n and $\bar{y}$ is the data mean, that is,

$$\bar{y} = \frac{1}{m} \sum_{i=1}^{m} y_i$$

The ratio R measures the variation of the fit about the data mean relative to the variation of the data about their mean. A value of $R \simeq 1$ is considered good. The quantity σ is the minimum value of E relative to the difference between the data sample size and the degree of polynomial fit. Note that, for $n = m - 1$, $\hat{f}$ is the polynomial interpolate of the data, in which case $R = 1$ but σ is indeterminate.

EXAMPLE 7.7

For $g(x) = \sin 4x$ on $[1, 3]$, we used polluted data:

$$y_i = g(x_i) + \varepsilon_i \qquad 1 \le i \le 51$$

where $x_i = 1 + 0.04(i - 1)$ and $|\varepsilon_i| \le 10^{-3}$ is random noise. Using polynomials of degree ≤ 6, we computed R, σ, and the maximum error

$$\max_{1 \le i \le 51} |g(x_i) - \hat{f}(x_i)|$$

where $\hat{f}$ is the polynomial least squares fit. The results are as follows:

```
F(X)=SIN(4X), EPS=1.E-3,POLYNOMIAL LEAST SQUARES FIT
```

N	R	SIGMA	MAX ERROR
1	0.00115	0.5383728E 00	0.1195053E 01
2	0.68495	0.1733440E 00	0.9561769E 00
3	0.70046	0.1683193E 00	0.8639690E 00
4	0.98971	0.5905960E-02	0.2086815E 00
5	0.99108	0.5232006E-02	0.1559816E 00
6	0.99992	0.4775166E-04	0.1854563E-01

As the degree N increases, the value of $R \to 1$ and $\sigma \to 0$, indicating a better and better fit. This is confirmed by the maximum errors.

EXERCISES

7.29 Show that the Chebyshev basis $\{T_k(x)\}_{k=0}^6$ and power basis $\{x^k\}_{k=0}^6$ are related by

$$x^0 = T_0(x)$$

$$x^1 = T_1(x)$$

$$x^2 = \tfrac{1}{2}[T_0(x) + T_2(x)]$$

$$x^3 = \tfrac{1}{4}[3T_1(x) + T_3(x)]$$

$$x^4 = \tfrac{1}{8}[3T_0(x) + 4T_2(x) + T_4(x)]$$

$$x^5 = \tfrac{1}{16}[10T_1(x) + 5T_3(x) + T_5(x)]$$

$$x^6 = \tfrac{1}{32}[10T_0(x) + 15T_2(x) + 6T_4(x) + T_6(x)]$$

7.30 Use exercise 7.29 to find $c_0, \ldots, c_6$ in terms of $a_0, \ldots, a_6$ where

$$\sum_{k=0}^6 a_k x^k = \sum_{k=0}^6 c_k T_k(x)$$

Conclude that every polynomial of degree ≤ 6 can be expressed as a linear combination of $T_0, T_1, \ldots, T_6$.

7.31 For data points $\{(i, y_i)\}_{i=1}^5$ compute the Gram matrix for the basis $\{x^k\}_{k=0}^3$. Determine the condition number of this matrix.

7.32 Repeat exercise 7.31 for the modified Chebyshev basis $\{\tilde{T}_k(x)\}_{k=0}^3$ on $[1, 5]$. Hint: See Example 7.5.

*7.33 Use the trigonometric identity $\cos(\theta \pm \phi) = \cos\theta\cos\theta \mp \sin\phi$ to show that $\cos n\theta = 2\cos\theta\cos(n-1)\theta - \cos(n-2)\theta$. From this conclude that the Chebyshev polynomials satisfy $T_n(x) = 2xT_{n-1}(x) - T_{n-2}(x), n \geq 2$.

7.34 Let $p(x) = 1 - 2x + 3x^2 - 4x^3 + 5x^4 - 6x^5 + 7x^6$ and evaluate p at 50 points chosen randomly in $[0, 2]$. Then pollute the data by adding a perturbation term $y_i = p(x_i) + \varepsilon_i$, where $|\varepsilon_i| \leq 0.05$. Use LSPOLY to "recover" the polynomial by a sixth-degree least squares fit. Compute the maximum and RMS errors.

*7.35 For the modified Chebyshev basis denote the normal equations as the matrix problem $G\hat{c} = r$, where $g_{ij} = (\tilde{T}_i, \tilde{T}_j)$, $\hat{c} = [\hat{c}_0, \ldots, \hat{c}_n]^T$, and $r = [(y, \tilde{T}_0), \ldots, (y, \tilde{T}_n)]^T$.

(a) Show that $\|y - \bar{y}\|^2 = \|y\|^2 - m(\bar{y})^2$, where $\bar{y}$ is the mean of the data values.
(b) Show that $\|\hat{f} - \bar{y}\|^2 = \hat{c}^T r - m(\bar{y})^2$.
(c) Conclude from parts (a) and (b) that

$$R = \frac{\|\hat{f} - \bar{y}\|^2}{\|y - \bar{y}\|^2} = \frac{\hat{c}^T r - m(\bar{y})^2}{\|y\|^2 - m(\bar{y})^2}$$

7.36 The following data were determined by adding random noise of magnitude $\leq 10^{-2}$ to values of some polynomial of degree 3. For $n = 1, \ldots, 6$, use LSPOLY to determine the least squares fit of degree n and print R, σ in each case. Based on the output, which degree polynomial best fits the data?

x_i	y_i	x_i	y_i
1.0	49.7861	2.3	13.4711
1.3	38.5901	2.5	10.3518
1.5	32.1196	2.6	8.9906
1.6	29.1902	2.75	7.1734
1.75	25.1376	2.8	6.6341
1.8	23.8741	3.0	4.7231
1.85	22.6384	3.1	3.9277
1.9	21.4692	3.3	2.6321
2.0	19.2307	3.45	1.8508
2.1	17.1448	3.6	1.2647

7.37 Subroutine LSPOLY does not generate the working array very efficiently. The Gram matrix is symmetric, and hence $W(I,J) = W(J,I)$ for $1 \leq I,J \leq N1$. Recode LSPOLY to take advantage of this symmetry.

7.38 Write a subroutine CHANGE(A,B,C) that accepts the coefficients $\{c_i\}_{i=0}^n$, $n \leq 6$, of a polynomial $p(x) = \sum_{i=0}^n c_i \tilde{T}_i(x)$ in array C and returns in C the coefficients $\{a_i\}_{i=0}^n$ for the same polynomial in power form, that is, $p(x) = \sum_{i=0}^n a_i x^i$. The parameters A and B give the endpoints of the interval for which $\{\tilde{T}_i\}$ are defined. Test your program on a polynomial of degree 6.

7.4 SPLINE APPROXIMATION

In this section we show how cubic splines can be used in the least squares problem. We consider only cubic splines with uniformly spaced knots and utilize the basic representation discussed in Section 4.3.

Given data $\{(x_i, y_i)\}_{i=1}^{m}$ with $a = x_1 < x_2 < \cdots < x_m = b$, we define, for $n > 2$, $h = (b - a)/(n - 2)$ and

$$\phi_k(x) = \Phi(h^{-1}(x - t_k)) \qquad 0 \le k \le n$$

where $t_k = a + (k - 1)h$ and Φ is as defined in Section 4.3 (sketched in Figure 4.3). Then $S(x) = \sum_{k=0}^{n} a_k \phi_k(x)$ is a cubic spline on $[a, b]$ with knots $t_1, \ldots, t_{n-1}$. The data points $\{x_i\}_{i=1}^{m}$ and knots need to be related in some way since Theorem 7.1 requires that

$$\|S\| = \left\| \sum_{k=0}^{n} a_k \phi_k \right\| = 0$$

if and only if $S(x) = 0$ for all x. This condition is satisfied if there is a data point between each adjacent pair of knots, that is, $t_j < x_i < t_{j+1}$ for some x_i. Since $m \gg n$, there are usually several data points between adjacent knots.

The problem is to determine $\hat{S} = \sum_{j=0}^{n} \hat{a}_j \phi_j$ such that $\|\hat{S} - y\|^2 = \min_S \|S - y\|^2$ or, equivalently, to solve the normal equations

$$\sum_{j=0}^{n} (\phi_j, \phi_k)\hat{a}_j = (y, \phi_k) \qquad 0 \le k \le n$$

In the following subroutine LSQSPL we solve the normal equations using Gaussian elimination via subroutines ELIMIN and BAKSUB from Chapter 3. The values of the basic splines $\{\phi_k\}$ at the data points are computed using subroutines SPLVAL and BASIC from Chapter 4. The values $\phi_k(x_1), \ldots, \phi_k(x_m)$ are stored in column $K + 1$ of array PHI, that is, values of ϕ_0 in column 1, values of ϕ_1 in column 2 and so forth. Array PHI is used to compute the entries of the Gram matrix and the right-hand side, which are stored in the working array W for use in ELIMIN. The number of basic spline functions $n + 1$ is N in LSQSPL, and M is the number of data points.

```
      SUBROUTINE LSQSPL(X,Y,M,N,B)
C   X:ARRAY OF DATA POINTS
C   Y:ARRAY OF DATA VALUES
C   M:NUMBER OF DATA POINTS
C   N:NUMBER OF BASIC CUBIC SPLINE FUNCTIONS
C   B:COEFFICIENTS IN BASIC REPRESENTATION ON RETURN
      REAL X(M),Y(M),B(N),PHI(100,25),W(13,14),S(13)
      DOUBLE PRECISION DD,DB,P,Q,R
      DO 1 K=1,N
        B(K)=0.0
1     CONTINUE
C   DETERMINE VALUES OF BASIC SPLINES AT DATA POINTS
C   AND STORE IN ARRAY PHI
      DO 3 J=1,N
        B(J)=1.0
        DO 2 I=1,M
          CALL SPLVAL(B,X(1),X(M),X(I),N-3,PHI(I,J))
2       CONTINUE
```

```
              B(J)=0.0
   3    CONTINUE
C    GENERATE ENTRIES IN NORMAL MATRIX
        DO 6 I=1,N
          DO 5 J=1,N
            DD=0.0
            DO 4 K=1,M
              P=PHI(K,I)
              Q=PHI(K,J)
              DD=DD+P*Q
   4        CONTINUE
            W(I,J)=DD
   5      CONTINUE
   6    CONTINUE
        N1=N+1
C    GENERATE RIGHT HAND SIDE AND
C    STORE IN LAST COLUMN OF W
        DO 8 I=1,N
          DB=0.D0
          DO 7 J=1,M
            P=PHI(J,I)
            R=Y(J)*P
            DB=DB+R
   7      CONTINUE
          W(I,N1)=DB
   8    CONTINUE
C    SOLVE THE NORMAL EQUATIONS TO FIND
C    THE COEFFICIENTS FOR BASIC REPRESENTATION
        CALL ELIMIN(W,N,N1,IFLG,S)
        IF(IFLG.LT.0) GO TO 10
        CALL BAKSUB(W,N,N1,B)
        RETURN
  10    WRITE(6,101)
 101    FORMAT(' ',2X,'THE NORMAL MATRIX IS SINGULAR')
        RETURN
        END
```

In order to minimize roundoff error, the inner products (ϕ_j, ϕ_k) and (y, ϕ_k) are computed in double precision and then stored in (single-precision) array W.

EXAMPLE 7.8

We perform an experiment for $g(x) = \sin 4x$ on $[1, 3]$ that is similar to Example 7.6. For the noisy data

$$y_i = g(x_i) + \varepsilon_i \qquad 1 \le i \le 51$$

where $x_i = 1 + 0.04(i - 1)$ and $|\varepsilon_i| \le 10^{-3}$, we compute the best cubic spline data fit with $h = 0.1$, using

$$\phi_k(x) = \Phi(10(x - t_k)) \qquad 0 \le k \le 12 = n$$

where $t_k = 1 + 0.1(k - 1)$, $0 \le k \le 12$. In addition to computing the maximum and RMS errors given by

$$\max_{1 \le i \le 201} |\hat{S}(z_i) - g(z_i)|$$

and

$$\left\{ \frac{1}{201} \sum_{i=1}^{201} [\hat{S}(z_i) - g(z_i)]^2 \right\}^{1/2}$$

we compute the derivative of $\hat{S}$ and its corresponding maximum error

$$\max_{1 \le i \le 201} |\hat{S}'(z_i) - g'(z_i)|$$

where $z_i = 1 + 0.01(i - 1)$. The main program and output are as follows:

```
      REAL X(51),Y(51),B(13)
      DOUBLE PRECISION DT
      F(X)=SIN(4.*X)
      DF(X)=4.*COS(4.*X)
      EPS=1.E-3
      DO 1 J=1,51
        T=1.+.04*FLOAT(J-1)
        DT=T
        X(J)=T
        CALL NOISE(DT,RAND)
        Y(J)=F(T)+(2.*RAND-1.0)*EPS
1     CONTINUE
      M=51
      N=13
      CALL LSQSPL(X,Y,M,N,B)
      DERERR=0.0
      ABSERR=0.0
      RMSERR=0.0
      DO 2 J=1,201
        Z=1.0+.01*FLOAT(J-1)
        CALL SPLVAL(B,X(1),X(M),Z,N-3,SVAL)
        CALL DSPVAL(B,X(1),X(M),Z,N-3,DVAL)
        DERR=DF(Z)-DVAL
        ERR=F(Z)-SVAL
        TERM=ERR**2
        RMSERR=RMSERR+TERM
        DIFF=ABS(ERR)
        DDIFF=ABS(DERR)
        IF(DDIFF.GT.DERERR) DERERR=DDIFF
        IF(DIFF.GT.ABSERR) ABSERR=DIFF
2     CONTINUE
      WRITE(6,88) ABSERR
88    FORMAT(' ',2X,'MAXIMUM ERROR IS ',E14.7/)
      RMSERR=SQRT(RMSERR/201.0)
      WRITE(6,77) RMSERR
77    FORMAT(' ',2X,'ROOT MEAN SQUARE ERROR= ',E14.7/)
      WRITE(6,99) DERERR
```

```
99   FORMAT(' ',2X,'DERIVATIVE MAX ERROR IS ',E14.7)
     STOP
     END
```

```
MAXIMUM ERROR IS   0.1614034E-02
```

```
ROOT MEAN SQUARE ERROR=   0.1055909E-02
```

```
DERIVATIVE MAX ERROR IS   0.1855087E-01
```

The maximum and RMS errors are of the same order of magnitude as the noise in the data, and the maximum error in the derivative is quite respectable. Note: We use subroutine DSPVAL as described in exercise 4.63.

EXAMPLE 7.9

In Example 6.11 the adaptive Runge-Kutta method, subroutine RK2ADP, was applied to the initial value problem

$$y'(t) = e^{-t} - y(t) \qquad y(0) = 1$$

whose solution is $y(t) = (1 + t)e^{-t}$. The output $\{(t_i, Y_i)\}_{i=1}^{27 = M}$ from this example is now used as data for a least squares cubic spline fit via LSQSPL. We have $t_1 = 0$ and $t_{27} = 1$ and use the cubic spline basis defined by

$$\phi_k(t) = \Phi(10(t - \tau_k)) \qquad 0 \le k \le 12$$

where $\tau_k = 0.1(k - 1)$, $0 \le k \le 12$. For the resultant cubic spline fit $\hat{S}$ of the discrete variable output we compute

$$\max_{1 \le i \le 201} |\hat{S}(z_i) - y(z_i)|$$

$$\left\{ \frac{1}{201} \sum_{i=1}^{201} [\hat{S}(z_i) - y(z_i)]^2 \right\}^{1/2}$$

and $\quad \max_{1 \le i \le 201} |\hat{S}'(z_i) - y'(z_i)|$

where $z_i = 0.005(i - 1)$, $1 \le i \le 201$. The results are as follows:

```
C   USING THE DATA FROM EXAMPLE 6.11 WHERE RK2ADP WAS
C   APPLIED TO Y'=EXP(-T)-Y WE COMPUTE THE LEAST
C   SQUARES CUBIC SPLINE FIT AND SAMPLE THE ERROR AT THE
C   201 POINTS Z(I)=.005*(I-1),I=1,201
C   THE RESULTS ARE AS FOLLOWS:
```

```
MAXIMUM ERROR IS   0.8374453E-04
```

```
ROOT MEAN SQUARE ERROR=   0.5415580E-04
```

```
DERIVATIVE MAX ERROR IS   0.3814697E-03
```

The estimated errors in Example 6.11 are all about 10^{-4}, and the least squares cubic spline fit gives a comparable maximum error on [0, 1]. In an adaptive routine like RK2ADP the output is generally not equally spaced. If values at equally spaced points are desired, a least squares spline fit can be utilized. Note also that $\hat{S}'(t)$ gives an accurate approximation to $y'(t)$ throughout [0, 1].

EXERCISES

7.39 Subroutine LSQSPL does not exploit the symmetry of the normal equations in generating working array W. Modify the code to generate W more efficiently.

7.40 Modify LSQSPL so that all computation is done in single precision. Repeat Example 7.8 using both versions of LSQSPL on your computer along with a suitable NOISE subroutine. Try various values of EPS and note any difference between the single-precision and partial double-precision versions.

7.41 Perform an experiment like Example 7.8 for $g(x) = \tan x$ on [0, 1.5]. Try several values of EPS, say 0.1, 0.05, 0.01, and 0.005.

7.42 Repeat exercise 7.41 for data determined by $y_i = (1 + \varepsilon_i)g(x_i)$.

7.43 Use the data in exercise 7.34 to compute, for $h = 0.1$, the least squares cubic spline fit. Compare the maximum and RMS errors with that of the sixth-degree least squares polynomial fit. Repeat for $h = 0.05$.

7.44 Repeat the computation of exercise 7.36 using LSQSPL for $h = 0.52$ and $h = 0.26$.

7.45 For $h = 1$, determine the least squares cubic spline fit of the data given in exercise 4.58. Use an automatic plotter to graph the data and spline fit. Compare with exercise 4.59.

7.46 Write a subroutine for least squares data fitting using linear splines with uniformly spaced knots. See exercises 4.70 to 4.72. Use the subroutine to perform an experiment like Example 7.8 for $g(x) = (1 + x^2)^{1/2}$ on [−1, 1].

7.5 EXTENSION TO OTHER PROBLEMS

In this section we indicate how the least squares principle can be extended to problems other than curve fitting of discrete data. Specifically we consider least squares approximation of data given on an interval (as opposed to discrete data) and the least squares solution of an overdetermined linear system of equations.

Suppose $g(x)$ is given on an interval $[a, b]$ and it is required to approximate g by a function of the form

$$f(x) = \sum_{j=0}^{n} a_j \phi_j(x) \tag{7.8}$$

where $\{\phi_j\}_{j=0}^n$ is a given set of linearly independent functions. The least squares approach consists in minimizing

$$e_2(a_0, \ldots, a_n) = \int_a^b [f(x) - g(x)]^2 \, dx = \int_a^b \left[\sum_{j=0}^n a_j \phi_j(x) - g(x) \right]^2 dx$$

by an appropriate choice of the coefficients $\{a_j\}_{j=0}^n$. In some situations it may be expedient to force a better approximation on some portion of $[a, b]$. This can be accomplished by introducing a positive weight function w into the integral; that is, we minimize instead

$$e_2(a_0, \ldots, a_n) = \int_a^b [f(x) - g(x)]^2 w(x) \, dx \tag{7.9}$$

Of course the choice $w(x) \equiv 1$ gives equal weight to all parts of the interval.

Let us show how the minimization of (7.9) fits into the theory of Section 7.2. All that is required is the definition of a suitable inner product and norm. For a given weight function w and interval $[a, b]$, define

$$\langle u, v \rangle = \int_a^b u(x)v(x)w(x) \, dx \tag{7.10}$$

and $\quad \|u\| = \sqrt{\langle u, u \rangle} \tag{7.11}$

If all functions considered (including w) are continuous on $[a, b]$, then (7.10) and (7.11) define an inner product and norm that satisfy properties (i) to (iv) and (i') to (iv'), respectively. Moreover all of the results of Section 7.2 are valid in the present context. Thus if $\hat{f}$ minimizes e_2, that is, $\|\hat{f} - g\|^2 = \min_f \|f - g\|^2$, then define, for f in the form (7.8),

$$F(t) = \|\hat{f} + tf - g\|^2$$

and proceed exactly as before to find

$$F(t) \geq F(0) = \|\hat{f} - g\|^2 \quad \text{all } t$$

Consequently $F'(0) = 0$ and there results the normal equation

$$\langle \hat{f} - g, \phi_i \rangle = 0 \quad 1 \leq i \leq n$$

which can be written as

$$\sum_{j=0}^n \langle \phi_i, \phi_j \rangle \hat{a}_j \equiv \langle \hat{f}, \phi_i \rangle = \langle g, \phi_i \rangle \quad 0 \leq i \leq n$$

Just as in the case of discrete least squares approximation, the normal equations give a symmetric matrix problem. The only change is the definition of norm and inner product.

EXAMPLE 7.10

If $[a, b] = [0, 1]$, $w(x) \equiv 1$, and the basis functions are chosen to be $\phi_k(x) = x^k$, $0 \leq k \leq n$, then

$$\langle g, \phi_j \rangle = \int_0^1 g(x)x^j \, dx$$

$$\langle \phi_i, \phi_j \rangle = \int_0^1 x^i x^j \, dx = \frac{1}{i + j + 1}$$

The coefficient matrix for the normal equations is the Hilbert matrix

$$\begin{bmatrix} 1 & \frac{1}{2} & \frac{1}{3} & \cdots & \frac{1}{n+1} \\ \frac{1}{2} & \frac{1}{3} & \frac{1}{4} & \cdots & \frac{1}{n+2} \\ \cdots & \cdots & \cdots & & \cdots \\ \frac{1}{n+1} & \frac{1}{n+2} & \frac{1}{n+3} & \cdots & \frac{1}{2n+1} \end{bmatrix}$$

Recall from exercises 3.62 and 3.63 that Hilbert matrices are notoriously ill-conditioned.

This example indicates that the power functions $\{x^k\}_{k=0}^n$ can lead to severe computational difficulties. Fortunately the solution of the normal equations can, in many instances, be simplified by a judicious choice of basis functions. The basis functions are said to be orthogonal with respect to w if

$$\langle \phi_i, \phi_j \rangle \begin{cases} = 0 & \text{for } i \neq j \\ \neq 0 & \text{for } i = j \end{cases}$$

(See also the discussion of orthogonal polynomials in Section 5.4.) For orthogonal basis functions the normal equations become

$$\sum_{j=0}^n \langle \phi_i, \phi_j \rangle \hat{a}_j = \langle \phi_i, \phi_i \rangle \hat{a}_i = \langle g, \phi_i \rangle \qquad 0 \leq i \leq n$$

and hence

$$\hat{f}(x) = \sum_{i=0}^n \frac{\langle g, \phi_i \rangle}{\|\phi_i\|^2} \phi_i(x)$$

EXAMPLE 7.11

The functions $\{\cos kx\}_{k=0}^n$ are orthogonal on $[0, \pi]$ with respect to the weight function $w(x) \equiv 1$.

EXAMPLE 7.12

The functions $\{1, \cos x, \ldots, \cos mx, \sin x, \sin 2x, \ldots, \sin mx\}$ are orthogonal on $[-\pi, \pi]$ with respect to the weight function $w(x) \equiv 1$.

EXAMPLE 7.13

The Legendre polynomials are orthogonal on $[-1, 1]$ with respect to the weight function $w(x) \equiv 1$. If the weight function is $w(x) = (1 - x^2)^{1/2}$, then the Chebyshev polynomials are orthogonal on $[-1, 1]$ (Section 5.4).

EXAMPLE 7.14

Consider the choice of basis consisting of the basic cubic splines, that is, for $h = (b - a)/(n - 2)$

$$\phi_k(x) = \Phi(h^{-1}(x - x_k)) \qquad 0 \le k \le n$$

where $x_k = a + (k - 1)h$. In this case the basis is not orthogonal, but it is nearly so. Consider the integral

$$\langle \phi_i, \phi_j \rangle = \int_a^b \phi_i(x)\phi_j(x)w(x) \, dx$$

and recall that $\phi_i(x) = 0$ if $|x - x_i| \ge 2h$. Since ϕ_i and ϕ_j are nonzero only on the intervals $[x_{i-2}, x_{i+2}]$ and $[x_{j-2}, x_{j+2}]$, respectively, it follows that $\langle \phi_i, \phi_j \rangle = 0$ for $|i - j| \ge 4$. The coefficient matrix for the normal equation is sparse with bandwidth ≤ 7. Special algorithms are available for economically solving sparse banded systems. Moreover it can be shown that the condition number of this coefficient matrix is bounded independent of n.

Finally we consider the least squares "solution" of a system of linear equations. Consider the matrix problem

$$\mathbf{Ax} = \mathbf{b} \tag{7.12}$$

with $\mathbf{A}$ an $m \times n$ matrix. If $m < n$, then solutions are not unique, whereas for $m > n$, the linear system is overdetermined and there may be no solution to the problem. In applications the latter is more common. In any case we are presented with a matrix problem for which unique solvability is not guaranteed. Instead of attempting to solve (7.12), we choose $\mathbf{x}$ to minimize the residual $\mathbf{r} = \mathbf{b} - \mathbf{Ax}$ in a least squares sense. Recall from Section 3.4 the vector norm

$$\|\mathbf{y}\|_2 = \left\{ \sum_{i=1}^m |y_i|^2 \right\}^{1/2}$$

and define the corresponding inner product

$$[\mathbf{y}, \mathbf{w}] = \sum_{i=1}^m y_i w_i$$

for vectors $\mathbf{y} = [y_1, \ldots, y_m]^T$, $\mathbf{w} = [w_1, \ldots, w_m]^T$. We "solve" (7.12) in a least squares sense by choosing $\hat{\mathbf{x}}$ such that

$$\|\mathbf{A}\hat{\mathbf{x}} - \mathbf{b}\|_2^2 = \min_{\mathbf{x}} \|\mathbf{Ax} - \mathbf{b}\|_2^2$$

Note that $\|A\hat{x} - b\|_2 = 0$ if and only if $\hat{x}$ is a solution of (7.12). Usually $\|A\hat{x} - b\|_2 \neq 0$, in which case we call $\hat{x}$ a **pseudosolution** of (7.12) or a **least squares solution** of (7.12).

The normal equations for $\hat{x}$ can be determined in a manner similar to that discussed in Section 7.2. Here we define, for an arbitrary but fixed vector x and real t,

$$F(t) = \|A(\hat{x} + tx) - b\|_2^2$$
$$= \|A\hat{x} - b + tAx\|_2^2$$

Since $[\cdot, \cdot]$ is an inner product, we can write $F(t)$ as

$$F(t) = [A\hat{x} - b + tAx, A\hat{x} - b + tAx]$$
$$= \|A\hat{x} - b\|_2^2 + 2t[A\hat{x} - b, Ax] + t^2\|Ax\|_2^2$$

It follows that $F(t) \geq F(0)$ and hence

$$F'(0) = 2[A\hat{x} - b, Ax] = 0$$

or, equivalently,

$$[A^T(A\hat{x} - b), x] = 0$$

Since x is arbitrary we must have

$$A^TA\hat{x} = A^Tb \tag{7.13}$$

which are the normal equations. It can be shown that the normal equations always have a solution (not necessarily unique) that minimizes $\|Ax - b\|_2$. In principle a least squares solution of (7.12) can be found by solving the (square) matrix problem (7.13); however, A^TA is frequently ill-conditioned, and special techniques have been devised for finding $\hat{x}$. Two such techniques involve the QR factorization and the singular value decomposition. A discussion of these methods is beyond the scope of the text, and we refer the reader to Noble and Daniel (1977) and Stewart (1973).

EXAMPLE 7.15

Suppose that y and t are known to be related by $y = a_0 + a_1 t$ for some unknown coefficients a_0 and a_1. Several experimental data points $\{(t_i, y_i)\}_{i=1}^m$ are obtained. Thus with no data error we should have

$$y_1 = a_0 + a_1 t_1$$
$$\cdots\cdots\cdots\cdots\cdots$$
$$y_m = a_0 + a_1 t_m$$

This overdetermined system of equations is "solved" in a least squares sense as follows. Define the 2-vector $x = [a_0, a_1]^T$, the m-vector $b = [y_1, \ldots, y_m]^T$, and the $m \times 2$ matrix

$$A = \begin{bmatrix} 1 & t_1 \\ 1 & t_2 \\ \cdots & \cdots \\ 1 & t_m \end{bmatrix}$$

For $\|A\hat{x} - b\|_2^2 = \min_x \|Ax - b\|_2^2$, the normal equations are $A^T A \hat{x} = A^T b$ with

$$A^T A = \begin{bmatrix} m & \sum_{i=1}^{m} t_i \\ \sum_{i=1}^{m} t_i & \sum_{i=1}^{m} t_i^2 \end{bmatrix} \qquad A^T b = \begin{bmatrix} \sum_{i=1}^{m} y_i \\ \sum_{i=1}^{m} y_i t_i \end{bmatrix}$$

Compare these normal equations with (7.2) and convince yourself that this is another (but equivalent) way of viewing the linear least squares polynomial.

EXERCISES

7.47 Suppose $\hat{f}(x) = \sum_{j=0}^{n} \hat{a}_j \phi_j(x)$ satisfies the normal equations $\langle \hat{f}, \phi_i \rangle = \langle g, \phi_i \rangle$, $0 \le i \le n$. Show that $\|\hat{f} - g\|^2 \le \|f - g\|^2$ for all f in the form (7.8). Hint: See exercise 7.25.

7.48 The Legendre polynomials $\{L_k(x)\}$ are orthogonal with respect to $w(x) \equiv 1$ on the interval $[-1, 1]$. For an interval $[c, d]$ let $x(t) = (2t - c - d)/(d - c)$ and define the modified Legendre polynomials by $\tilde{L}_k(t) = L_k(x(t))$. Show that $\{\tilde{L}_k(t)\}$ are orthogonal with respect to $w(t) \equiv 2/(d - c)$ on $[c, d]$.

7.49 Find the polynomial of degree 2 that best approximates $g(x) = \sin x$ on $[0, \pi]$ in the least squares sense. Hint: Use the basis of modified Legendre polynomials given in exercise 7.48.

7.50 Show that for $m, n \ge 0$

$$\int_0^\pi \cos mx \cos nx \, dx = \begin{cases} 0 & \text{for } m \ne n \\ \pi/2 & \text{for } m = n \end{cases}$$

7.51 Define $g(x)$ on $[0, \pi]$ by

$$g(x) = \begin{cases} 1 & \text{for } 0 \le x \le \pi/2 \\ -1 & \text{for } \pi/2 < x \le \pi \end{cases}$$

Find $\{\hat{a}_k\}_{k=0}^n$ that minimize

$$\int_0^\pi \left[\sum_{k=0}^n a_k \cos kx - g(x) \right]^2 dx$$

7.52 Show that for $m, n \ge 1$

$$\int_0^\pi \sin mx \sin nx \, dx = \begin{cases} 0 & \text{for } m \ne n \\ \pi/2 & \text{for } m = n \end{cases}$$

7.53 Show that for $m, n \ge 1$

$$\int_{-\pi}^{\pi} \sin nx \cos mx \, dx = 0$$

7.54 Let $h = (b - a)/(n - 1)$ and $x_i = a + (i - 1)h$. Define $\psi_i(x) = L(h^{-1}(x - x_i))$,

$1 \leq i \leq n$, where L is as defined in exercise 4.70. Then $\{\psi_i(x)\}_{i=1}^n$ are the basic linear splines. If $w(x)$ is a continuous weight function on $[a, b]$ show that

$$\langle \psi_i, \psi_j \rangle \equiv \int_a^b \psi_i(x)\psi_j(x)w(x)\,dx = 0 \qquad \text{if } |i - j| \geq 2$$

Conclude that the basis $\{\psi_i\}_{i=1}^n$ for least squares approximation results in a symmetric tridiagonal system of normal equations. For $w(x) \equiv 1$, show that

$$\langle \psi_i, \psi_i \rangle = \langle \psi_{i+1}, \psi_{i+1} \rangle \qquad 2 \leq i \leq n - 2$$

$$\langle \psi_i, \psi_{i+1} \rangle = \langle \psi_{i+1}, \psi_{i+2} \rangle \qquad 2 \leq i \leq n - 3$$

Conclude that, except for the first and last rows, the main, super, and subdiagonals are constant in the coefficient matrix of the normal equations [for $w(x) \equiv 1$].

7.55 Show that $[\mathbf{y}, \mathbf{w}] = \mathbf{y}^T\mathbf{w}$ for m-vectors $\mathbf{y}$ and $\mathbf{w}$.

7.56 Show that $[\mathbf{x}, \mathbf{Ay}] = [\mathbf{A}^T\mathbf{x}, \mathbf{y}]$, where $\mathbf{A}$ is $m \times n$, $\mathbf{y}$ is an n-vector, and $\mathbf{x}$ is an m-vector.

7.57 Fill in all of the details for the derivation of (7.13).

7.58 If $[\mathbf{A}\hat{\mathbf{x}} - \mathbf{b}, \mathbf{Ax}] = 0$ for all n-vectors $\mathbf{x}$, show that $\|\mathbf{A}\hat{\mathbf{x}} - \mathbf{b}\|_2^2 \leq \|\mathbf{Ax} - \mathbf{b}\|_2^2$ for all $\mathbf{x}$. Hint: First show that $\|\mathbf{Ax} - \mathbf{b}\|_2^2 = \|\mathbf{Ax} - \mathbf{A}\hat{\mathbf{x}}\|_2^2 + \|\mathbf{A}\hat{\mathbf{x}} - \mathbf{b}\|_2^2$.

7.59 Find the least squares solution of the overdetermined system

$$x - y = 2$$
$$2x + 3y = -1$$
$$x + y = 3$$

7.60 Determine the normal equations for the underdetermined system

$$x - y + z = 1$$
$$2x + y - z = 2$$

Is $\mathbf{A}^T\mathbf{A}$ for this problem invertible? Solve the normal equations. How many solutions are there? Do any least squares solutions solve the original system?

7.61 Can you generalize Example 7.15 to the least squares polynomial of degree n for discrete data $\{(t_i, y_i)\}_{i=1}^m$? Hint: Consider the system

$$\sum_{k=0}^n a_k t_i^k = y_i \qquad 1 \leq i \leq m$$

and determine the normal equations $\mathbf{A}^T\mathbf{A}\hat{\mathbf{x}} = \mathbf{A}^T\mathbf{b}$.

NOTES AND COMMENTS

Sections 7.1 to 7.3

We first expose the reader to the least squares problem for linear functions—a problem of considerable importance in statistics. A readable undergraduate text on statistics is Walpole and Myers (1978). By proceeding from the concrete to the abstract, we attempt to develop the theory as a natural and understandable progression. The more knowledgeable reader will recognize the collection of all functions of the form (7.5) as a vector space and that $\hat{f}$ is the projection of y onto this space with respect to the inner product $(\cdot, \cdot)$.

The usual approach for the derivation of the normal equations in Theorem 7.1 is to set the partial derivatives of $E(a_0, \ldots, a_n)$ equal to zero, as is done in Section 7.1. While this is more consistent with the preliminary discussion, we thought it worthwhile to present an alternate (but equivalent) point of view.

Section 7.4

The use of cubic splines with uniform knot spacing is somewhat limited. The book by de Boor (1978) discusses least squares spline approximation in a more general setting. The subroutine L2APPR given therein provides for nonuniform knot spacing with splines of order k and requires a user-supplied weighting of the data points. In Lemma XIV.2 necessary and sufficient conditions on the knots and data points are given in order that property (iv') be valid for splines.

The least squares solution of a system of linear equations is discussed in more detail in Strang (1980). This book also develops the notion of a pseudoinverse of a matrix and its relation to least squares problems. A similar but more advanced treatment is given in Noble and Daniel (1977).

Finally we mention that the least squares problem can be considered a special case of Fourier analysis. The book by Churchill (1969) develops the theory of orthogonal functions and Fourier series at a level appropriate for advanced undergraduate students.

BIBLIOGRAPHY

Andrews, J. G., and R. R. McLone: *Mathematical Modelling*, Butterworths, London, 1976.

Atkinson, K. E.: *An Introduction to Numerical Analysis*, Wiley, New York, 1978.

Birkhoff, G., and G. C. Rota: *Ordinary Differential Equations*, Blaisdell, New York, 1969.

Boyce, W. E., and R. C. DiPrima: *Elementary Differential Equations*, Wiley, New York, 1977.

Braun, M.: *Differential Equations and Their Applications*, Springer-Verlag, New York, 1975.

Brent, R. P.: *Algorithms for Minimization without Derivatives*, Prentice-Hall, Englewood Cliffs, NJ, 1973.

Bronson, R.: *Matrix Methods*, Academic Press, New York, 1970.

Burden, R. L., J. D. Faires, and A. C. Reynolds: *Numerical Analysis*, Prindle, Weber & Schmidt, Boston, 1981.

Cheney, E. W.: *Introduction to Approximation Theory*, McGraw-Hill, New York, 1966.

Cheney, W., and D. Kincaid: *Numerical Mathematics and Computing*, Brooks/Cole, Monterey, CA, 1980.

Churchill, R. V.: *Fourier Series and Boundary Value Problems*, McGraw-Hill, New York, 1969.

Coddington, E. A., and N. Levinson: *Theory of Ordinary Differential Equations*, McGraw-Hill, New York, 1955.

Conte, S. D., and C. de Boor: *Elementary Numerical Analysis: An Algorithmic Approach*, McGraw-Hill, New York, 1980.

Dahlquist, G., and A. Björck: *Numerical Methods*, Prentice-Hall, Englewood Cliffs, NJ, 1974.

Davis, P. J.: *Interpolation and Approximation*, Ginn Blaisdell, Waltham, MA, 1963.

———, and P. Rabinowitz: *Methods of Numerical Integration*, Academic Press, New York, 1975.

De Boor, C.: *A Practical Guide to Splines*, Springer-Verlag, New York, 1978.

Donagarra, J., J. Bunch, C. Moler, and G. Stewart: *LINPACK Users Guide*, SIAM, Philadelphia, 1978.

Fike, C. T.: *Computer Evaluation of Mathematical Functions*, Prentice-Hall, Englewood Cliffs, NJ, 1968.

Forsythe, G. E., M. A. Malcolm, and C. B. Moler: *Computer Methods for Mathematical Computations*, Prentice-Hall, Englewood Cliffs, NJ, 1977.

Forsythe, G. E., and C. Moler: *Computer Solution of Linear Algebraic Systems*, Prentice-Hall, Englewood Cliffs, NJ, 1967.

Gear, C. W.: *Numerical Initial Value Problems in Ordinary Differential Equations*, Prentice-Hall, Englewood Cliffs, NJ, 1971.

George, A., and J. Liu: *Computer Solution of Large Sparse Positive Definite Systems*, Prentice-Hall, Englewood Cliffs, NJ, 1980.

Goldstine, H. H.: *A History of Numerical Analysis from the 16th through the 19th Century*, Springer-Verlag, New York, 1977.

Haberman, R.: *Mathematical Models*, Prentice-Hall, Englewood Cliffs, NJ, 1977.

Henrici, P.: *Discrete Variable Methods in Ordinary Differential Equations*, Wiley, New York, 1962.

———: *Elements of Numerical Analysis*, Wiley, New York, 1964.

Hildebrand, F. B.: *Introduction to Numerical Analysis*, McGraw-Hill, New York, 1956.

Householder, A. S.: *The Numerical Treatment of a Single Nonlinear Equation*, McGraw-Hill, New York, 1970.

Johnson, L. W., and R. D. Riess: *Numerical Analysis*, Addison-Wesley, Reading, MA, 1982.

Knuth, D. E.: *The Art of Computer Programming*, vol. 2, Addison-Wesley, Reading, MA, 1969.

Lambert, J. D.: *Computational Methods in Ordinary Differential Equations*, Wiley, New York, 1973.

Lapidus, L., and J. H. Seinfeld: *Numerical Solution of Ordinary Differential Equations*, Academic Press, New York, 1971.

Lawson, C. L., and R. J. Hanson: *Solving Least Squares Problems*, Prentice-Hall, Englewood Cliffs, NJ, 1974.

Maki, D. P., and M. Thompson: *Mathematical Models and Applications*, Prentice-Hall, Englewood Cliffs, NJ, 1973.

Malkevitch, J., and W. Meyer: *Graphs, Models, Finite Mathematics*, Prentice-Hall, Englewood Cliffs, NJ, 1974.

Noble, B.: *Applications of Undergraduate Mathematics in Engineering*, Macmillan, New York, 1967.

———, and J. W. Daniel: *Applied Linear Algebra*, Prentice-Hall, Englewood Cliffs, NJ, 1977.

Ortega, J. M., and W. C. Rheinboldt: *Iterative Solution of Nonlinear Equations in Several Variables*, Academic Press, New York, 1970.

Parlett, B.: *The Symmetric Eigenvalue Problem*, Prentice-Hall, Englewood Cliffs, NJ, 1980.

Prenter, P. M.: *Splines and Variational Methods*, Wiley-Interscience, New York, 1975.

Ralston, A., and P. Rabinowitz: *A First Course in Numerical Analysis*, McGraw-Hill, New York, 1978.

Rheinboldt, W. C.: "Algorithms for Finding Zeros of Functions," *UMAP* **2** (1):43 (1981).

Rice, J. R.: *Matrix Computations and Mathematical Software*, McGraw-Hill, New York, 1981.

——— (ed.): *Mathematical Software*, Academic Press, New York, 1971.

Shampine, L. F., and R. C. Allen: *Numerical Computing: An Introduction*, Saunders, Philadelphia, 1973.

———, and M. K. Gordon: *Computer Solution of Ordinary Differential Equations*, Freeman, San Francisco, 1975.

Sterbenz, P.: *Floating-Point Computation*, Prentice-Hall, Englewood Cliffs, NJ, 1974.

Stewart, G. W.: *Introduction to Matrix Computations*, Academic Press, New York, 1973.

Strang, G.: *Linear Algebra and Its Applications*, Academic Press, New York, 1976.

———, and G. Fix: *An Analysis of the Finite Element Method*, Prentice-Hall, Englewood Cliffs, NJ, 1973.

Stroud, A. H., and D. Secrest: *Gaussian Quadrature Formulas*, Prentice-Hall, Englewood Cliffs, NJ, 1966.

Traub, J. F.: *Iterative Methods for the Solution of Equations*, Prentice-Hall, Englewood Cliffs, NJ, 1964.

Varga, R. S.: *Matrix Iterative Analysis*, Prentice-Hall, Englewood Cliffs, NJ, 1962.

Walpole, R. E., and R. H. Myers: *Probability and Statistics for Engineers and Scientists*, Macmillan, New York, 1978.

Wilkinson, J. H.: *Rounding Errors in Algebraic Processes*, Prentice-Hall, Englewood Cliffs, NJ, 1963.

———: *The Algebraic Eigenvalue Problem*, Clarendon Press, Oxford, 1965.

ANSWERS TO SELECTED EXERCISES

CHAPTER 1

1.5 $n \geq 5$

1.6 $\cos \dfrac{5\pi}{16} \simeq \dfrac{\sqrt{2}}{2}\left[1 - \dfrac{\pi}{16} - \dfrac{\pi^2}{2(16)^2}\right] = .55464$

1.11 $n \geq 8$

1.15   Measurement errors in data, blunders like incorrect data statements

1.16   A priori

1.22 $1\frac{2}{3} = (.16666) \times 10^1$; $-\frac{3}{100} = -(.30000) \times 10^{-1}$

1.27 $\dfrac{|x - \tilde{x}|}{|x|} \leq \dfrac{\beta^{t-r}u}{2}$ $\dfrac{\beta^{t-r}}{2} \quad \dfrac{\beta^{1-t}}{2}$

1.31 $w = -(.3000) \times 10^{-1}$; $z = -(.3600) \times 10^{-1}$

1.33   **(a)** $(101)_8 = 17_{10}$

       (c) $(1101101)_2 = 109_{10}$

1.34 $(431)_{10} = (657)_8$

1.43   **(a)**   Well-conditioned

       (d)   Ill-conditioned

       (e)   Well-conditioned

1.44   Part (d) of exercise 1.43

CHAPTER 2

2.2 $(0, 1)$ contains a root

2.7   Root is approximately .7034674

2.14   No

2.15   BYSEKT uses TOL as an absolute error tolerance. For $x \simeq 50$ the specification TOL $= 10^{-6}$ is too small.

2.28 $x^* = 0$ has multiplicity 3.

2.31 g is better conditioned

2.43   (b) $n(n + 1)/2$ multiplications and n additions.

2.56   This polynomial has no real roots.

2.57   The convergence is only linear.

CHAPTER 3

3.1 $$\begin{bmatrix} 4 & 6 & 3 \\ 2 & 1 & 0 \\ 1 & -1 & 1 \\ 0 & 1 & -1 \end{bmatrix} \begin{bmatrix} x_2 \\ x_2 \\ x_3 \end{bmatrix} = \begin{bmatrix} 5 \\ 4 \\ 7 \\ 9 \end{bmatrix}$$

3.5 $$\begin{bmatrix} .5 & 0 & 0 \\ 0 & .25 & 0 \\ 0 & 0 & -1 \end{bmatrix}$$

3.7 $$\begin{bmatrix} -\frac{1}{11} \\ \frac{3}{11} \\ 1 \end{bmatrix}$$

3.9 $\mathbf{A}x$ requires n^2; $\mathbf{AB}$ requires n^3

3.12   (b) $$\begin{bmatrix} 14 & -7 \\ 2 & -3 \\ 3 & -1 \end{bmatrix}$$ (d) $$\begin{bmatrix} 6 & -2 \\ -2 & 17 \end{bmatrix}$$

3.14 $\mathbf{A}^3 = \mathbf{0}$

3.16   No

3.24   If $b_3 = 6$ there is no solution. By Theorem 3.2, $\mathbf{A}$ is not invertible.

3.27 $\mathbf{P} = \mathbf{P}^{-1}$

3.31   The number of operations is $\displaystyle\sum_{k=1}^{n-1} k^2 + 2k = \frac{n^3}{3} + \frac{n^2}{2} - \frac{5n}{6}$.

3.43   Cost is approximately \$0.44 by Algorithms 3.3 to 3.5.

3.44 $x_1 = \dfrac{2}{2 - \varepsilon}$; $x_2 = \dfrac{2(1 - \varepsilon)}{2 - \varepsilon}$

3.51 $\|\mathbf{x}\|_1 = 21$, $\|\mathbf{x}\|_2 = \sqrt{812} = 28.4956$

3.56 $K(\mathbf{A}) = 12.007 \times 10^4$

3.69   No row interchanges are necessary.

CHAPTER 4

4.1 $P_3(x) = -116.22(x - 1.2)(x - 1.3)(x - 1.45)$
$\qquad\qquad + 685.92(x - 1.05)(x - 1.3)(x - 1.45)$
$\qquad\qquad - 960.56(x - 1.05)(x - 1.2)(x - 1.45)$
$\qquad\qquad + 549.21(x - 1.05)(x - 1.2)(x - 1.3)$

4.4 $P_3(x) = 1.7433 + 5.526(x - 1.05) + 19.092(x - 1.05)(x - 1.2)$
$\qquad\qquad + 158.35(x - 1.05)(x - 1.2)(x - 1.3)$

4.14 $f'(0) \simeq 0.69329$

4.16 $P_3(x) = 0.90125 + 0.064686\tau + 0.004643 \dfrac{\tau(\tau - 1)}{2}$

$$+ 0.000332 \dfrac{\tau(\tau - 1)(\tau - 2)}{6}$$

4.22 $e_n(f, x) = 0$ for all x

4.28 $P_1(0.55) = 1.14178$ based on 0.5, 0.58.

$P_2(0.6) = 1.16618$ based on 0.5, 0.58, 0.66

$P_3(0.46) = 1.04403$ based on 0.3, 0.42, 0.5, 0.58

4.34 $P_6(3\pi/8) = 0.3827$ is accurate to within 0.165×10^{-4}.

4.39 $n \geq 3$

4.42 Yes

CHAPTER 5

5.5 $N \geq 1059$

5.11 (a) $k = 1$ (b) $k = 3$

5.12 Yes, since $I_N(a_0 + a_1 x + \cdots + a_n x^n) = \displaystyle\sum_{k=0}^{n} a_k I_N(x^k)$.

5.15 $-w_1 = w_4 = (b - a)^2/12$, $w_2 = w_3 = (b - a)/2$

5.21 $f(x) = \cos x$ on $[0, 2\pi]$

5.33 The T table has seven rows and columns. The Romberg method requires 130 function evaluations.

5.35 $T_{44} = T_{33}$

5.42 (a), (c), (d), (f)

5.48 NOFUN increases with k.

5.49 Yes

5.50 The integrand has the same value at 0, $\pi/4$, $\pi/2$, $3\pi/4$, and π. Consequently ADPSIM would return the value 3π after the first stage.

5.53 (c) gives result more efficiently.

5.58 This is Gauss–Legendre rule $I_3(f)$.

5.63 $x = t/\sqrt{3}$.

5.67 $I = \int_{-1}^{1} (1 + t)/\sqrt{1 - t^2} \, dt$ and Gauss–Chebyshev gives exact results for $N = 1$.

5.68 Let $x(t) = t^2$.

5.72 $\displaystyle\int_{-\infty}^{\infty} \dfrac{1}{1 + x^2} \, dx = 4 \int_0^1 \dfrac{1}{1 + x^2} \, dx$

5.76 $\displaystyle\int_0^2 \dfrac{e^x}{\sqrt{|x - 1|}} \, dx = 2e \int_0^1 (e^{t^2} + e^{-t^2}) \, dt$

5.80 $f'(1.4) \simeq 35.092 = D_h f_3$, and estimated error is -0.5043.

5.86 $f'(1.4) \simeq 34.5877$ is more accurate.

CHAPTER 6

6.1 (a) 3 (b) 2 (c) 1

6.2 (c) $f(t, y) = e^{t+y}$

6.4 (a) 5568 years

6.6 (a) Arbitrary rectangle
(b) Arbitrary rectangle
(d) Any rectangle that excludes the line $y = -t$.

6.7 (a) y large, say $y \geq 100$
(c) $|y| \leq 1$ and t large, say $|t| \geq 100$

6.10 (a) $L = 10$ for $|y| \leq 10$ and all t
(e) $L = 10$ for $|t| \leq 10$ and all y

6.15 No, because $f(0, 1)$ is undefined

6.17 $Y_N^\delta = (1 + \delta)(1 + h/N)^N$

6.25 Want $h(\partial f/\partial y)$ to be in I_A.

6.26 For $y' = \lambda y$ the method gives $Y_{n+1} = Y_n/(1 - \lambda h)$. Thus for absolute stability we need $|1 - \lambda h| \geq 1$.

6.28 (b) is stiff and hence requires small h for absolute stability.

6.29 For $y' = \lambda y$, Heun's method gives $Y_{n+1} = (1 + \lambda h)Y_n$.

6.36 Require $h(\partial f/\partial y)$ to be in $I_A = [-2.78, 0]$.

6.46 Caution: When TOLR $= 0$, the test for an increase in step size should be modified.

6.59 For $y' = \lambda y$ the one-step Adams–Moulton formula gives

$$Y_{n+1} = \left(\frac{1 + h\lambda/2}{1 - h\lambda/2} \right) Y_n$$

Thus for absolute stability we require $|1 + \lambda h/2| \leq |1 - \lambda h/2|$.

6.70 Let $x_1 = y, x_2 = y', x_3 = y''$. Then

$$\begin{aligned}
x_1' &= x_2 & x_1(0) &= 0 \\
x_2' &= x_3 & x_2(0) &= 1 \\
x_3' &= x_1 + x_2 - x_3 + \cos t & x_3(0) &= 0
\end{aligned}$$

CHAPTER 7

7.9 $y = 0.989537 + 0.509128x$ rounded to six digits

7.10 $y = 1.0004 + 0.47929x$ rounded to five digits

7.12 $y = (-2.0536 + 3.0267x)^{-1}$ rounded to five digits

7.13 $f(x) = 1.5575 \, e^{1.9379x}$ rounded to five digits

7.21 Yes

7.22 No

7.30 $C_0 = \frac{1}{16}(16a_0 + 8a_2 + 6a_4 + 5a_6)$
$C_1 = \frac{1}{8}(8a_1 + 6a_3 + 5a_5)$
$C_2 = \frac{1}{32}(16a_2 + 16a_4 + 15a_6)$
$C_3 = \frac{1}{16}(4a_3 + 5a_5)$
$C_4 = \frac{1}{16}(2a_4 + 3a_6)$
$C_5 = \frac{1}{16}a_5$
$C_6 = \frac{1}{32}a_6$

7.31
$$G = \begin{bmatrix} 5 & 15 & 55 & 216 \\ 15 & 55 & 216 & 979 \\ 55 & 216 & 979 & 4425 \\ 216 & 979 & 4425 & 8015 \end{bmatrix} \qquad K(G) = 22057.5$$

7.49 $\quad p = \dfrac{6}{\pi} \tilde{L}_0 + \dfrac{45}{71} \dfrac{48 - 2\pi^2}{\pi^3} \tilde{L}_2$

7.51 $\quad a_k = \dfrac{4}{k\pi} \sin\left(\dfrac{k\pi}{2}\right)$

7.59 $\quad x = \frac{3}{2}, y = -1$

7.60
$$A^T A = \begin{bmatrix} 5 & 1 & -1 \\ 1 & 2 & -2 \\ -1 & -2 & 2 \end{bmatrix}$$

Solutions of normal equations are given by

$$\mathbf{x} = \begin{bmatrix} 1 \\ 0 \\ 0 \end{bmatrix} + \alpha \begin{bmatrix} 0 \\ 1 \\ 1 \end{bmatrix}$$

for any scalar α.

INDEX